Autonomous Control of Unmanned Aerial Vehicles

Autonomous Control of Unmanned Aerial Vehicles

Special Issue Editor

Victor Becerra

MDPI • Basel • Beijing • Wuhan • Barcelona • Belgrade

Special Issue Editor
Victor Becerra
University of Portsmouth
UK

Editorial Office
MDPI
St. Alban-Anlage 66
4052 Basel, Switzerland

This is a reprint of articles from the Special Issue published online in the open access journal *Electronics* (ISSN 2079-9292) from 2018 to 2019 (available at: https://www.mdpi.com/journal/electronics/special_issues/Unmanned_Aerial_Vehicles)

For citation purposes, cite each article independently as indicated on the article page online and as indicated below:

LastName, A.A.; LastName, B.B.; LastName, C.C. Article Title. *Journal Name* **Year**, *Article Number*, Page Range.

ISBN 978-3-03921-030-5 (Pbk)
ISBN 978-3-03921-031-2 (PDF)

Cover image courtesy of Victor M. Becerra.

Contents

About the Special Issue Editor

Victor Becerra is currently a Professor of Power Systems Engineering at the University of Portsmouth, United Kingdom, since his appointment in December 2015. Between 2000 and 2015, he was an academic at the School of Systems Engineering, University of Reading, United Kingdom, where he became Professor of Automatic Control in 2012. He obtained his PhD in Control Engineering in 1994 from City, University of London for his work on nonlinear optimal control. He has published over 140 research papers, one research monograph, and one edited volume. His current research interests include computational optimal control, autonomous systems, control and optimisation of power systems and smart grids, and control of energy storage. His research has been funded by the EPSRC, the Knowledge Transfer Partnership Programme, the EC, the European Space Agency, the Royal Academy of Engineering, and various UK industries. He is a Fellow of the IET, a Senior Member of the IEEE, a Senior Member of the AIAA, and a Chartered Engineer in the United Kingdom. He is an Associate Editor of the *IMA Journal of Mathematical Control and Information and the International Journal of Automation and Computing.*

electronics

MDPI

Editorial

Autonomous Control of Unmanned Aerial Vehicles

Victor M. Becerra

School of Energy and Electronic Engineering, University of Portsmouth, Anglesea Road, Portsmouth PO1 3DJ, UK; victor.becerra@port.ac.uk; Tel.: +44-23-9284-2393

Received: 12 April 2019; Accepted: 15 April 2019; Published: 22 April 2019

1. Introduction

Unmanned aerial vehicles (UAVs) are being increasingly used in different applications in both military and civilian domains. These applications include, for example, surveillance, reconnaissance, remote sensing, target acquisition, border patrol, infrastructure monitoring, aerial imaging, industrial inspection, and emergency medical aid.

Vehicles that can be considered autonomous must be able to make decisions and react to events without direct intervention by humans [1]. There are some fundamental aspects that are common to all autonomous vehicles. These aspects include the abilities of sensing and perceiving the environment, analyzing the sensed information, communicating, planning and decision making, as well as acting using control algorithms and actuators.

Although some UAVs are becoming able to perform increasingly complex autonomous maneuvers, most UAVs are not fully autonomous; instead, they are mostly operated remotely by humans [2]. To make UAVs fully autonomous, many technological and algorithmic developments are still needed. For instance, UAVs will need to improve their sensing of obstacles and subsequent avoidance. This becomes particularly important as autonomous UAVs start to operate in a civil air space that is used by other aircraft.

Operating unmanned flying vehicles is useful yet it can be challenging when the vehicle interacts with the environment [3]. This interaction could be, for instance, in the form of landing on ground or landing pads, docking into a station, approaching terrain for inspection, or approaching another aircraft for refueling purposes. Such tasks can often be solved when the vehicle is remotely piloted, especially when the pilot has a first-person view of the environment. However, human control may not always be possible, for instance due to the unavailability of a suitable data link, or because of the precision and/or speed that is required for the maneuver, which may be outside human capabilities. Thus, it is important to find effective and flexible strategies to enable vehicles to perform such tasks autonomously.

Well-developed features of autonomous UAV control include, for instance, stability enhancement and waypoint flight [4]. However, new developments in the design of UAVs and the emergence of new application areas demand robust and adaptive control techniques for different flight conditions, aggressive maneuvering flight, robust disturbance rejection, obstacle avoidance, fault tolerance, formation flying, and the use of new sensing and perception paradigms, such as computer vision. Even when the vehicle performs tasks autonomously, the efficiency and reliability of the communication link to the ground station or other aerial vehicles is important, as the autonomous UAV may need to send information about itself or its environment to the ground station or other vehicles, or it may need to receive updated mission parameters from the ground station, or information from other vehicles. To achieve all the ambitious requirements that autonomous operation brings about, systematic and innovative methods for planning, navigation, decision making, control, sensing and communications are needed.

The aim of this Special Issue is to bring together researchers and practitioners in the field of unmanned aerial systems, with a common interest in their autonomy. The contributions that are part of this special Special Issue present key challenges associated with autonomous control of unmanned aerial vehicles, propose solution methodologies to address such challenges, analyse the proposed methodologies and evaluate their performance.

2. The Present Special Issue

This special issue consists of thirteen selected articles covering different aspects of autonomous aerial vehicles, including 3D path planning with obstacle avoidance, visual control of near ground maneuvers, visual inspection, vision-based safe emergency landing, control strategies for robust disturbance rejection, efficient communication links, autonomous decision making in automated air confrontation systems, remote sensing using multi-UAV systems, ground vehicle detection, and novel autonomous UAV designs, such as flying wings and coaxial rotor UAVs.

The ability to plan colision-free paths in complex environments is an important element of UAV autonomy. In [5], Samaniego and co-workers present a computationally efficient method for 3D path planning of UAVs using an adaptive discrete mesh. The proposed method explores and decomposes the 3D environment under a recursive reward cost paradigm, resulting in an efficient and simple 3D path detection. Their algorithm saves computational time and memory compared with classical techniques.

The ability of some vehicles to transition from hover to lift-based forward flight and vice-versa brings the possibility for an autonomous flying vehicle to perform complex missions where the two different flight modes are needed. The work by Garcia-Nieto et al. [6] presents the design, implementation, harware-in-the-loop simulation and prototype testing of a control system that allows an unmanned flying-wing to perform vertical take-off and landing (VTOL) maneuvers using two tilting rotors. This work is considered by the authors as a first step towards the development of an autonomous flying-wing with VTOL capabilities.

Complex near-ground maneuvers, such as landing and capturing moving pray, are performed by flying animals with ease. These animals perform such maneuvres by exclusively using the information from their vision and vestibular system. It has been suggested that flying insects and birds use a particular visual strategy described by Tau theory to perform maneoeuvres that involve closing gaps with objects. Inspired by flying animals, the article by Armendariz and co-authors [7] describes and evaluates a visual approach that uses optic flow and Tau theory to perform autonomous near-ground manoeuvres involving vertical and horizontal motion relative to a moving target, without knowledge of height and velocity of the flying vehicle or the velocity of the target.

A coaxial rotor UAV uses a pair of coaxial reversing rotors which compensate for each other's torque, instead of balancing the yaw moment of the aircraft with a tail rotor. Therefore, a coaxial rotor aircraft has a compact structure, a small radial size, and a higher power efficiency. In their contribution, Li and co-authors [8] propose a decoupling algorithm to improve the reliability of the attitude control for the longitudinal motion stability of a coaxial rotor UAV. Based on a dynamic model that describes the vehicle's longitudinal motion, an under-actuated controller is designed using the fuzzy sliding-mode approach. The study provides simulation results showing that the position and attitude performance of the coaxial rotor UAV can be improved with the proposed control methods.

Visual inspection of aircraft is another application area where autonomous aerial systems are being used. The work by Papa and Ponte [9] describes the preliminary design of a general visual inspection system onboard a commercial quadrotor UAV. A high-definition camera is used to detect visual damage on the inspected aircraft caused by hail or lightning strikes, which are among the most dangerous threats for the airframe. Preliminary experimental results obtained from initial test flights are given, showing the performance of the ultrasonic distance keeping system and of the image acquisition/processing module for damage detection.

Because of their nature, autonomous flying vehicles must be able to reject disturbances in a robust manner. The article by Song et al. [10] presents a fixed-time active disturbance rejection control approach for the attitude control problem of a quadrotor UAV. The authors consider the presence of dynamic wind, mass eccentricity and actuator faults. The work is based on the feedback linearisation technique, along with a sliding mode feedback law and an extended state observer. The work provides mathematical proofs of convergence of the proposed extended state observer and feedback laws, along with simulation and experimental results that demonstrate the robustness and capabilities of the proposed control approach.

The efficiency of the communication link between a UAV and the groud control station is a key aspect in military applications, delivery services as well as search and rescue operations. In their contribution, Atoev et al. [11] investigate the single-carrier frequency division multiplexing modulation technique as a means to achieve high efficiency in the communications link between the UAV and the ground control station. The authors provide experimental results and compare the performance of their proposed approach with a commonly used modulation method.

The demand for autonomous decision-making algorithms to support automated air confrontation systems is growing. The work by Zhang et al. [12] addresses such demand by presenting the development of a super-horizon air confrontation training environment. The authors employ computational intelligence approaches, including reinforcement learning and neural networks, to create a self-learning air confrontation maneouver decision making system, which is tested by means of complex simulations of different air confrontation situations.

Agricultural applications of UAVs have mainly focused on a few areas, such as pest control and crop monitoring. However, agricultural UAVs are expected to be used for many other useful purposes such as field surveys, sowing, spraying, and remote sensing. In their article, Ju and Son [13] describe the development of a multi-UAV system for remote sensing in agriculture using a distributed swarm control algorithm. The authors show through their extensive experimental work and thorough analysis that their developed agricultural multi-UAV system solves the problem of battery shortage and reduces working time and control effort.

Due to their small size, autonomous UAVs are often sensitive to environmental disturbances such as wind gust. The contribution by Shi et al. [14] deals with high precision attitude control for a quadrotor UAV subject to wind gust and actuator faults. Their control strategy is based on the online disturbance uncertainty estimation and attenuation method. The authors propose and analyse state observer and sliding mode control laws based on the super-twisting algorithm, which is used to mitigate the chattering effects that often occur in sliding mode control and estimation methods. The effectiveness of their approach is demonstrated by means of simulations and real-time experiments.

The presence of a slung load attached to an autonomous helicopter exerts a swing effect on the system which significantly changes the dynamics of the vehicle and can threaten the stability of the attitute control system. Aiming to address this problem, the work by Shi and co-workers [15] proposes a high precision disturbance compensation method for a quadrotor. The authors model the quadrotor-slung load system, representing the slung load as a disturbance, and propose a harmonic state observer, along with an attitude tracking controller based on backstepping. The control system is tested by means of simulations and real-time experiments, showing improvements in the robustness of a quadrotor subject to a slung load.

An important task for some autonomous aerial systems involves the detection of vehicles and other objects on the ground. The work by Liu et al. [16] presents a method for ground vehicle detection in aerial infrared images based on a convolutional neural network. The proposed method is able to detect both stationary and moving vehicles in real urban environments. As part of their research, the authors created and have publicly shared a database of aerial vehicle imagery that can be used for research in vehicle detection. Their tests demonstrate that the proposed method is effective and efficient in recognizing ground vehicles, and is suitable for real-time application.

A current area of research of clear importance to the operation of autonomous aerial vehicles is their safe landing and recovery. As most UAV navigation methods rely on global positioning system (GPS) signals, many drones cannot land properly in the absence of such signals. Given that with the use of vision and image recognition technology the position and posture of the UAV in three dimensions can be estimated, and the environment where the drone is located can be perceived, the contribution by Yang and co-workers [17] proposes a monocular autonomous landing system that utilizes vision-based simulteaneous localization and mapping (SLAM) algorithms for use in emergencies and in unstructured environments. Experiments carried out by the authors with multiple sets of real scenes are reported and demonstrate the effectiveness of their proposed methods.

3. Future Possibilities

The UAV market is growing at a fast pace and in 2017 it was expected to triple from the the annual value of $4 billion to $14 billion in 2027 [18]. Although the market is still dominated by military applications, commercial aapplications are increasing their market share, with commercial UAV production expected to grow from $4.1 billion worldwide in 2018 to $13.1 billion in 2027 [19]. Moreover, the size of UAV-based solutions and services has been estimated to have a potential value of over $127 billion [20].

With the demand for autonomous features in UAVs growing alongside the UAV market as a whole, it can only be expected that the future research activity in the area of autonomous control of unmanned aerial vehicles will be very active, with commercial R&D aimed at enriching the technological capabilities of products to better compete in a growing and demanding market, but also with universities supported by their own funds and by government funding agencies, which see great future potential in autonomous systems. In [20], market analysts have identified the following key areas for R&D in unmanned aerial vehicles: artificial intelligence, drone detection and avoidance technology, control and communications, image processing, and battery capacity. All of these key areas are fundamental to UAV autonomy and are reflected in different ways in the contributions that are part of this Special Issue.

Funding: This research received no external funding.

Acknowledgments: The Guest Editor and the Editor-in-Chief MDPI Electronics journal wish to thank all authors who submitted their excellent research work to this special issue. We are grateful to all reviewers who contributed evaluations and views of the merits and quality of the manuscripts, and provided valuable suggestions and comments to improve their quality and the overall scientific value. Special thanks go to the Editorial Board of MDPI Electronics journal for the opportunity to edit this special issue, and to the Electronics Editorial Office staff for the hard and attentive work in keeping a rigorous peer-review schedule and timely publication process.

Conflicts of Interest: The author declares no conflict of interest.

Abbreviations

The following abbreviations are used in this manuscript:

MDPI	Multidisciplinary Digital Publishing Institute
UAV	Unmanned aerial vehicle
SLAM	Simultaneous localization and mapping
3D	Three-dimensional
VTOL	Vertical take-off and landing
R&D	Research and development

References

1. Sebbane, Y.B. *Smart Autonomous Aircraft: Flight Control and Planning for UAV*, 1st ed.; CRC Press: Boca Raton, FL, USA, 2015.
2. Grifantini, K. How to Make UAVs Fully Autonomous. Available online: https://www.technologyreview.com/s/414363/how-to-make-uavs-fully-autonomous/ (accessed on 20 April 2019).
3. Alkowatly, M.T.; Becerra, V.M.; Holderbaum, W. Bioinspired Autonomous Visual Vertical Control of a Quadrotor Unmanned Aerial Vehicle. *J. Guid. Control. Dyn.* **2014**, *38*, 249–262. [CrossRef]
4. Zhang, R.; Zhang, J.; Yu, H. Review of modeling and control in UAV autonomous maneuvering flight. In Proceedings of the 2018 IEEE International Conference on Mechatronics and Automation (ICMA), Changchun, China, 5–8 August 2018; pp. 1920–1925.

5. Samaniego, F.; Sanchis, J.; García-Nieto, S.; Simarro, R. Recursive Rewarding Modified Adaptive Cell Decomposition (RR-MACD): A Dynamic Path Planning Algorithm for UAVs. *Electronics* **2019**, *8*, 306. [CrossRef]
6. Garcia-Nieto, S.; Velasco-Carrau, J.; Paredes-Valles, F.; Salcedo, J.V.; Simarro, R. Motion Equations and Attitude Control in the Vertical Flight of a VTOL Bi-Rotor UAV. *Electronics* **2019**, *8*, 208. [CrossRef]
7. Armendariz, S.; Becerra, V.; Bausch, N. Bio-Inspired Autonomous Visual Vertical and Horizontal Control of a Quadrotor Unmanned Aerial Vehicle. *Electronics* **2019**, *8*, 184. [CrossRef]
8. Li, K.; Wei, Y.; Wang, C.; Deng, H. Longitudinal Attitude Control Decoupling Algorithm Based on the Fuzzy Sliding Mode of a Coaxial-Rotor UAV. *Electronics* **2019**, *8*, 107. [CrossRef]
9. Papa, U.; Ponte, S. Preliminary Design of an Unmanned Aircraft System for Aircraft General Visual Inspection. *Electronics* **2018**, *7*, 435. [CrossRef]
10. Song, C.; Wei, C.; Yang, F.; Cui, N. High-Order Sliding Mode-Based Fixed-Time Active Disturbance Rejection Control for Quadrotor Attitude System. *Electronics* **2018**, *7*, 357. [CrossRef]
11. Atoev, S.; Kwon, O.H.; Lee, S.H.; Kwon, K.R. An Efficient SC-FDM Modulation Technique for a UAV Communication Link. *Electronics* **2018**, *7*, 352. [CrossRef]
12. Zhang, X.; Liu, G.; Yang, C.; Wu, J. Research on Air Confrontation Maneuver Decision-Making Method Based on Reinforcement Learning. *Electronics* **2018**, *7*, 279. [CrossRef]
13. Ju, C.; Son, H.I. Multiple UAV Systems for Agricultural Applications: Control, Implementation, and Evaluation. *Electronics* **2018**, *7*, 162. [CrossRef]
14. Shi, D.; Wu, Z.; Chou, W. Super-Twisting Extended State Observer and Sliding Mode Controller for Quadrotor UAV Attitude System in Presence of Wind Gust and Actuator Faults. *Electronics* **2018**, *7*, 128. [CrossRef]
15. Shi, D.; Wu, Z.; Chou, W. Harmonic Extended State Observer Based Anti-Swing Attitude Control for Quadrotor with Slung Load. *Electronics* **2018**, *7*, 83. [CrossRef]
16. Liu, X.; Yang, T.; Li, J. Real-Time Ground Vehicle Detection in Aerial Infrared Imagery Based on Convolutional Neural Network. *Electronics* **2018**, *7*, 78. [CrossRef]
17. Yang, T.; Li, P.; Zhang, H.; Li, J.; Li, Z. Monocular Vision SLAM-Based UAV Autonomous Landing in Emergencies and Unknown Environments. *Electronics* **2018**, *7*, 73. [CrossRef]
18. Canetta, L.; Mattei, G.; Guanziroli, A. Exploring commercial UAV market evolution from customer requirements elicitation to collaborative supply network management. In Proceedings of the 2017 International Conference on Engineering, Technology and Innovation (ICE/ITMC), Funchal, Portugal, 27–29 June 2017; pp. 1016–1022.
19. Finnegan, P. *2018 World Civil Unmanned Aerial Systems Market Profile & Forecast*; Technical Report; Teal Group Corporation: Fairfax, VA, USA, 2018.
20. Mazur, M.; Wiśniewski, A. *Clarity from Above—PwC Global Report on the Commercial Applications of Drone Technology*; Technical Report; PwC Polska: Warszawa, Poland, 2016.

electronics

MDPI

Article

Monocular Vision SLAM-Based UAV Autonomous Landing in Emergencies and Unknown Environments

Tao Yang [1,2,*], Peiqi Li [1], Huiming Zhang [3], Jing Li [4,*] and Zhi Li [1]

1 SAIIP, School of Computer Science, Northwestern Polytechnical University, Xi'an 710072, China; page_7026@mail.nwpu.edu.cn (P.L.); zLeewack@mail.nwpu.edu.cn (Z.L.)

2 Research & Development Institute of Northwestern Polytechnical University in Shenzhen, Shenzhen 518057, China

3 National Key Laboratory for Novel Software Technology, Nanjing University, Nanjing 210023, China; mf1733077@smail.nju.edu.cn

4 School of Telecommunications Engineering, Xidian University, Xi'an 710071, China

* Correspondence: tyang@nwpu.edu.cn (T.Y.); jinglixd@mail.xidina.edu.cn (J.L.); Tel.: +86-150-0291-9079 (T.Y.); +86-139-9132-0168 (J.L.)

Received: 17 April 2018; Accepted: 11 May 2018; Published: 15 May 2018

Abstract: With the popularization and wide application of drones in military and civilian fields, the safety of drones must be considered. At present, the failure and drop rates of drones are still much higher than those of manned aircraft. Therefore, it is imperative to improve the research on the safe landing and recovery of drones. However, most drone navigation methods rely on global positioning system (GPS) signals. When GPS signals are missing, these drones cannot land or recover properly. In fact, with the help of optical equipment and image recognition technology, the position and posture of the drone in three dimensions can be obtained, and the environment where the drone is located can be perceived. This paper proposes and implements a monocular vision-based drone autonomous landing system in emergencies and in unstructured environments. In this system, a novel map representation approach is proposed that combines three-dimensional features and a mid-pass filter to remove noise and construct a grid map with different heights. In addition, a region segmentation is presented to detect the edges of different-height grid areas for the sake of improving the speed and accuracy of the subsequent landing area selection. As a visual landing technology, this paper evaluates the proposed algorithm in two tasks: scene reconstruction integrity and landing location security. In these tasks, firstly, a drone scans the scene and acquires key frames in the monocular visual simultaneous localization and mapping (SLAM) system in order to estimate the pose of the drone and to create a three-dimensional point cloud map. Then, the filtered three-dimensional point cloud map is converted into a grid map. The grid map is further divided into different regions to select the appropriate landing zone. Thus, it can carry out autonomous route planning. Finally, when it stops upon the landing field, it will start the descent mode near the landing area. Experiments in multiple sets of real scenes show that the environmental awareness and the landing area selection have high robustness and real-time performance.

Keywords: UAV automatic landing; monocular visual SLAM; autonomous landing area selection

1. Introduction

Unmanned aerial vehicles (UAVs) are non-manned aircraft that are operated by radio remote control equipment or a self-contained program control device. Drones have wide applicability in military and civilian areas because they have advantages of simple and practical structure, convenient and flexible operation, and low cost. Furthermore, users do not have to worry about casualties that drones may cause. They are widely used in military missions such as tactical reconnaissance and territorial surveillance, target positioning, and so on. In civil use, drones can be used for field monitoring, meteorological exploration, highway inspection, etc. With the popularization and wide application of drones in military and civilian fields, drones' safety issues must be considered. Relevant data show that the number of failures in the recycling process of drones accounts for more than 80% of the total number of failures of drones. Therefore, research on the safe landing and recovery of drones has become an urgent task.

However, due to the complex application environment of drones (especially in the context of war), drones' landing research needs to consider many factors and versatility, with improved practicality. To be more specific, the main challenges cover the following points: (1) **Autonomous control without GPS signal**. The anti-jamming capability of GPS is extremely weak. If the on-board GPS signal receiver of drones malfunctions due to electronic interference, drones will lose their navigation and positioning function, and thereby fail to land safely. In the natural state, GPS signals can be easily interfered. The influencing factors are mainly divided into four categories: (a) weather factors and sunspots may reduce signal strength, but generally do not affect positioning; (b) electromagnetic interference, radio, and strong magnetic fields all generate different levels of interference; (c) GPS signals will decrease under shelters, such as buildings, vehicles, insulation paper, trees, and metal components; and (d) high-rise buildings and dense high-rise buildings will affect GPS signals. Therefore, it is very important to study the autonomous positioning and flight control of drones without GPS signals; (2) **Passive landing in an emergency**. Since the drone's compensation mechanism does not allow the failed drone to continue flying for a long time, it should begin to select a site for emergency landing. Although this is a helpless move, it is also an important measure to prevent the drone from falling into densely populated areas. The Federal Aviation Administration of the United States believes that, in the future, drones must not only guarantee their own secure flight, but also have the ability to interact safely with a variety of aircraft in their airspace in the event of an emergency. Such regulations still assume the ability of drones to maintain communication between the air and ground during emergencies. In fact, when some more serious failures occur, a drone is likely to completely lose contact with the ground. At that point, drones' abilities to autonomously plan routes, autonomously search for landing sites, and autonomously land become the last resort to save themselves; (3) **Autonomous landing in an unknown environment**. In the military field or in disaster relief situations, the place where drones need to perform tasks is mostly an unknown environment or a variegated environment. It is essential that drones can choose landing sites with proper strategies and land safely.

To address these problems, researchers have made their contributions to drones' autonomous flight and secure landing. Jung et al. [1] propose a four-rotor drone guided landing algorithm. This paper presents a framework for the utilization of low-cost sensors for precise landing in moving targets. Based on the previous paper, authors in [2] describes the tracking guidance for autonomous drone landing and the vision-based detection of the marker on a moving vehicle with a real-time image processing system. Falanga et al. [3] presents a quadrotor system capable of autonomously landing on a moving platform using only onboard sensing and computing. This paper relies on computer vision algorithms, multi-sensor fusion for localization of the robot, detection and motion estimation of the moving platform, and path planning for fully autonomous navigation. Authors in [4,5] propose drone landing technology by identifying a sign and then landing the drone on the marker. Therefore, the drone needs to place the landing mark in the landing area before landfall. Vlantis et al. [6] studies the problem of landing a quadrotor on an inclined moving platform. The aerial robot employs a forward-looking on-board camera to detect and observe the landing platform, which is carried by a mobile robot moving independently on an inclined surface. Kim et al. [7] propose a vision-based target following and landing system for a quadrotor vehicle on a moving platform. The system employs a vision-based landing site detection and locating algorithm using an omnidirectional lens. Measurements from the omnidirectional camera are combined with a proper dynamic model in order to estimate the position and velocity of the moving platform. Forster et al. [8] proposes a resource-efficient system for real-time three-dimensional terrain reconstruction and landing spot detection for micro aerial vehicles. This paper uses the semi-Direct monocular visual odometry (SVO) algorithm to extract the key points to create the terrain map. However, SVO is a visual odometer based on the semi-direct method, which inherits some drawbacks of the direct method and discards optimization and loop detection. Authors in [9] propose a fixed-wing drone landing method based on optical guidance, using the ground landing guidance system to optically guide the landing. A measuring camera is arranged on both sides of the runway, and a marker light is installed in front of the drone, and the drone is spatially positioned by binocular stereo vision. This method has many outstanding advantages, being a self-contained system with high measurement accuracy that is low-cost and has low power consumption. Furthermore, it is less susceptible to interference without time accumulation errors. However, it is a ground guidance system and is not suitable for fully-autonomous landing of quadrotor UAVs in emergencies and unknown environments.

Despite the above, these methods have their own drawbacks and limitations: (1) most existing methods focus on landing the drone on a marker; and (2) some methods use model-based approaches to deal with missing visual information. Alternative solutions are realized with the use of additional sensors attached to landing area. Among many, these sensors include inertial measurement units (IMUs), GPS receivers, or infrared markers; and (3) previous research has only been able to accomplish landing in a given environment. Additionally, GPS [7,10–12] or motion capture systems [13,14] are often used for state estimation, either only while patrolling or throughout the entire task. Conversely, this paper relies only on visual–inertial odometry for state estimation. These approaches do not work in many cases, such as an emergency landing of a UAV or landing in a stricken area. There are also many emergency situations during the flight of UAVs, such as low battery power, machine malfunctions, or some unexpected conditions where drones need to be landed in unmarked areas. Therefore, we need UAVs to be able to land autonomously in unstructured and natural environments.

In order to land in unknown environments, our approach is to use visual landing technology. The simultaneous localization and mapping (SLAM) algorithm is a hotspot of robot and computer vision research, and is considered as one of the key technologies for automatic navigation in unknown environments. In 2007, Professor Davison presented MonoSLAM [15], which is the first real-time monocular vision SLAM system. MonoSLAM was designed to expand the Kalman filter for the back-end, tracking very sparse feature points. Parallel tracking and mapping (PTAM) [16] is a well-known single-SLAM algorithm that proposes and implements the tracking and mapping into two separate thread modules, and greatly improves the efficiency of the algorithm so that the algorithm can run in real time. PTAM combined with augmented reality (AR) is used in augmented reality software. Nevertheless, it also has its own limitations. For example, it can only be applied in a relatively small working environment. Forster et al. propose a semi-direct monocular visual odometry (SVO) based on a semi-direct method [17]. It uses pixel brightness to estimate pose, resulting in the ability to maintain pixel-level precision in high-frame-rate video. However, SVO abandons the optimization and loop detection part in order to improve speed and make the system lightweight, which results in increased calculation error and inaccurate posture estimation under a long running time, and the loss is not easy to reposition. LSD-SLAM (large-scale direct monocular SLAM) [18] is an algorithm based on features and direct methods, proposed by Engel et al. It applies the direct method to semi-dense monocular SLAM, which can realize semi-dense scene reconstruction on a CPU. Since LSD-SLAM uses direct methods to track, it also inherits the disadvantages of direct methods. For example, LSD-SLAM is very sensitive to the intrinsic camera parameters and exposure, and it fails very easily in the process of fast motion. Mur-Artal et al. propose a feature-based monolithic SLAM system, ORB_SLAM2 [19], which can be applied to all scenes in real-time. The algorithm is divided into four modules: tracking, building, relocating, and closed-loop detection. The system is divided into three separate threads, which can successfully track and build a map. With the advent of the new sensor event-based camera, many SLAM studies based on event cameras [20,21] have emerged in recent years. However, the event-based camera is expensive and has a low spatial resolution, which limits the performance of the application.

In this paper, a vision-based UAV landing method is used. With the help of optical equipment and image recognition technology, the UAV is capable of autonomously identifying the landing zone and reconstructing three-dimensional terrain to accomplish automatic return and route planning. Before preparing for an autonomous landing, the drones scan the scene, analyze the appropriate location, and initiate an autonomous calibration of the initial landing site. Then, it carries out autonomous route planning for this point. When the appropriate landing path is calculated, the autonomous landing control mode is automatically switched on and the proper solution is adopted to approach the landing site. After reaching the landing field along with a correct landing route, it starts the descent mode. As drones continue to decline, the control system systematically identifies information such as altitude change rate and pose of the drone, and adjusts the altitude of the drone at any time until it lands in a predetermined landing zone. As shown in Figure 1, the experiments were performed in multiple scenarios. The results of the experiments show that the environmental awareness and the landing area selection have high robustness and real-time performance.

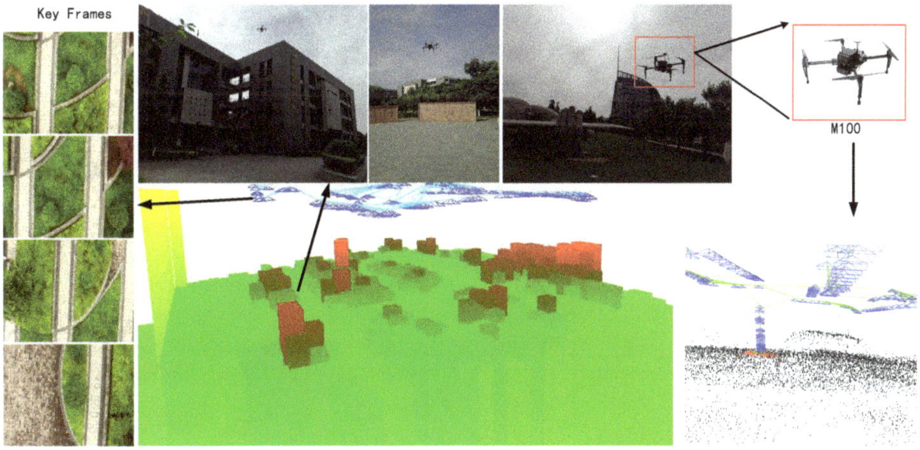

Figure 1. The M100, also known as the Matrice M100, is a quadrocopter drone for developers, released by DJI (Shenzhen, China), the world's largest consumer drone maker. The drone uses a monocular camera to scan the ground for key frames and a three-dimensional point cloud map. Then, it converts the three-dimensional point cloud map into a grid map, detects suitable landing areas, and carries out a landing. The map in the lower center of the picture shows a flat area in green and an area higher than the horizon in red. The depth of red indicates the height. The blue lines above the map show the keyframes created by the drone during the flight. The left column of the picture is part of the key frames.

When an airport fails to receive a control signal in some case (e.g., system failure or signal interference), the ability to land at the airport with a good strategy and land safely will greatly reduce the damage of the unmanned drone to the ground personnel. Advanced autonomous visual landing control technology can avoid the danger of drones facing emergency landings. With the use of optical and other detection equipment, the autonomous sensing capabilities of drones will be greatly improved, and evasion will be implemented before ground controllers are put in danger. An anti-collision algorithm that takes both safety and economy into consideration will automatically re-plan the route after the UAV has implemented collision avoidance maneuvers to continue the task.

This method can be applied not only to the passive landing of drones in complex scenarios or in emergency situations and the active landing of drones, but also to many areas such as the automatic driving of unmanned vehicles, augmented reality, and the autonomous positioning of robots.

The main contributions of this work can be summarized as follows:

- This paper proposes and implements the vision-based drone autonomous landing system in an unstructured environment. By combining existing technologies, they are improved to better meet the requirements of the system proposed herein.
- This paper proposes a novel map representation approach that combines three-dimensional features and a mid-pass filter. Each visible feature is converted into a grid map by utilizing the mid-pass filter to remove noise that is too high and too low in each grid. After constructing a grid map, feature points of different heights can be visualized as grids of different heights.
- This paper recommends a region segmentation to detect the edge of different-height grid areas. It smooths the areas with the same height based on a mean shift algorithm. An edge detector is used to identify obstacles and flat areas. By region segmentation, the speed and accuracy of the subsequent landing area selection are substantially improved.
- Based on a grid map and region segmentation, we present a visual landing technology to explore a suitable landing area for drones in emergencies and unknown environments. Furthermore, with the pose calculated by SLAM, drones can autonomously fulfill path planning and implement landing. To evaluate the proposed algorithm, we apply it in multiple sets of real scenes. Experimental results demonstrate that the proposed method achieves encouraging results.

The remainder of this paper is organized as follows. In Section 2, we propose a UAV autonomous landing approach based on monocular visual SLAM. The experimental results are presented in Section 3. Finally, we conclude the paper in Section 4.

2. The Approach

An overview of the proposed algorithm for the detection of landing sites is shown in Figure 2. When the drone begins the landing procedure, the approach can estimate the position and posture of the drone, build the grid map of the environment, and select the most suitable area for landing via the filtering algorithm. A selecting landing area and vision navigation method is demonstrated, which uses SLAM to estimate the current pose of the drone.

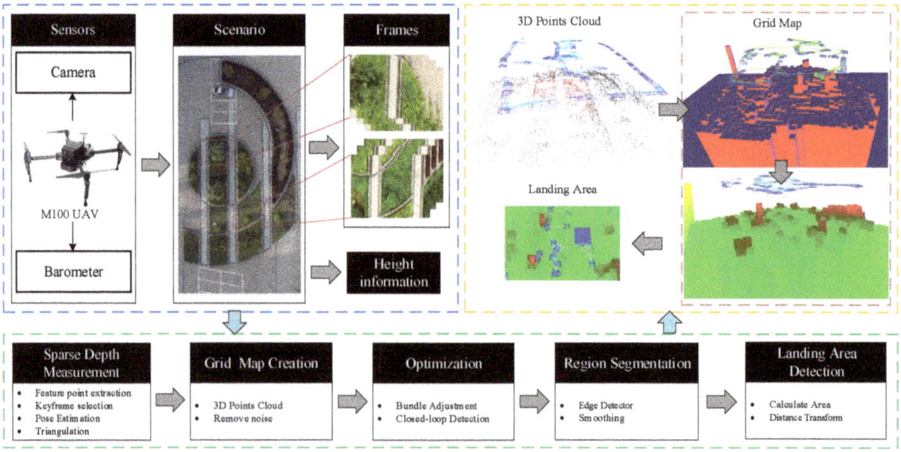

Figure 2. Overview of the main components and connection in the proposed approach.

This paper establishes a three-dimensional point cloud map of the environment by visual SLAM. Then, a two-dimensional grid map is set up by the three-dimensional point cloud of the feature points proposed by the SLAM algorithm. The height of each grid is calculated by projecting the map points of the graph into the corresponding grids. Then, the mean shift-based image segmentation algorithm is used to smooth the height of the grid map, divide the obstacles and ground, and combine the highly similar image blocks together. By calculating the space distance between the landing area and the obstacle, the algorithm selects the region which is the farthest from the obstacle as the filtered landing area. In this way, a suitable area for UAV landing is selected. The UAV finally lands on the safe area by following the descent program.

2.1. Sparse Depth Measurement

The camera pose is forwarded to the on-board computer, which associates the camera pose with the corresponding images based on the pre-calibrated camera. These camera pose estimates are used as priors in the bundle adjustment if an area is marked as a potential landing spot. Additionally, feature tracks are generated by the provided framerate. The ORB (oriented FAST and rotated BRIEF) [22] feature tracker is used to generate coarse depth measurements for region tracking in unknown terrain and in the bundle adjustment of the backend thread. The ORB feature tracker is made up of the oriented FAST (Features from Accelerated Segment Test) [23] corner detector, which detects corners with a description of scale and rotation, and BRIEF (Binary Robust Independent Elementary Features) [24], an efficient feature point descriptor.

Monocular vision SLAM is a feature-based system that can be applied to all scenes in real time. The algorithm divides the algorithm into four modules: tracking, building, relocating, and closed-loop detection. It divides the system into three separate threads, which can track well and build a map. The ORB_SLAM2 [19] algorithm can guarantee the global consistency of trajectory and map through its optimization and closed-loop detection. If the camera returns to the same scene as before, the algorithm can optimize posture and map by conducting closed-loop detection.

If the scene is a plane, or is approximated as a plane, or when the parallax is small, the motion estimation can be done by homography. The motion is restored by the basic matrix F in a non-planar scene with large parallax. Although the camera is facing the ground, the data captured by the drone may involve rugged terrain. The basic matrix F indicates the relationship between any two images of the same scene that constrains where the projection of points from the scene can occur in both images. At the same time, in order to improve the robustness of the system, the basic matrix F and the homography matrix H are estimated at the same time when the real data always contains some noise. The homography matrix H describes the mapping relation between two images of the same planar surface in space. Then, the smaller one is chosen as the motion estimation matrix by comparing the weight projection error. The method in the ORB_SLAM2 algorithm is to calculate the SH value and select the corresponding model according to the RH worth.

(1) Extract reference frame and current frame features p_r, p_c, and then match features between two frames $p_r \leftrightarrow p_c$. If the number of matching features are not enough, the reference frame is reset.

(2) Calculate the homography matrix H: $p_c = Hp_r$, and then calculate the fundamental matrix F: $p_c^T F p_r = 0$, $F = K^{-T}EK^{-1}$, where K is the internal matrix of the camera and E is the essential matrix. The essential matrix E can be seen as a precursor to the fundamental matrix, and its relationship with F is as above. The degrees of freedom of the homography matrix is 8, which can be computed by four pairs of matching features. The fundamental matrix F can be calculated by the classical eight-point-algorithm [25]. It is unavoidable that there is a large number of mismatches in feature matching, and we use random sampling consistency (RANSAC) to solve them.

(3) Restore motion from the fundamental matrix F or the homography matrix H.

We can get the motion T_{12} of the camera by the polar constraint. The depth information of the map points can be estimated by the motion of the camera by triangulation. p_1, p_2 are set to represent the coordinates after the features are normalized on two frames,

$$d_1 p_1 = d_2 T_{12} p_2, \tag{1}$$

where d_1, d_2 represent the depth of two-frame features, p_1, p_2 are the three-dimensional coordinates of the current frame and reference frame features, and T_{12} comprised of rotation matrix R and translation matrix t is the transformation matrix of the first graph to the second graph.

However, when solving the pose of the camera, because of the scale equivalence of the essential matrix E itself, there is also a scale equivalence of the t, R obtained by decomposing E. The normal method is to normalize the scale of t, which leads directly to the scale uncertainty of monocular vision. For t, after it is multiplied by an arbitrary constant, the polar constraint is always established. In order to solve this scale uncertainty, we compared the height information measured by the barometer with the flight altitude calculated by the SLAM system to obtain the scale factor.

The Matrice M100 comes with a barometer module. The barometer is based on the experimental principle of Evangelista Torricelli for measuring atmospheric pressure. Most of the aircraft's altitude measurement is achieved through a barometer, and GPS-equipped aircraft also generally have a barometer as a backup. For every 12 m of height raised, the mercury column is lowered by about 1 millimeter, so the height of the aircraft when it is flying in the air can be measured. In the experimental scenario, the drone's flying height is usually no more than 30 m, so the error of the height measured by the barometer is not very large and satisfies the experimental requirements. The experiment in this article has the feature that the monocular visual SLAM application environment is oriented to the ground. In addition, the camera ZENMUSE X3 (DJI, Shenzhen, China) used in the experiment has a PTZ (Pan/Tilt/Zoom) self-balancing function, which can ensure that the camera maintains its ground-facing state. A PTZ camera is a camera that supports all-around (up, down, left, and right) movement and lens zoom control. Because of the scale problem of monocular vision SLAM and the characteristics of the experiments in this paper, the height information obtained by the barometer measurement is sufficient to restore the SLAM scale factor on the z-axis:

$$\triangle H = \triangle h_v * s, \tag{2}$$

where $\triangle h_v$ represents the height difference in the z-axis measured by monocular visual SLAM, $\triangle H$ represents the height variation in world coordinate system, and s represents the scale factor. The value of s can be calculated by replacing $\triangle H$ with the height variation measured by a barometer. It is thus possible to continue to obtain true height information by scaling the visual pose.

After the success of the map initialization, PNP (Perspective-N-Point) is used to transform the three-dimensional motion into two-dimensional point pairs. Therefore, the position posture of the current frame can be obtained by using a PNP solution to calculate the three-dimensional motion to two-dimensional points by using the three-dimensional map point P in the reference frame and the two-dimensional keypoints p on the current frame.

Given a three-dimensional map point set P and two-dimensional matching on the set of points p, we can calculate the pose by minimizing the re-projection error pose:

$$\xi^* = argmin_{\xi} \frac{1}{2} \sum_{i=1}^{n} \left\| u_i - \frac{1}{s_i} K T P_i \right\|_2^2 . \tag{3}$$

The error is obtained by comparing the pixel coordinates (i.e., the observed projected position) with the position of the three-dimensional point projected according to the currently-estimated pose. This error is called the re-projection error. We minimize the re-projection error of the matching point by constantly optimizing the pose in order to obtain the optimal camera pose.

For each frame, when the map is initialized, the system estimates the position of the current frame in accordance with the previous frame. Hence, with successful tracking, it is relatively easy to get the posture information of each frame.

However, a sparse feature point map does not meet the requirements of the screening landing area. Thus, it is imperative to integrate other methods to optimize the map.

2.2. Grid Map Creation

First of all, as shown in Figure 3, we divide the plane into small grids. The size of the grid can be adjusted according to the actual situation. Then, the SLAM algorithm is applied to calculate the three-dimensional location of feature point in the world coordinate system and the pose of each key frame. Then, we will convert the three-dimensional point clout into a two-dimensional grid map. Furthermore, this article sets the point to be projected into the grid only when it is observed by multiple frames. If it is observed by merely one frame, the point will not be projected into the grid, which can avoid points that are noise. Thus, each grid has a pile of two-dimensional points with the height information. There is one final step needed to determine which grids are suitable for drones to land on based on these points.

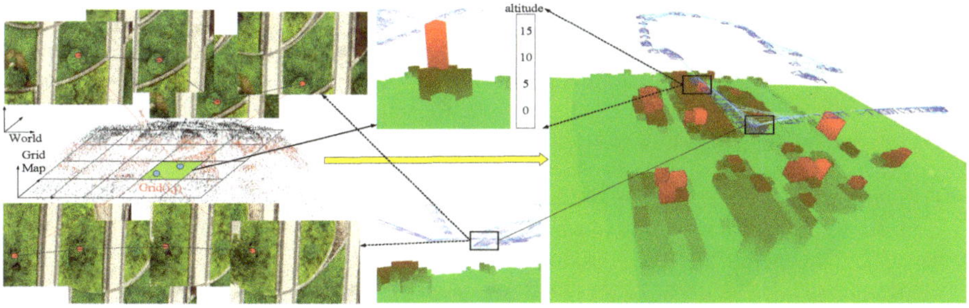

Figure 3. A feature point in real-world coordinates is observed by multiple key frames. Through triangulation, its three-dimensional position is obtained and converted into a three-dimensional point cloud. Then, the three-dimensional point cloud is projected into the two-dimensional grid map, and the height of the grid is obtained by calculating the height of all of the filtered three-dimensional points that fall in each grid. The premise of triangulation is to know the pose of each key frame, as shown in the blue block diagram.

First, we define the height of each grid:

$$h(i,j) = \frac{\sum_{k \epsilon N(i,j)} -h_{min} - h_{max}}{N(i,j) - 2}, \tag{4}$$

where h_k represents the map points in the grid $Grid(i,j)$, and the map points h_{min} and h_{max} represent the maximum and minimum values in the grid. The highest and lowest points in the grid are removed, and then the mean of the map points is computed to assign the value to the grid height $h(i,j)$.

The following formula defines whether the grid is suitable for landing:

$$T(i,j) = \sum_{m,n \epsilon R(i,j,r)} \|h(m,n) - h(i,j)\|^2, \tag{5}$$

where $h(m,n)$ is the height value of the map point (m,n) on the two-dimensional grid map and $h(i,j)$ represents the height of the grid (i,j) in the grid coordinates. r is the radius of the search, and is adjusted according to the size of the UAV. Through traversal of each grid, the drone can search for the landing area. Grids that do not have a projection point are regarded as unreliable and marked as non-landing areas. Finally, the threshold of the formula $T(i,j)$ is set according to the actual application to determine the grid suitable for the UAV landing, and at the same time the grid is marked out.

2.3. Pose and Map Optimization

There will be errors when the camera is calibrated and tracked, so it is necessary to do some optimization after the pose estimation. The estimate of the pose is obtained by tracking frames. By using this estimate as an initial value, we can model the optimization problem as a least-squares graph optimization problem and then use g2o (General Graph Optimization) [26] to optimize poses and maps. Even after optimization, there will be errors, and these tracking errors will continue to accumulate, which may lead to an increasingly growing rear frame pose estimation error, and eventually deviate from reality. Thus, long-term estimates of the results will be unreliable. Considering this, closed-loop testing, which is related to the correctness of estimated trajectory and maps after a long time, is particularly important.

Because the pose of key frames is estimated based on the previous reference frame, the error will be accumulated and result in increasingly inaccurate posture estimates. Therefore, we optimize the position and orientation using a closed-loop detection. When the camera captures the previously captured image, we can correct the position of the camera by detecting the similarity between the images. Closed-loop detection can be achieved through the word pocket model DBow3 [27]. DBoW3 is an open source C++ library for indexing and converting images into a bag-of-word representation. It implements a hierarchical tree for approximating nearest neighbours in the image feature space and creating a visual vocabulary. DBoW3 also implements an image database with inverted and direct files to index images and enabling quick queries and feature comparisons.

1. Feature extraction: select features based on the data set and then describe them to form feature data. For example, the sift key points in the image are detected, and then the feature descriptor is calculated to generate a 128-dimensional feature vector;
2. Learning the word bag: merge all of the processed feature data. Then, the feature words are divided into several classes by means of clustering. We set the number of these classes, and each class is equivalent to a visual word;
3. Use of a visual bag to quantify the image features: each image consists of many visual words. It can use statistical word frequency histograms to indicate which category an image belongs to.

With the dictionary, given any feature, the corresponding word can be found by looking up the dictionary tree layer-by-layer. When the new key frame is inserted, the distribution of the image in the word list or the histogram can be computed. This allows us to use the text search algorithm TF-IDF (term frequency-inverse document frequency) [28] and the approach in [29] to calculate the similarity between the two images. After detecting the closed loop, BA (bundle adjustment) is used to optimize some of the previous reference frames.

2.4. Region Segmentation-Based Landing Area Detection

It is necessary to divide the map according to the height before the screening of the landing area suitable for the UAV. The grid map of precise region segmentation based on height is helpful to improve the speed and accuracy of the subsequent landing area selection. In this paper, a method based on image segmentation to divide the height region of the grid map is proposed. This section introduces the algorithm flow in detail.

According to the experimental requirements, an image segmentation method based on mean shift [30] is used to segment the mesh map. In accordance with image segmentation based on the mean shift principle, the grid map obtained in the second chapter is smoothed and divided. Firstly, the size of the grid map and the height of each grid are input. Each grid is regarded as the smallest unit. Secondly, the mean shift algorithm is used to cluster the height of the grid to determine the total number of categories and the center of each category. Then, using these statistics as input, the final division of the grid map via the mean shift algorithm is obtained. Specific steps are shown in Algorithm 1.

Algorithm 1 Image Segmentation-Based Grid Map Partitioning Algorithm.

Input: grid map
use the mean shift algorithm to smooth the created grid map. For each grid, initialize $j = 1$, $y_{i,1} = x_i$.
while modulus point non-convergence **do**
  calculate $y_{i,j+1}$
  $z_i = (x_i^{(s)}, y_{i,c}^{(r)})$
end while
the grid map is smoothed with mean shift, and the convergence result is stored in z_i, $z_i = y_{i,c}$
for $i = 0, 1, 2 \cdots N$ z_i **do**
  if grid space distance $< h_s$ && height distance $< h_r$ **then**
    divided into different categories $C_{p_{p=1,\cdots m}}$
  end if
end for
Output: for each grid $i = 1, 2, \cdots, n$, the category logo $L_i = p | z_i \in C_p$.

After clustering the grid map, the ground condition without a priori environment information can be obtained. Therefore, the system gains understanding of the height distribution of the ground and the obstacle information to a certain extent, and is able to select an area suitable for the UAV to land. Then, the world coordinates in this district can be output to guide the landing of the UAV. Due to the skewing that may occur during the drone's landing, the UAV landing point needs to avoid obstacles in order to ensure a safe of landing. For the grid map, the algorithm sets the districts of all the altitude, except the landing height H, as obstacles. After districts with matching height and area are selected, it is necessary to calculate the integrated distance between the districts and all the obstacles. Specific steps are shown in Algorithm 2.

Algorithm 2 Choose the Best Landing Spot.

Input: the previous grid height categories C_i, the appropriate landing height H, the appropriate landing area S.
for $i = 0, 1, 2 \cdots N$ grid categories $C_{i_{i=1,\cdots N}}$ **do**
  if the height h of $C_i = H$ && the area $s = S$ **then**
    add grid g_i to the landing zone candidate set and number a_i
  else
    add grid g_i to the obstacle set and number b_j
  end if
end for
for $i = 0, 1, 2 \cdots N$ the landing zone candidate set a_i **do**
  for $j = 0, 1, 2 \cdots N$ the obstacle set b_j **do**
    calculate the distance d_{ij} of each area a_i from each obstacle b_j; that is, the distance from the candidate area to the nearest edge of the obstacle area.
    calculate the overall distance $d_i = d_i + d_{ij}$ of each area a_i from all obstacles.
  end for
end for
Output: The area a_i with the largest d_i is the landing point, which makes it possible to stay away from existing obstacles.

The appropriate landing height H is selected from the previous grid height categories and the appropriate landing area S is set according to the size of the UAV. Then, the system can determine the best landing location through these two screening approaches.

3. Experiments

3.1. Experimental Platform

This paper chose the commercial UAV DJI Matrice 100 (M100) as a platform for the data acquisition offline processing stage and the UAV autonomous real-time control stage experiment. It includes a flight controller, power system, barometer module, GPS module, and other modules. This paper used the monocular camera, ZENMUSE X3, as a visual sensor to carry out the experimental data collection. The small ZENMUSE X3 monocular camera can guarantee high-quality video during high-speed movement with its wide-angle fixed focus lens, powerful performance, a shooting screen without distortion, and clear images. At the same time, we used the camera's PTZ self-balancing function to make the camera face the ground. We also used the barometer module on the UAV to measure the flight altitude for the monocular SLAM scale correction. In this paper, the image data resolution was 640×480. The experimental configuration environment was an Intel Core i7-8700K CPU@3.70 GHz processor, 16.0 GB memory, and 64-bit operating system.

If the camera used does not have a PTZ self-balancing function, the height difference on the z-axis cannot be used directly when restoring the scale factor. In the calculation of the scale factor stage, the aircraft only moves in height, and the difference in the three-dimensional space of the pose in this time period is calculated to replace the difference in the z-axis. In addition, during the initial phase of the aircraft, it is necessary to ensure that the camera is parallel to the ground. After this, the PTZ self-balancing function no longer affects system functionality.

The experimental platform interacted with the M100 through the image acquisition card and the wireless serial port to simulate the on-board processing, as shown in Figure 4. The laptop captured the image stream taken by the drone in real-time through the image capture card, and sent the control command to the drone in real-time through the wireless serial port.

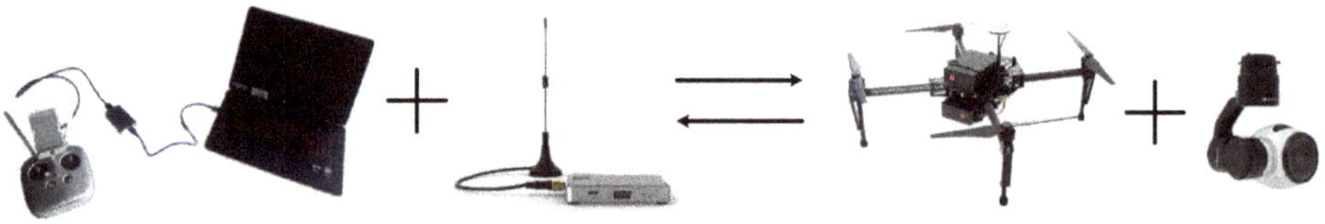

(a) (b) (c) (d)

Figure 4. The image (**a**) is the experimental configuration environment. One end of the image acquisition card is connected to the computer, and the other end is connected to the remote control of the drone. The image capture card transmits the image to the computer via a remote control; The image (**b**) shows the wireless serial port module; The image (**c**) shows the M100 drone; and the image (**d**) shows the ZENMUSE X3 camera.

3.2. Real-Time Control Experiments

In Section 2, we can get the key frame data of monocular image sequence processing, and obtain the UAV pose information and the three-dimensional point cloud data according to the visual SLAM system. Then, we convert the three-dimensional point cloud into a two-dimensional grid. Next, the image segmentation algorithm based on mean shift is used to deal with the two-dimensional grid map. Finally, the two-dimensional grid map is filtered and the appropriate landing area is obtained.

At the beginning of the experiment, we first turned on the computer and started the program. The drone flew over the scene and began to simulate entering the fault state. It was necessary to start the autonomous landing procedure. The startup program controlled the drone to scan the landing environment. When the UAV started the data acquisition, the first step is to initialize. At the initial location, the aircraft pose must have a translation instead of a rotation change. In order to reduce the error of the pose information and the 3D point cloud data estimated by the SLAM system, after completing the initialization, the UAV started the closed-loop flight, followed by some closed loops of smaller radius. By forming a closed loop, the construction and pose estimation results are optimized. The candidate landing area is selected from within the construction result. The drone moved to the top of the candidate area with the shortest path, and focused on the candidate landing area (i.e., movement with small displacement). Finally, if the program still confirmed that the area was the final landing area, the landing mode was started. Otherwise, the candidate area was re-determined and scanned.

Figure 5 shows the creation of a two-dimensional height grid map and the specific meaning of each part of the grid map. The color represents the depth of the map. Green represents the lowest point, red represents the highest point, and the middle height is represented by a gradient color. Dark blue indicates a suitable landing site. Light blue indicates the flight path of the drone. The experimental scene is a circular flower bed surrounded by large semi-circular flower beds. There are three cars parked in front of the flower bed. There are big trees, shrubs and weeds on the circular flower beds and the semicircular flower beds on the periphery. The other half of the flower bed is an empty square which is suitable for landing.

In order to prove the accuracy of the proposed UAV autonomous landing system, we carried out a real-time control experiment of UAV autonomous landing. The landing trajectory and the specific process are shown in Figure 6. It presents our study of the UAV autonomous landing area screening and the entire implementation process of the UAV autonomous landing system. When the UAV was ready to land, flight control switched from normal flight to autonomous landing. The autonomous landing system was started. Then, the system was initialized and the flight began. After the initialization was completed, the pose was estimated, and the closed-loop flight was carried out. Then, the area to be measured was screened. After the selection of the UAV landing area, it began to fall on the selected area.

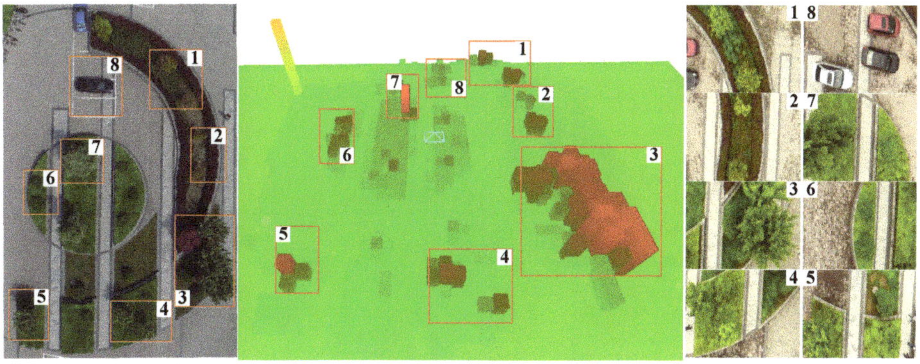

(a) Scenario overview (b) Grid map (c) Part of the scenario

Figure 5. *Cont.*

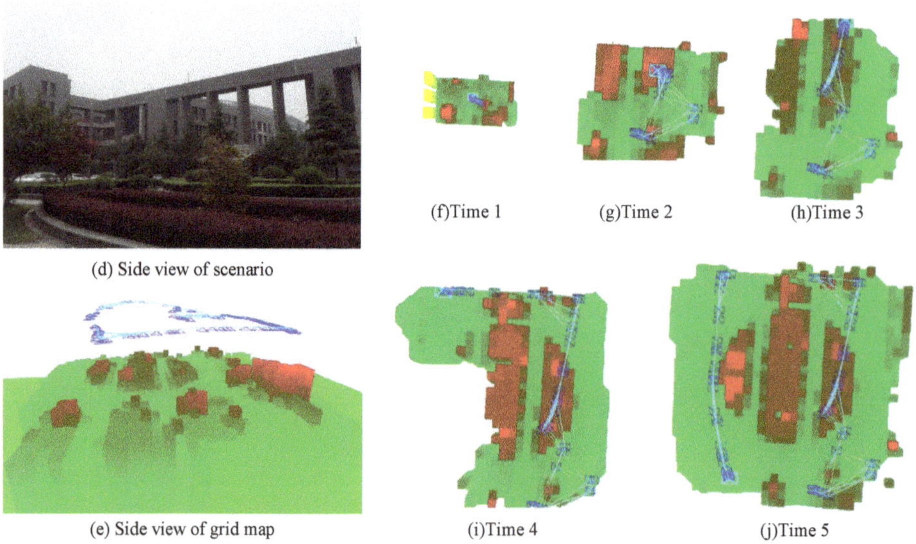

Figure 5. The images of (**a–c**) detail the correspondence between each part of the two-dimensional height map and the real scene; the image (**d**) is a real scene with different angles; The image (**e**) is the side view of the grid map; the images of (**f–j**) are the construction process of the grid map. The blue box above the map shows the trajectory of the drone and the pose of the drone when the key frame is generated.

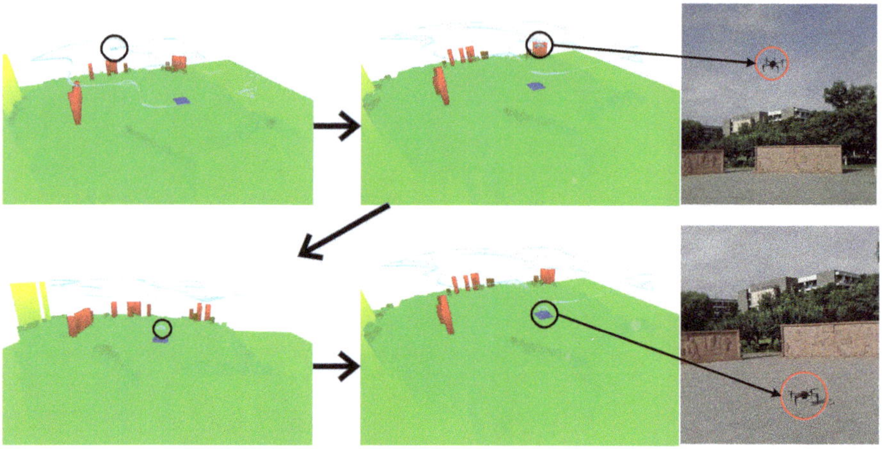

Figure 6. The landing process of a quadrotor and the real-world scene of the process. The color represents the depth of the map. Green represents the lowest point, red represents the highest point, and the middle height is represented by a gradient color. Dark blue indicates a suitable landing site. Light blue indicates the flight path of the aircraft. The black circle refers to the current position of the drone.

3.3. Landing Area Detection in Multi-Scenario

In this paper, several groups of data experiments and analysis were conducted. Five sets among the experimental results are shown in Figure 7. All experimental data were collected from the Northwestern Polytechnical University Chang'an campus. The first scene was a small forest with a suitable landing zone in the middle. The scene for the second set of data acquisition was a gentle slope. Surrounded by trees, a larger humanoid sculpture was located in the middle of the slope. The front part of the slope was relatively flat and suitable for landing. Several trees and shrubs were scattered in the third scene. The middle of the scene was flat and suitable for landing. The scene of the fourth set of data collection was a square with five highly-visible obstacles evenly distributed within it. Three of the obstacles were carved walls, while the others were long stone benches. The middle area of the square was flat and suitable for landing, without any obstacles.

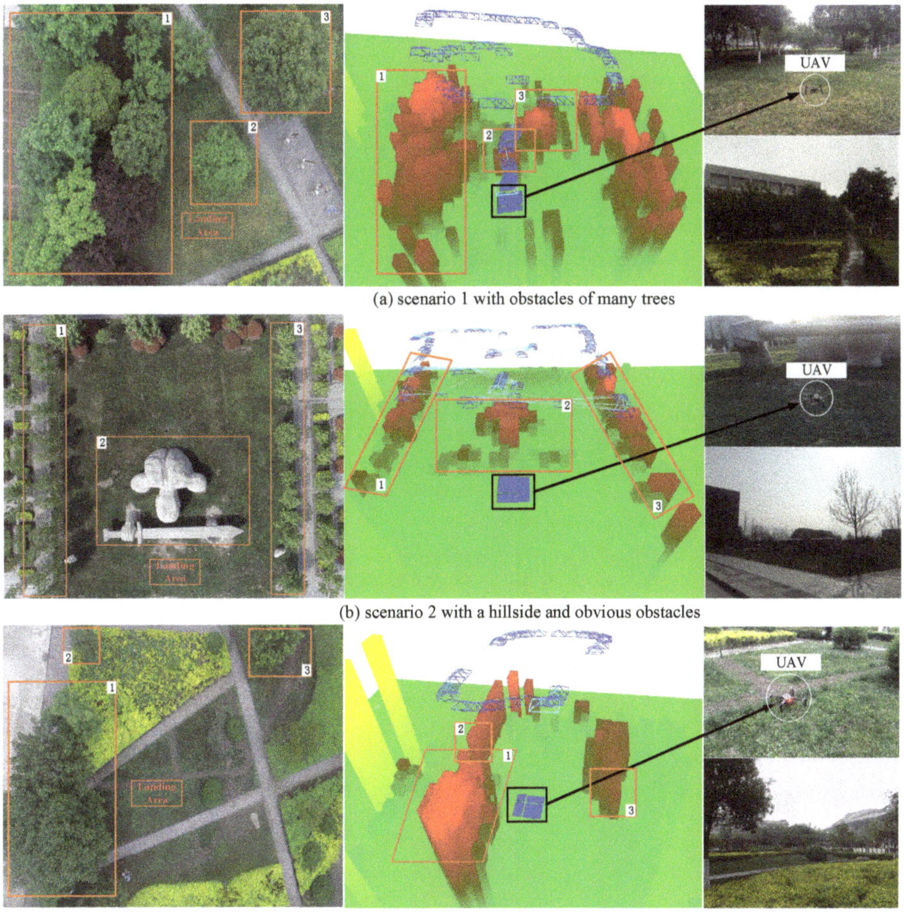

(a) scenario 1 with obstacles of many trees

(b) scenario 2 with a hillside and obvious obstacles

(c) scenario 3 with obstacles of trees and shrubs

Figure 7. *Cont.*

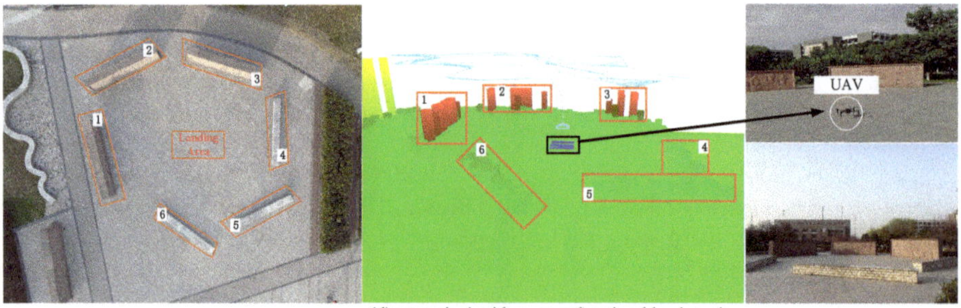

(d) scenario 4 with even and noticeable obstacles

Figure 7. Two-dimensional grid height map creation and landing site selection result in five different scenarios. The left column is the real scene, the middle is a two-dimensional grid height map, and the dark blue area is a suitable landing spot. The blue trajectory above the map is the drone flight trajectory, and the right column is the real scene where the drone landed accurately at the landing site.

Figure 7 shows the specific experimental locations and experimental results of the proposed method. The robustness of the proposed system is demonstrated by experimental results in different conditions. The characteristics of the four experimental sites in Figure 7 are different and they simulate different actual application environments.

Scenarios 1, 2, and 3 were experiments simulating field environments. Scenario 1 simulated a forest landscape, with a large areas of trees. There was only a small piece of land suitable for landing in the middle, which our method chose. Scenario 2 simulated wild hillside and boulders. The system chose the flat ground between the hillside and the flat land and avoided obstacles well. Scenario 3 simulated a farm field. Although the shrub area was relatively flat, our method chose a flatter grassland. Scenario 4 simulated the urban environment with regular tall buildings and flat squares. There were even and obvious obstacles distributed in Scenario 4, and the experimental results can be intuitively understood. The landing site selected by our algorithm had the largest integrated distance and was located in the center of the five obstacles.

Take scenario 1 as an example. The drone has a flying height of 20 m. The entire landing process took 1 min and 52 s. Firstly, the drone scans the scene and builds a map. However, due to the high flying height, high scene complexity and the existence of many empty areas (area with large height differences but small areas), the sparse point cloud map constructed by monocular SLAM can not meet the requirements for detecting the landing area of drones. At this point, the grid map proposed in this paper shows its practicality. Based on the three-dimensional point cloud map, the grid map expands the space, fills in the area around the feature point, and realizes the perception of the overall environment. Only in the presence of a coherent flat area, the drone can land. All the revised parts are shown by blue font in the revision.

Experiments showed that, although the three-dimensional point cloud of visual SLAM was estimated to be sparse, the two-dimensional height grid map reconstructed supplied excellent scene information. The map was accurate enough to meet the needs of UAV landing. In various simulation environments and simulation field experiments, the landing sites selected by the proposed method were the safest places in the scenes, and the drone landed accurately in these areas.

For more insights, we invite the reviewer to take a look at the multimedia demonstration UAV_AutoLanding_Demo.mp4 of our system in the Supplementary Materials. The demo is included in the submitted zip file, or you can click here directly (https://page0607.github.io/UAV_Landing/).

4. Conclusions

This paper proposes a novel UAV autonomous landing approach based on monocular visual SLAM. In the proposed approach, we exploit a feature-based method to estimate the drone's pose and attain the fixed point in three-dimensional space. Regarding grid map construction, we first establish an image stream with a corresponding time stamp and barometric altitude information. Then, through the extraction of robust feature points and descriptors, the motion of the drone, the flight path, and the visible three-dimensional point cloud map are obtained by matching and tracking the features. A visible grid map can be built by putting the three-dimensional point cloud to the grid map and calculating the height of each grid after removing noise. After that, a method based on image segmentation is used to divide the height region of the grid map. On the basis of the divided grid map, the system can obtain the appropriate landing place after decision optimization. Finally, the drone takes the shortest path to reach the destination and starts the descent mode near the landing area. After finishing this, the UAV autonomous landing system is successfully executed. Extensive experimental results on multiple real scenes confirmed that the landing area selection and navigation based on visual technology is effective and efficient. In addition, it can be used to find a suitable path in the field of automated driving where unmanned vehicles and UAVs are used in combination. The drone circles in the air to detect the environment, and then a three-dimensional map is constructed to select the appropriate driving path for the unmanned vehicle. Furthermore, the current work shows great potential in various visual domains. In future work, we will consider extending our work to multiple fields and develop it to the general level.

Supplementary Materials: The following are available online at VideoS1:UAV_AutoLanding_Demo.mp4.

Author Contributions: T.Y. and P.L. conceived and designed the algorithm and wrote the paper; H.Z. performed the experiments; J.L. and Z.L. analyzed the data.

Acknowledgments: This work was supported by the National Natural Science Foundation of China under Grant 61672429 and the ShenZhen Science and Technology Foundation under Grant JCYJ20160229172932237.

Conflicts of Interest: The authors declare no conflict of interest.

References

1. Jung, Y.; Cho, S.; Shim, D.H. A trajectory-tracking controller design using L1 adaptive control for multi-rotor UAVs. In Proceedings of the 2015 International Conference on Unmanned Aircraft Systems (ICUAS), Denver, CO, USA, 9–12 June 2015; pp. 132–138.
2. Lee, H.; Jung, S.; Shim, D.H. Vision-based UAV landing on the moving vehicle. In Proceedings of the 2016 International Conference on Unmanned Aircraft Systems (ICUAS), Arlington, VA, USA, 7–10 June 2016; IEEE: Piscataway, NJ, USA, 2016; pp. 1–7.
3. Falanga, D.; Zanchettin, A.; Simovic, A.; Delmerico, J.; Scaramuzza, D. Vision-based Autonomous Quadrotor Landing on a Moving Platform. In Proceedings of the 2017 IEEE International Symposium on Safety, Security and Rescue Robotics (SSRR), Shanghai, China, 11–13 October 2017.
4. Kim, J.; Jung, Y.; Lee, D.; Shim, D.H. Outdoor autonomous landing on a moving platform for quadrotors using an omnidirectional camera. In Proceedings of the 2014 International Conference on Unmanned Aircraft Systems (ICUAS), Orlando, FL, USA, 27–30 May 2014; IEEE: Piscataway, NJ, USA, 2014; pp. 1243–1252.
5. Jung, Y.; Lee, D.; Bang, H. Close-range vision navigation and guidance for rotary UAV autonomous landing. In Proceedings of the 2015 IEEE International Conference on Automation Science and Engineering (CASE), Gothenburg, Sweden, 24–28 August 2015; IEEE: Piscataway, NJ, USA, 2015; pp. 342–347.
6. Vlantis, P.; Marantos, P.; Bechlioulis, C.P.; Kyriakopoulos, K.J. Quadrotor landing on an inclined platform of a moving ground vehicle. In Proceedings of the 2015 IEEE International Conference on Robotics and Automation (ICRA), Seattle, WA, USA, 26–30 May 2015; pp. 2202–2207.
7. Kim, J.W.; Jung, Y.D.; Lee, D.S.; Shim, D.H. Landing Control on a Mobile Platform for Multi-copters using an Omnidirectional Image Sensor. *J. Intell. Robot. Syst.* **2016**, *84*, 1–13. [CrossRef]

8. Forster, C.; Faessler, M.; Fontana, F.; Werlberger, M. Continuous on-board monocular-vision-based elevation mapping applied to autonomous landing of micro aerial vehicles. In Proceedings of the 2015 IEEE International Conference on Robotics and Automation (ICRA), Seattle, WA, USA, 26–30 May 2015; pp. 111–118.
9. Yang, T.; Li, G.; Li, J.; Zhang, Y.; Zhang, X.; Zhang, Z.; Li, Z. A ground-based near infrared camera array system for uav auto-landing in GPS-denied environment. *Sensors* **2016**, *16*, 1393. [CrossRef] [PubMed]
10. Saripalli, S.; Montgomery, J.F.; Sukhatme, G. Vision-based autonomous landing of an unmanned aerial vehicle. In Proceedings of the IEEE International Conference on Robotics and Automation (ICRA '02), Washington, DC, USA, 11–15 May 2002; pp. 2799–2804.
11. Richardson, T.S.; Jones, C.G.; Likhoded, A.; Sparks, E.; Jordan, A.; Cowling, I.; Willcox, S. Automated Vision-based Recovery of a Rotary Wing Unmanned Aerial Vehicle onto a Moving Platform. *J. Field Robot.* **2013**, *30*, 667–684. [CrossRef]
12. Muskardin, T.; Balmer, G.; Wlach, S.; Kondak, K.; Laiacker, M.; Ollero, A. Landing of a fixed-wing UAV on a mobile ground vehicle. In Proceedings of the 2016 IEEE International Conference on Robotics and Automation (ICRA), Stockholm, Sweden, 16–21 May 2016; pp. 1237–1242.
13. Lee, D.; Ryan, T.; Kim, H.J. Autonomous landing of a VTOL UAV on a moving platform using image-based visual servoing. In Proceedings of the 2012 IEEE International Conference on Robotics and Automation (ICRA), Saint Paul, MN, USA, 14–18 May 2012; pp. 971–976.
14. Ghamry, K.A.; Dong, Y.; Kamel, M.A.; Zhang, Y. Real-time autonomous take-off, tracking and landing of UAV on a moving UGV platform. In Proceedings of the 2016 24th Mediterranean Conference on Control and Automation (MED), Athens, Greece, 21–24 June 2016.
15. Davison, A.J.; Reid, I.D.; Molton, N.D.; Stasse, O. MonoSLAM: Real-Time Single Camera SLAM. *IEEE Trans. Pattern Anal. Mach. Intell.* **2007**, *29*, 10–52. [CrossRef] [PubMed]
16. Kameda, Y. Parallel Tracking and Mapping for Small AR Workspaces (PTAM) Augmented Reality. *J. Inst. Image Inf. Telev. Eng.* **2012**, *66*, 45–51. [CrossRef]
17. Forster, C.; Pizzoli, M.; Scaramuzza, D. SVO: Fast semi-direct monocular visual odometry. In Proceedings of the 2014 IEEE International Conference on Robotics and Automation (ICRA), Hong Kong, China, 31 May–7 June 2014; pp. 15–22.
18. Engel, J.; Schops, T.; Cremers, D. LSD-SLAM: Large-Scale Direct Monocular SLAM. In Proceedings of the European Conference on Computer Vision, Munich, Germany, 8–14 September 2014; pp. 834–849.
19. Mur-Artal, R.; Tardos, J.D. ORB-SLAM2: An Open-Source SLAM System for Monocular, Stereo, and RGB-D Cameras. *IEEE Trans. Robot.* **2016**, *33*, 1255–1262. [CrossRef]
20. Weikersdorfer, D.; Adrian, D.B.; Cremers, D.; Conradt, J. Event-based 3D SLAM with a depth-augmented dynamic vision sensor. In Proceedings of the 2014 IEEE International Conference on Robotics and Automation (ICRA), Hong Kong, China, 31 May–7 June 2014; pp. 359–364.
21. Rebecq, H.; Horstschaefer, T.; Gallego, G.; Scaramuzza, D. EVO: A Geometric Approach to Event-Based 6-DOF Parallel Tracking and Mapping in Real Time. *IEEE Robot. Autom. Lett.* **2017**, *2*, 593–600. [CrossRef]
22. Rublee, E.; Rabaud, V.; Konolige, K.; Bradski, G. ORB: An efficient alternative to SIFT or SURF. In Proceedings of the 2011 IEEE International Conference on Computer Vision (ICCV), Barcelona, Spain, 6–13 November 2011; pp. 2564–2571.
23. Rosten, E.; Drummond, T. Fusing points and lines for high performance tracking. In Proceedings of the Tenth IEEE International Conference on Computer Vision (ICCV 2005), Beijing, China, 17–21 October 2005; Volume 2, pp. 1508–1515.
24. Calonder, M.; Lepetit, V.; Strecha, C.; Fua, P. BRIEF: Binary robust independent elementary features. In Proceedings of the 11th European Conference on Computer Vision, Crete, Greece, 5–11 September 2010; pp. 778–792.
25. Hartley, R.I. In defense of the eight-point algorithm. *IEEE Trans. Pattern Anal. Mach. Intell.* **1997**, *19*, 580–593. [CrossRef]
26. Kummerle, R.; Grisetti, G.; Strasdat, H.; Konolige, K.; Burgard, W. G^2o: A general framework for graph optimization. In Proceedings of the IEEE International Conference on Robotics and Automation (ICRA), Shanghai, China, 9–13 May 2011; pp. 3607–3613.
27. Dorian, G.L.; Tardos, J.D. Bags of Binary Words for Fast Place Recognition in Image Sequences. *IEEE Trans. Robot.* **2012**, *28*, 1188–1197.

28. Robertson, S. Understanding inverse document frequency: On theoretical arguments for IDF. *J. Doc.* **2004**, *60*, 503–520. [CrossRef]
29. Nister, D.; Stewenius, H. Scalable recognition with a vocabulary tree. In Proceedings of the IEEE Computer Society Conference on Computer Vision and Pattern Recognition, New York, NY, USA, 17–22 June 2006; Volume 2, pp. 2161–2168.
30. Comaniciu, D.; Meer, P. Mean shift: A robust approach toward feature space analysis. *IEEE Trans. Pattern Anal. Mach. Intell.* **2002**, *24*, 603–619. [CrossRef]

electronics

MDPI

Article

Bio-Inspired Autonomous Visual Vertical and Horizontal Control of a Quadrotor Unmanned Aerial Vehicle

Saul Armendariz *, Victor Becerra and Nils Bausch

School of Energy and Electronic Engineering, University of Portsmouth, Portsmouth PO1 3DJ, UK; victor.becerra@port.ac.uk (V.B.); nils.bausch@port.ac.uk (N.B.)

* Correspondence: saul.armendarizpuente@myport.ac.uk

Received: 29 December 2018; Accepted: 1 February 2019; Published: 5 February 2019

Abstract: Near-ground manoeuvres, such as landing, are key elements in unmanned aerial vehicle navigation. Traditionally, these manoeuvres have been done using external reference frames to measure or estimate the velocity and the height of the vehicle. Complex near-ground manoeuvres are performed by flying animals with ease. These animals perform these complex manoeuvres by exclusively using the information from their vision and vestibular system. In this paper, we use the Tau theory, a visual strategy that, is believed, is used by many animals to approach objects, as a solution for relative ground distance control for unmanned vehicles. In this paper, it is shown how this approach can be used to perform near-ground manoeuvres in a vertical and horizontal manner on a moving target without the knowledge of height and velocity of either the vehicle or the target. The proposed system is tested with simulations. Here, it is shown that, using the proposed methods, the vehicle is able to perform landing on a moving target, and also they enable the user to choose the dynamic characteristics of the approach.

Keywords: UAV; bio-inspiration; autonomous control; horizontal control; vertical control

---

1. Introduction

Unmanned Aerial Vehicle (UAV) usage and applications, specially those performed by Micro Aerial Vehicles (MAV), has increased. Now, more than ever they are being used in tasks such as inspection, surveillance, reconnaissance, and search and rescue [1]. This increased use demands for better navigation strategies to tackle more challenging approaches. To successfully accomplish this, UAV technologies need to be further advanced.

Navigation in unmanned vehicles is commonly performed using an external reference frames, such as global positioning systems and other sensors. This reliance on external reference frames severely hinders their autonomy. Constant changes in the mission context make it difficult for an autonomous vehicle to adapt to its changing environment. Near ground manoeuvrers are vital to complete any flight mission successfully. Accurate velocity control of the vehicle at touchdown is critical. A combination of positioning systems, range finding sensors and image sensors have been popular tools in navigation strategies to accomplish autonomous landing [2].

Biologically inspired controllers in robots, unlike traditional controllers, emulate animals to achieve complex tasks. Flying animals control mechanisms have been optimized through millions of years of natural evolution, allowing them to navigate complex environments with ease, without relying on any external reference frame.

Tau theory, as the base of a bio-inspired controller, has been used in [3] to generate trajectories during UAV perching using information from external reference sensors, such as Global Positioning System (GPS). Landing on a moving platform without knowledge of the vehicle's height or velocity has been achieved in [4] where the previously known size of the landing platform is used to estimate the position of the quadrotor body frame and generate an adequate landing trajectory.

The key contribution of this paper is a novel bio-inspired vertical and horizontal control system on-board the UAV to achieve near-ground manoeuvres on a moving target. This paper is organized as follows: The basics of Tau theory and its variants are described in Section 2. A body-centric control model is presented in Section 3 that it is complemented with a high-level control system described in Section 4. The estimation of visual motion is described in Section 5, followed by the objective tracking description in Section 6. Finally we perform simulations in Section 7, that we discuss in Section 8 and provide conclusions in Section 9.

2. Tau Theory

2.1. Flying Navigation Strategies in Nature

Flying insects have captured the attention of visual navigation researchers due to their ability to navigate complex and changing environments. Their large eyes with wide Field-of-View (FoV) suggest that they use optic flow to regulate motor actions. Flying bees, despite having two eyes, are not believed to use any depth perception information as their eyes separation does not allow them to capture this information [5]. This means that bees navigate using exclusively the optic flow patterns generated from their own motion. In [6], it was proposed that bees use a measure of image angular velocity ω_z, named the ventral flow, given by:

$$\omega_z = \frac{v_z(t)}{z(t)} \tag{1}$$

where $v_z(t)$ is the velocity and $z(t)$ is the distance to the objective at a given moment in time. When performing landing, it has been found that bees always land with a zero horizontal velocity at touch down [7]. This is achieved without knowledge of height or forward velocity, rather using their ratio, which is the image angular velocity in the vertical direction. While it descends towards the objective, the ventral flow increases due to the decrease in height. By holding constant the ventral flow while performing landing ($\omega_z = C$), both the velocity and the height decease, until zero forward velocity is achieved at touch down. This has been named as the constant ventral flow strategy [6].

2.2. Biological Evidence of Tau Theory

When flying animals approach an object to land, capture or perch, as if they use predictive timing information linked to visual cues of their surrounding to guide and adjust their actions. Time-to-contact (TTC), sometimes refereed to as time-to-collision, is defined as the remaining time before an anticipated contact between the approaching animal and the target. Based on the TTC, Lee introduced Tau theory [7]. He proposed that the variable Tau could be used to represent the TTC in the animals' visual systems. It is defined as the inverse of the target's relative rate of expansion on the animal's retina. In addition, Lee also proposed a general Tau theory, which states that the information from Tau is used in the guidance of the general movements of animals, not only on their perceptual mechanisms. This theory has been verified mathematically and experimentally, inspiring robotics researchers to apply Tau theory. In this project we use Tau theory to perform near-ground manoeuvrers in a MAV.

Lee proposed that the animal movement is goal-directed. If a motion gap is defined as the difference between the animal's current motion state and its target state, then all the intended control actions are made for the purpose of closing the motion gap. If an object is at a distance $z > 0$ along some axis, then the Time-to-Contact to the object is defined as

$$TTC(t) = -\frac{z(t)}{\dot{z}(t)} \tag{2}$$

This can only be true when $\dot{z} \neq 0$. As the subject moves towards the target, the retinal image of the object in the subject's eyes will dilate and the features of the target inside the subject's retina will move radially. This image dilation is caused by the reduction of the relative distance between the subject and the target. It has been demonstrated [8] that the time-to-contact is the reciprocal of the image dilation and can be registered optically from the targets's image features in the subject's retina, such that:

$$TTC(t) = -\frac{\Phi(t)}{\dot{\Phi}(t)} \tag{3}$$

where $\Phi(rad)$ is the angle in the object's retinal image. This shows that the time-to-contact can be registered optically without knowledge of the distance to the object or the relative velocity. The Time-to-Contact and Tau (τ) are connected as follows:

$$\tau = \frac{z(t)}{\dot{z}(t)} = -TTC(t) \tag{4}$$

2.3. Basic Tau Strategies

Assuming that the UAV has arrived at the desired location for landing and it is ready to descend, with Tau it is possible to initiate a descending trajectory, starting from the initial location at non-zero speed and ending right upon the target with a zero speed for a no impact landing. The only information needed to control an on-going descent action is the time rate of tau. It has been observed that animals tend to keep the time rate of tau constant as they close the gap towards their target [7].

$$\dot{\tau}(t) = k \tag{5}$$

where k is a constant. Integrating the previous equation we obtain

$$\tau(t) = kt + \tau_0 \tag{6}$$

where τ_0 is the initial constant value, which is:

$$\tau_0 = x_0/\dot{x}_0 < 0 \tag{7}$$

where x_0 and $\dot{x}_0$ are the initial position and velocity of the vehicle, respectively. Substituting, we obtain:

$$x(t)/\dot{x}(t) = kt + \tau_0 \tag{8}$$

solving for $x(t)$, $\dot{x}(t)$ and $\ddot{x}(t)$ we obtain:

$$\begin{aligned} x(t) &= x_0(1 + kt\dot{x}_0/x_0)^{\frac{1}{k}} \\ \dot{x}(t) &= \dot{x}_0(1 + kt\dot{x}_0/x_0)^{\frac{1-k}{k}} \\ \ddot{x}(t) &= \frac{\dot{x}_0^2}{x_0}(1-k)\dot{x}_0(1 + kt\dot{x}_0/x_0)^{\frac{1-2k}{k}} \end{aligned} \tag{9}$$

To visualize the effects of k independently from initial conditions, namely position, velocity and acceleration, each of the equations are normalized and the results are displayed in Table 1.

Table 1. Motion with different constant k values.

k	t	x	$\dot{x}$	$\ddot{x}$	**Final Goal**
$k < 0$	$\to t_d$	$\to \infty$	$\to \infty$	$\to \infty$	Gap not closed
$k = 0$	$\to t_d$	$= x_0$	$= \dot{x}_0$	$= 0$	Gap not closed
$0 < k < 0.5$	$\to t_d$	$\to 0$	$\to 0$	$\to 0$	Zero Touchdown
$k = 0.5$	$\to t_d$	$\to 0$	$\to 0$	$= C$	Slight Collision
$0.5 < k < 1$	$\to t_d$	$\to 0$	$\to 0$	$\to \infty$	Slight Collision
$k = 1$	$\to t_d$	$\to 0$	$= C$	$\to \infty$	Collision
$k > 1$	$\to t_d$	$\to 0$	$\to \infty$	$\to \infty$	Strong Collision

Table 1 and Figure 1 show the values of x, $\dot{x}$ and $\ddot{x}$ with different k values. We can see that only the case with $0.5 \leq k < 1$ achieves a slight collision.

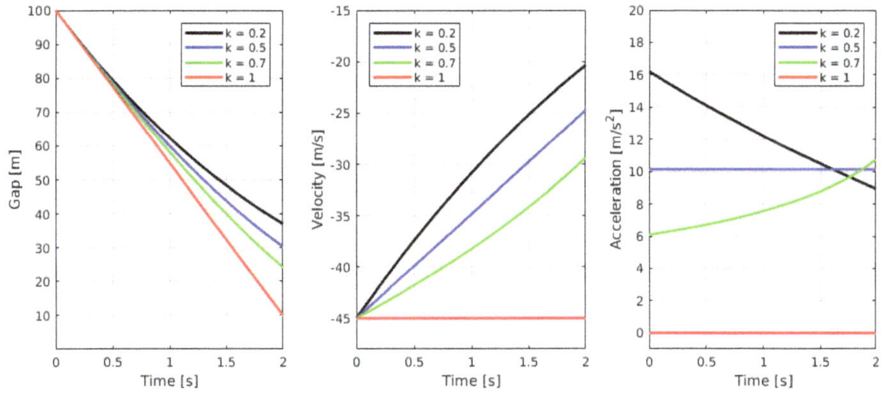

Figure 1. Values of x, $\dot{x}$ and $\ddot{x}$ with different values of k ($k = 0.2, 0.5, 0.7, 1.0$).

2.4. Tau Coupling

In a more realistic scenario, multiple gaps exist when approaching an objective and they all need to be closed simultaneously. Tau coupling [9] can be used for such situations. For example, if we need to close two translational gaps, $\alpha(t)$ and $\beta(t)$, the two corresponding tau variables will be linked by a constant ratio of $k_{\alpha\beta}$ during the course of the approach.

$$\tau_\beta = k_{\alpha\beta}\tau_\alpha \tag{10}$$

Taking this into consideration, we can rewrite Equation (9):

$$\begin{aligned} \beta &= C\alpha^{1/k_{\alpha\beta}} \\ \dot{\beta} &= \frac{C}{k_{\alpha\beta}}\alpha^{\frac{1}{k_{\alpha\beta}}-1}\dot{\alpha} \\ \ddot{\beta} &= \frac{C}{k_{\alpha\beta}}\alpha^{\frac{1}{k_{\alpha\beta}}-2}\left(\frac{1-k_{\alpha\beta}}{k_{\alpha\beta}}\dot{\alpha}^2 + \alpha\ddot{\alpha}\right) \end{aligned} \tag{11}$$

where the constant C is defined as $C = \beta_0/{\alpha_0}^{1/k_{\alpha\beta}}$. Similarly to the previous case, we can find the motion caused by different values of $k_{\alpha\beta}$.

These results indicate that when $0 < k_{\alpha\beta} \leq 0.5$ or $k_{\alpha\beta} = 1$, the distance, velocity and acceleration of the gap $\beta(t)$ will become zero in parallel to the closure of gap $\alpha(t)$, as seen in Table 2. Just as in the previous case, the gap closure can be modified with constant $k_{\alpha\beta}$ to perform different strategies, such as: landing with zero velocity at touchdown, never closing the gap or achieving an aggressive gap closure.

Table 2. Motion with different constant $k_{\alpha\beta}$ values in coupling movement.

$k_{\alpha\beta}$	t	α	β	$\dot{\alpha}$	$\ddot{\beta}$	**Final Goal**
$k_{\alpha\beta} < 0$	$\to t_d$	$\to 0$	$\to \infty$	$\to \infty$	$\to \infty$	Gap y not closed
$k_{\alpha\beta} = 0$	$\to t_d$	$\to 0$	$= 0$	?	?	Error
$0 < k_{\alpha\beta} < 0.5$	$\to t_d$	$\to 0$	$\to 0$	$\to 0$	$\to 0$	Zero Touchdown
$0.5 \leq k_{\alpha\beta} < 1$	$\to t_d$	$\to 0$	$\to 0$	$\to 0$	$\to \infty$	Slight Collision
$k_{\alpha\beta} = 1$	$\to t_d$	$\to 0$	$\to 0$	$\to 0$	$\to 0$	Collision
$k_{\alpha\beta} > 1$	$\to t_d$	$\to 0$	$\to 0$	$\to \infty$	$\to \infty$	Strong Collision

2.5. Gravity Guidance Strategy

Previous examples had the disadvantages of requiring a downward velocity in order to be usable for landing. This can be achieved easily when the vehicle is in motion and the near-ground manoeuvre is initialized, but it will not initialize if the vehicle starts with a zero downward velocity. To solve this problem a method called "intrinsic Tau gravity guidance" was developed [7]. This is a special instance of Tau coupling where the $\alpha(t)$ gap is guided by the gravity's constant vertical acceleration. This manoeuvre can be expressed as:

$$\tau_\alpha(t) = k_{\alpha g}\tau_g(t) \tag{12}$$

where the constant $k_{\alpha g}$ will determine the movement characteristics, and $\tau_g(t)$ specifies the time of the gap to be closed with gravity's constant acceleration. The gap $x_g(t)$ makes use of $\tau_g(t)$, which can be derived from the free-fall equations under gravitational acceleration:

$$\begin{aligned} x_g(t) &= \frac{1}{2}gt_d^2 - \frac{1}{2}gt^2 \\ \dot{x}_g(t) &= -gt \end{aligned} \tag{13}$$

$$\tau_g(t) = \frac{x_g(t)}{\dot{x}_g(t)} = \frac{1}{2}\left(t - \frac{t_d^2}{t}\right) \tag{14}$$

where t_d is the time duration of the entire operation. Using Tau coupling, we can find the solution for $\alpha(t)$ as follows:

$$\begin{aligned} \alpha(t) &= \frac{\alpha_0}{t_d^{2/k_{\alpha g}}}(t_d^2 - t^2)^{\frac{1}{k_{\alpha g}}} \\ \dot{\alpha}(t) &= \frac{-2\alpha_0 t}{k_{\alpha g} t_d^{2/k_{\alpha g}}}(t_d^2 - t^2)^{\frac{1}{k_{\alpha g}}-1} \\ \ddot{\alpha}(t) &= \frac{2\alpha_0}{k_{\alpha g} t_d^{2/k_{\alpha g}}}\left(\frac{2t^2}{k_{\alpha g}} - t^2 - t_d^d\right)\left(t_d^2 - t^2\right)^{\frac{1}{k_{\alpha g}}-2} \end{aligned} \tag{15}$$

Table 3 and Figure 2 show the motion of gap closure on α, $\dot{\alpha}$ and $\ddot{\alpha}$ for different values of $k_{\alpha g}$.

Table 3. Motion with different constant $k_{\alpha g}$ values in during intrinsic Tau gravity movement.

$k_{\alpha g}$	t	α	$\dot{\alpha}$	$\ddot{\alpha}$	Final Goal
$k_{\alpha g} < 0$	$\rightarrow t_d$	$\rightarrow \infty$	$\rightarrow \infty$	$\rightarrow \infty$	Gap not closed
$k_{\alpha g} = 0$	$\rightarrow t_d$	$= 0$	?	?	Error
$0 < k_{\alpha g} < 0.5$	$\rightarrow t_d$	$\rightarrow 0$	$\rightarrow 0$	$\rightarrow 0$	Zero Touchdown
$0.5 \leq k_{\alpha g} < 1$	$\rightarrow t_d$	$\rightarrow 0$	$\rightarrow 0$	$\rightarrow \infty$	Slight Collision
$k_{\alpha g} = 1$	$\rightarrow t_d$	$\rightarrow 0$	$\rightarrow 0$	$\rightarrow 0$	Collision
$k_{\alpha g} > 1$	$\rightarrow t_d$	$\rightarrow 0$	$\rightarrow \infty$	$\rightarrow \infty$	Strong Collision

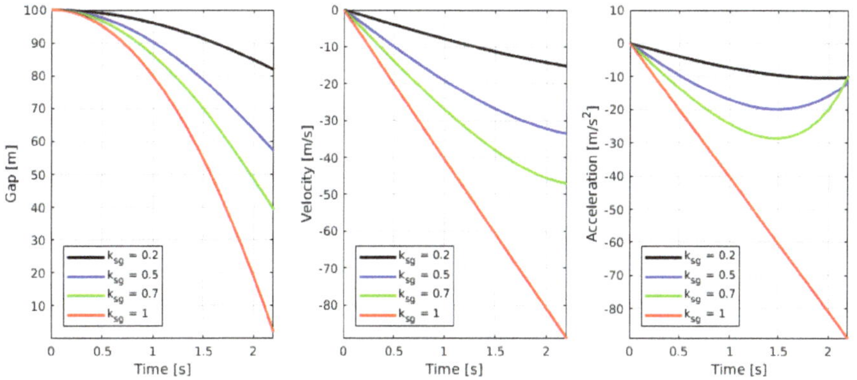

Figure 2. Values of α, $\dot{\alpha}$ and $\ddot{\alpha}$ with different values of $k_{\alpha g}$ ($k_{\alpha g} = 0.2, 0.5, 0.7, 1.0$).

2.6. Tau Theory Link to Constant Optic Flow Approach

Tau strategies have also been found in more developed species, such as birds and mammals, which require more complex visual locomotion strategies than insects with their constant optic flow approach. During vertical landing, using the constant dilation approach [10] for asymptotic closure of vertical gaps, the image dilation ω_z is given by:

$$\omega_z = -\frac{\dot{z}}{z} \tag{16}$$

which is held constant during the execution of the constant dilation strategy. Since the image dilation is the reciprocal of τ:

$$\tau = -\frac{1}{\omega_z} \tag{17}$$

This means that $\dot{\tau} = 0$, making the constant dilation strategy an implementation of the tau control strategy with a constant value of $k = 0$. This creates a soft touch landing with constant deceleration. The constant dilation strategy is a special case of the tau theory.

3. Body-Centric Quadrotor Model

The quadrotor model presented here is similar to the one developed in [11] and taken from [12]. For the purpose of modelling the quadrotor, two Cartesian coordinate frames are defined. The Earth-surface fixed frame, with axes 1_x^e, 1_y^e and 1_z^e aligned with north, east and down directions in the Earth frame. The second body frame is a body-fixed frame with its origin at the body centre of mass, and axes 1_x, 1_y and 1_z aligned with the forward, starboard (right), and down body orientations. The Earth and body coordinate frames, motor numbering and the rotation directions are illustrated in Figure 3.

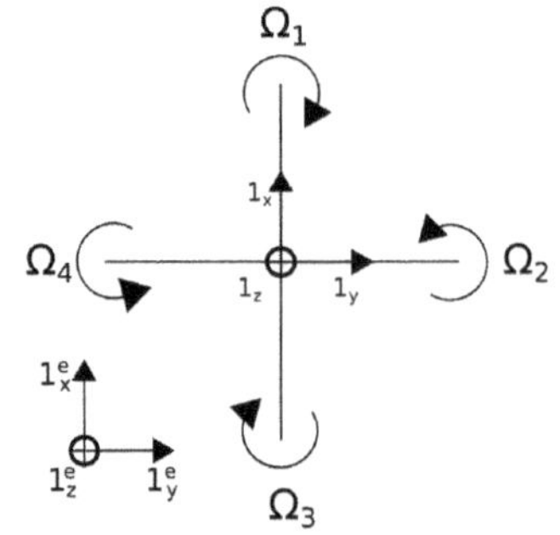

Figure 3. Top view of the quadrotor with the defined coordinate frames, motor numbering and positive motor rotation directions.

3.1. Attitude and Rotation Representation

The body attitude is represented, relative to the Earth frames, by the right-handed rotation sequence (yaw, pitch, roll) with angles ψ, θ, and ϕ about 1_z, 1_y and 1_x axes respectively. These three rotations define the transformation matrix $R_{b/e}$. Consequently, the quadrotor angular velocity in the Earth frame $\omega^e_{b/e} = [\dot{\psi}, \dot{\theta}, \dot{\phi}]$ and in the body frame $\omega^b_{b/e} = [p, q, r]$ are related as follows [13]:

$$\begin{bmatrix} mI_{3x3} & 0_{3x3} \\ 0_{3x3} & I_q \end{bmatrix} \begin{bmatrix} \dot{V}^b \\ \dot{w}^b_{b/e} \end{bmatrix} + \begin{bmatrix} \omega^b_{b/e} & \times & mV^b \\ \omega^b_{b/e} & \times & I_q\omega^b_{b/e} \end{bmatrix} = \begin{bmatrix} F^b \\ \tau^b \end{bmatrix} \quad (18)$$

3.2. Quadrotor Body Dynamics

Using Newton's Euler formalism, the boy dynamics are expressed in the body-fixed frame as:

$$\omega^e_{b/e} = \begin{bmatrix} 1 & \tan(\theta)\sin(\phi) & \tan(\theta)\cos(\phi) \\ 0 & \cos(\phi) & -\sin(\phi) \\ 0 & \sin(\phi)/cos(\theta) & \cos(\phi)/\cos(\theta) \end{bmatrix} \omega^b_{b/e} \quad (19)$$

We assume that the quadrotor is symmetric about its body principal axes, which coincide with the body frame. This assumption cancels all products of inertia and the inertial matrix becomes a diagonal matrix $I_q = diag(I_{xx}, I_{yy}, I_{zz})$.

The external forces acting on the quadrotor body are the weight force mg and the thrust forces generated be the four propellers T_i. Each thrust force is modelled as:

$$T_i = n\Omega_i^2, \qquad i = 1,2,3,4 \quad (20)$$

and the total thrust force $T_a = T_1, T_2, T_3, T_4$ is always aligned with the body 1_z axis in the negative direction. The total torque acting on the quadrotor is composed of the control torques and gyroscopic effect torque. Control torques τ_x and τ_y, which generate a positive rolling and pitching moment, can be expressed as

$$\tau_x = \ell(T_4 - T_2)1_x, \qquad \tau_y = \ell(T_1 - T_3)1_y \quad (21)$$

The aerodynamic drag torque Q_i acting on a propeller i is modelled as

$$Q_i = d\Omega_1^2, \qquad i = 1,2,3,4 \quad (22)$$

The total drag torque that generates a positive yawing moment is expressed as

$$\tau_z = d(Q_2^2 + Q_4^2 - Q_1^2 - Q_3^2)1_z \quad (23)$$

Body angular rates induce a gyroscopic effect torque τ_j on each of the rotating propellers due to rotor inertia J and the total imbalance Ω_{res} in the propeller angular velocities; τ_j can be expressed as

$$\tau_J = J(\omega_{b/e}^{b} \times 1_z)\Omega_{res} = \begin{bmatrix} Jq\Omega_{res} \\ -Jp\Omega_{res} \\ 0 \end{bmatrix} \tag{24}$$

where

$$\Omega_{res} = \Omega_2 + \Omega_4 - \Omega_1 - \Omega_3 \tag{25}$$

By defining the following variables

$$\begin{aligned} U_1 &= (\Omega_1^2 + \Omega_2^2 + \Omega_3^2 + \Omega_4^2) \\ U_2 &= (\Omega_4^2 - \Omega_2^2) \\ U_3 &= (\Omega_1^2 - \Omega_3^2) \\ U_4 &= (\Omega_2^2 + \Omega_4^2 - \Omega_1^2 - \Omega_3^2) \end{aligned} \tag{26}$$

the quadrotor model dynamic equations $(\dot{p}, \dot{q}, \dot{r}, \dot{v}_x, \dot{v}_y, \dot{v}_z)$ expressed in the body-fixed coordinates frame as well as the local Earth attitude kinematics $(\dot{\psi}, \dot{\theta}, \dot{\phi})$ can be written as

$$\begin{aligned} \dot{p} &= [qr(I_{yy} - I_{zz}) + Jq\Omega res + \ell n U_2]/I_{xx} \\ \dot{q} &= [pr(I_{zz} - I_{xx}) - Jp\Omega res + \ell n U_3]/I_{yy} \\ \dot{r} &= [pq(I_{xx} - I_{yy}) + dU_4]/I_{zz} \\ \dot{v}_x &= rv_y - qv_z - g\sin(\theta) \\ \dot{v}_y &= pv_z - rv_x + g\cos(\theta)\sin(\phi) \\ \dot{v}_z &= qv_x - pv_y + g\cos(\theta)\cos(\phi) - nU_1/m \\ \dot{\phi} &= p + q\tan(\theta)\sin(\phi) + r\tan(\theta)\cos(\phi) \\ \dot{\theta} &= q\cos(\phi) - r\sin(\phi) \\ \dot{\psi} &= q\sin(\phi)/\cos(\theta) + r\cos(\phi)/\cos(\theta) \end{aligned} \tag{27}$$

4. Control Scheme

The quadrotor is an open-loop unstable system with fast rotational dynamics. The proposed control scheme has two parts: a low-level stabilizing controller and a high-level bio-inspired controller in charge of near ground manoeuvrers.

4.1. Low-Level Controller

For the low-level controller a discrete time linear regulator with a direct feed-through matrix [14] is selected to perform stabilizing control on the quadrotor. The controller takes as input a vector of references

$$y_r = [\psi_r, a_{xr}, a_{yr}, a_{zr}]^T \tag{28}$$

and a state vector

$$x = [\phi, \theta, \psi, p, q, r]^T \tag{29}$$

Finally, it outputs a control vector

$$u = [\Omega_1, \Omega_2, \Omega_3, \Omega_4]^T \tag{30}$$

The controller is designed with basis on the previous Jacobian linearised dynamic model (27), about the equilibrium point $x_{eq} = [0,0,0,0,0,0]^T$ and $u_{eq} = [\Omega_h, \Omega_h, \Omega_h, \Omega_h]^T$, where Ω_h is the necessary speed in rad/s to maintain hover. The low-level control method is taken from [12] and uses a linear quadratic tracker approach. The control is given by

$$u(n) = -Kx(n) + Fy_r(n+1) \tag{31}$$

where matrices K and F are the state feedback and reference feed-forward gains, respectively. The purpose of the low-level controller is to stabilize the quadrotor's fast rotational dynamics by tracking a body acceleration and heading reference signal (28). This complements the high-level controller whose purpose is to use Tau theory to command the low-level controller with a suitable reference signal. The values of matrices K and F can be found on Appendix A.

4.2. High-Level Controller

The high-level controller will be in charge of supplying the low-level controller with suitable reference signals based on Tau theory. This can be achieved knowing that the vertical image dilation ω_z is equal to the reciprocal of Lee's basic Tau law [10]:

$$\omega_{zr}(t) = -\frac{1}{\tau(t)} \tag{32}$$

Substituting Tau, we obtain:

$$\omega_{zr}(t) = -\frac{1}{kt + \tau_0} \tag{33}$$

This means that regulating the visually registered image dilation to track $\omega_{zr}(t)$ becomes equivalent to enforcing the original Tau theory [7] with a constant k value that reflects the manoeuvrer we wish to accomplish. Looking at Equation (7) we can see that, in order for this implementation to be viable, a downward vertical (negative) velocity is necessary be properly initialized. This limits τ_0 to only negative values, otherwise the control law will cause the quadrotor to open the gap and fly away from the ground. Additionally, t needs to satisfy

$$t < \frac{\tau_0}{k} \quad \text{if} \quad k < 0 \quad \text{and} \quad t > -\frac{\tau_0}{k} \quad \text{if} \quad k > 0 \tag{34}$$

A simple solution to this problem, to perform near ground manoeuvres from hover, is to substitute the basic Tau implementation with its intrinsic tau gravity guidance counterpart [15]:

$$\omega_{zr}(t) = -\frac{1}{k_{\alpha g}\tau_g(t)} \tag{35}$$

Two values need to be defined for this implementation, the constant $k_{\alpha g}$ and t_d. Constant $k_{\alpha g}$ will dictate the approach that the quadrotor will take regarding the manoeuvre. As indicated in Table 3, the constant chosen will modify the way the action will be performed, from a zero velocity manoeuvre at touchdown, to a strong collision. The choice of t_d will dictate the manoeuvre execution time.

4.3. High-Level Control for Horizontal Manoeuvres

When multiple gaps need to be closed simultaneously, namely, $\alpha(t)$ and $\beta(t)$, the Tau coupling strategy [15] can be implemented. During this operation, the two corresponding Tau will be kept at a constant ratio $k_{\alpha\beta}$. This can be used to close vertical and horizontal gaps simultaneously. As previously discussed, the value of the $k_{\alpha\beta}$ constant will dictate the characteristics of the manoeuvre.

4.4. High-Level Vertical and Horizontal Control Implementation

Tau is controlled by tracking a time-varying image dilation reference signal obtained from the intrinsic tau gravity $\omega_{zr}(t)$ using Equation (35). The visual on-board processing system registers the value of the image dilation $\omega_z(n)$ with a sampling time T_s at a discrete time step n. A PI controller is used to regulate the dilation error e_{ω_z}

$$e_{\omega_z}(n) = \omega_{zr}(n) - \omega_z(n) \tag{36}$$

by providing a suitable reference signal (a_{zr})

$$a_{zr}(n) = K_P e_{\omega_z}(n) + K_I \sum_{i=0}^{n} e_{\omega_z}(i) \tag{37}$$

to the low level controller, where K_I and K_P are the PI controller gains. The values of the control gains can be found on Appendix A.

To achieve horizontal landing and tracking, the tau coupling strategy is used. If we have three gaps that need to be closed simultaneously, namely $z(t)$, $x(t)$ and $y(t)$, that coincide with the quadrotor 1_z, 1_x and 1_y respectively, it is possible to use the Tau coupling strategy in Equation (11) to find a suitable $\ddot{x}(t)$ and $\ddot{y}(t)$ linked to a_{zr} that will provide body acceleration reference signals, a_{xr} and a_{yr} respectively, for the low-level controller to be input into the reference vector (28) as follows

$$\begin{aligned} a_{xr} &= \ddot{x}(t) \\ a_{yr} &= \ddot{y}(t) \end{aligned} \tag{38}$$

The previous reference values will only be useful when they align with the vehicle reference frames 1_x and 1_y respectively. In order for the reference signals to point towards a target, they would need to be updated. As explained on Section 6.1: Equation (38) would need to be updated as Equation (60). The heading reference component ψ_r in the reference input vector (28) is set to a predefined constant value to hold the heading value while the manoeuvre is performed.

5. Estimation of Visual Motion Parameters

Optic flow corresponds to the image velocities (u, v) in the patterns of apparent motion of objects on frame caused by the relative motion between the subject and a scene. This includes three translational velocities (v_x^c, v_y^c, v_z^c) and three angular velocities (p^c, q^c, r^c), the depth of the observed objective Z, and the cameras focal length f. This can be expressed as:

$$\begin{aligned} u &= -f\left(\frac{v_x^c}{Z} + q^c\right) + x\frac{v_z^c}{Z} + yr^c - x^2\frac{q^c}{f} + xy\frac{p^c}{f} \\ v &= -f\left(\frac{v_y^c}{Z} - p^c\right) + y\frac{v_z^c}{Z} - xr^c + y^2\frac{p^c}{f} - xy\frac{q^c}{f} \end{aligned} \tag{39}$$

The translational and angular velocities are given in the camera frame, rigidly attached to the camera where its 1_x^c and 1_y^c axes are aligned with the image horizontal and vertical frames, and the 1_z^c axis is aligned with the optical axis flow towards the scene. This estimation will find the visual motion parameters, namely, the Focus of Expansion (FOE), the camera frame dilation ω_z^c and the ventral flows ω_x^c and ω_y^c. This will be used to control the quadrotor during the near ground manoeuvrers. The system implemented here is taken from [16].

5.1. Simultaneous Visual Motion Parameters Estimation

By removing the rotational component of the optic flow from Equation (39), the translational components of the optic flow, u_T, v_T can be expressed as

$$u_T = -f\frac{v_x^c}{Z} + x\frac{v_z^c}{Z}$$
$$v_T = -f\frac{v_y^c}{Z} + y\frac{v_z^c}{Z} \tag{40}$$

We rewrite the previous equation in terms of the visual motion parameters $\omega_x^c, \omega_y^c, \omega_z^c$, keeping in mind that $v_z^c = \dot{Z}$, the image dilation can be described as

$$\omega_z^c = \frac{v_z^c}{Z} \tag{41}$$

Using Equations (40) and (41), the translational optic flow components can be rewritten as:

$$u_T = -f\omega_x^c + x\omega_z^c$$
$$v_T = -f\omega_y^c + x\omega_z^c \tag{42}$$

In addition, the image frame coordinates of the Focus of Expansion (FOE), x_{FOE}, y_{FOE} can be calculated as

$$x_{FOE} = \frac{v_x^c}{v_z^c}$$
$$y_{FOE} = \frac{v_y^c}{v_z^c} \tag{43}$$

Note that the FOE can only exist when $v_z^c \neq 0$.

Due to the high number of points were the optic flow can be evaluated, a parametric model can be used to simultaneously calculate the visual motion parameters. The translational components (42) can be represented using the following model:

$$u_T = a_1 + a_2 x$$
$$v_T = a_3 + a_2 y \tag{44}$$

Then, the optic flow calculations are used to form a least-square regression problem. In this way, Equation (44) can be rewritten as

$$\begin{bmatrix} u_{T1} \\ v_{T1} \\ u_{T2} \\ v_{T2} \\ \vdots \\ u_{Tn} \\ v_{Tn} \end{bmatrix} = \begin{bmatrix} 1 & x_1 & 0 \\ 0 & y_1 & 1 \\ 1 & x_2 & 0 \\ 0 & y_2 & 1 \\ \vdots & \vdots & \vdots \\ 1 & x_n & 0 \\ 0 & y_n & 1 \end{bmatrix} \begin{bmatrix} a_1 \\ a_2 \\ a_3 \end{bmatrix} \tag{45}$$

This can be solved using least squares to find the estimated model parameters $\hat{a}_1, \hat{a}_2$ and $\hat{a}_3$. Then, the image dilation in camera frame, ventral flows and FOE can be found with:

$$\omega_z^c = \hat{a}_2$$
$$\omega_x^c = -\frac{\hat{a}_1}{f}$$
$$\omega_y^c = -\frac{\hat{a}_3}{f}$$
$$x_{FOE} = -\frac{\hat{a}_1}{\hat{a}_2}$$
$$y_{FOE} = -\frac{\hat{a}_3}{\hat{a}_2} \tag{46}$$

We need to consider that the camera is attached to the quadrotor in such way that the 1_z^c axis coincides with the body 1_z axis, while the camera 1_x^c and 1_y^c axes are rotated with and angle ψ_c about the body 1_z axis with respect to the body axes 1_x and 1_y, respectively. This means that while the image dilation in the camera and body frame are equal, the ventral flows need to be adjusted. This can be done as follows:

$$\begin{bmatrix} \omega_x \\ \omega_y \end{bmatrix} = \begin{bmatrix} \cos(\psi_c) & -\sin(\psi_c) \\ \sin(\psi_c) & \cos(\psi_c) \end{bmatrix} \begin{bmatrix} \omega_x^c \\ \omega_y^c \end{bmatrix} \tag{47}$$

5.2. Outlier Rejection

The proposed method for visual motion parameters estimation has been shown to produce accurate results [16], however, the raw estimates obtained from Equation (47) can exhibit outliers, caused by the the temporary violation of assumptions made by the optic flow method and due to the noisy nature of digital visual information. To deal with this issue, the outliers need to be eliminated in real time. A median filter is a good robust statistical filter that can be used for outlier rejection. The running median filter presented in [17] is used to reject outlines over a window of previous values.

5.3. Sensor Fusion: IMU Aided Estimation of Visual Motion Parameters (VMP)

Images captured from cameras are naturally noisy, reducing their accuracy during its processing. Even with the fast real-time method used here, image capture update rate is low compared to the dynamics of aerial vehicles. In order to use this information in a control system with a higher sampling rate, it is necessary to estimate the visual information between sampling instants. Even after the application of the outlier rejection filter, the resulting estimates will contain noise. Inter-sampling estimation can be achieved using a stochastic model-based estimation algorithm, such as a Kalman Filter. A dynamic filter of the visual motion parameters is derived for this purpose.

If a downward-looking camera is rigidly mounted on a quadrotor, the height of the quadrotor z and the scene depth at the centre of the image Z are related by their attitude angles. If we assume small attitude angles, it is possible to use the approximation $z \approx Z$.
Defining x_d as:

$$x_d = \frac{1}{z} \tag{48}$$

Taking its time derivative yields

$$\dot{x}_d = -\frac{\dot{z}}{z^2} \tag{49}$$

using a previously defined definition of image dilation (41), the derivative can be rewritten as

$$\dot{x}_d = \omega_z x_d \tag{50}$$

Taking the time derivative of (16) and assuming $z \neq 0$ give

$$\dot{\omega}_x = \frac{\dot{v}_x}{z} - \frac{\dot{z}}{z}\frac{v_x}{z} \tag{51}$$

Revising Equation (27), the acceleration component in the 1_x axis of the body frame is $a_x = -g\sin(\theta)$, and $\dot{\omega}_z$ can be rewritten as

$$\dot{\omega}_x = r\frac{v_y}{z} - q\frac{v_z}{z} + \frac{a_x}{z} - \frac{\dot{z}}{z}\frac{v_x}{z} \tag{52}$$

which can be also be written in the following form, taking into consideration Equations (1) and (48)

$$\dot{\omega}_x = r\omega_y - q\omega_z + \omega_x\omega_y + a_x x_d \tag{53}$$

Given that the body-frame starboard and downward accelerations are $a_y = g\cos(\theta)\sin(\phi)$ and $a_z = g\cos(\theta)\cos(\phi) - T_a/m$, the equations for $\dot{\omega}_y$ and $\dot{\omega}_z$ can be derived as well

$$\begin{aligned}\dot{\omega}_y &= p\omega_z - r\omega_x + \omega_y\omega_z + a_y x_d \\ \dot{\omega}_z &= q\omega_x - p\omega_y + \omega_z^2 + a_z x_d\end{aligned} \tag{54}$$

Using Equations (50), (53) and (54), the dynamic system for the visual motion parameters defined by the state vector $x_f = [\omega_x, \omega_y, \omega_z, x_d]^T$ which will be used in the Kalman Filter. The filter will predict the visual motion parameters at a higher rate to allow a more responsive high-level control.

Similar examples of data fusion can be observed in nature. Visual and non visual cues, such as gravito-inertial senses and efferent copies all play a collaborative role in forming the perception of motion [18]. With this information the brain is capable of build an estimate based on the information available. This is further supported by [19], where a mismatch between the expected and received motion cues can trigger motion sickness in humans.

The Cubature Kalman Filter [20], a variation of the Unscented Kalman Filter (UKF) [21] with a spherical-radial cubature rule is used here. It has been proved to be superior to the Extended Kalman Filter (EKF) [22] and, in some cases, superior to the UKF [23].

The CKF will output a state vector $x_f = [\omega_x, \omega_y, \omega_z, x_d]^T$, while the input vector $u_f = [p, q, r, a_x, a_y, a_z]^T$ is provided by the IMU, and the measurement vector $y_f = [\omega_x, \omega_y, \omega_z]$ is provided by the visual system. This system will be implemented by discretizing Equations (50), (53) and (54). The CKF produces estimates of $\hat{x}_f$ at the same rate as the IMU readings, to be used by the high-level controller, to enable a smoother and more responsive control at a higher rate. The CKF uses a constant process covariance matrix Q_f that can be chosen manually to take into account the unmodelled input noise. Additionally, a time variant noise covariance R_f is defined as

$$R_f = diag\left(\frac{1}{\sigma_v^2}, \frac{1}{\sigma_v^2}, \frac{1}{\sigma_v^2}\right) \tag{55}$$

where σ_v is calculated from the root mean square of the optic flow residuals in the fitting process described in Equation (45).

6. Objective Tracking

In order to perform near ground manoeuvrers on a moving target, first we need to be able to accurately detect and track the object. For this work we have decided to use AprilTags [24]. AprilTags are open source fiducial marks; artificial visual features designed for automatic detection. Initially used for augmented reality applications, they have since been widely adopted by the robotics community for uses such as: ground truth, pose estimation, and object detection and tracking. AprilTags are black-and-white square tags with an encoded binary payload.

In experiments [24], these tags have probed to have high accuracy, low false positive rate and inexpensive computation time. The main drawback of using any fiducial mark is the need to perform camera calibration to take into consideration the camera's focal length, principal point and radial distortion coefficients for each camera model.

6.1. Adjust Body Reference to Target Location

To use the body accelerations as reference signals a_{xr} and a_{yr} from the high-level into the low-level controller to point towards out target, we need to know the objective's quadrant in the camera's Cartesian system. Using the AprilTags, we can extract the target coordinates in the camera frame in pixels, namely u^c and v^c. If we consider the centre of the camera frame, u_0^c and v_0^c in pixels, as the origin of the Cartesian system, then we can calculate the distance from the origin to the objective as:

$$\begin{aligned}u_\Delta &= u_0^c - u^c \\ v_\Delta &= v_0^c - v^c\end{aligned} \tag{56}$$

Then, we can proceed to determine the angle of the objective from the centre of the camera in polar coordinates with

$$\lambda^c = \tan^{-1}\left(\frac{v_\Delta}{-u_\Delta}\right) \tag{57}$$

Finally, we separate the angle into its x and y components

$$\begin{aligned} u_\lambda &= -cos(\lambda^c) \\ v_\lambda &= -sin(\lambda^c) \end{aligned} \tag{58}$$

Each of the components will have a value that will range from -1 to 1, depending on the position of the target in the camera Cartesian system. This value will be multiplied by the reference body acceleration, this way the reference body acceleration signal will always move the vehicle, in the body 1_x and 1_y axes to the location of the objective (See Figure 4).

Since the position of the camera does not align with the body 1_x and 1_y axes, we need to rotate the camera Cartesian system with an angle ψ_c. Subtracting ψ_c from Equation (58), we obtain:

$$\begin{aligned} u_\lambda &= -cos(\lambda^c - \psi_c) \\ v_\lambda &= -sin(\lambda^c - \psi_c) \end{aligned} \tag{59}$$

Finally, the acceleration reference signal that will be feed into the input vector (28) to control the horizontal movement of the vehicle towards the objective is

$$\begin{aligned} a_{xr} &= \ddot{x}(t)u_\lambda \\ a_{yr} &= \ddot{y}(t)v_\lambda \end{aligned} \tag{60}$$

Equation (60) is an update on Equation (38) that takes into consideration the position of the camera in the vehicle and corrects the reference signals to point towards the objective.

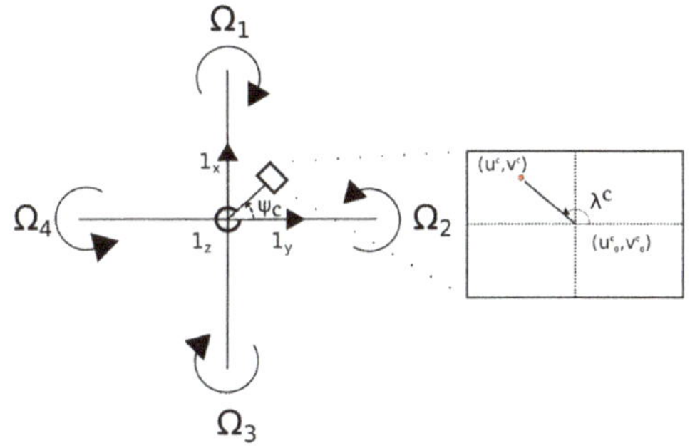

Figure 4. Top view of the quadrotor with the defined coordinate frames, camera location and camera frame conventions.

7. Simulations

7.1. Simulation Environment

Simulations are performed using the Robot Operating System (ROS) [25], a flexible framework for writing robot software. ROS includes a collection of tools, libraries, and conventions to design complex and robust robots. ROS is used in conjunction with Gazebo [26] a simulation environment to rapidly test algorithms using realistic scenarios. The algorithms prototyping will be tested using RotorS [27], a modular gazebo MAV simulator framework. The estimation of motion parameters makes use of the OpenCV [28] software library and the Eigen [29] C++ template library for linear algebra.

RotorS provides us with multi-rotor models such as the AscTec Hummingbird, the AscTec Pelican, and the AscTec Firefly, with the possibility to build custom multi-rotors and even fixed-wing unmanned vehicles. For our experiments we will be using the AscTec Pelican. The Pelican is a flexible research UAV platform that allows us to perform all the computer vision and high-level processes on-board, without the need of any external computing units.

The simulated Pelican incorporates a variety of sensors, including the IMU (three-axes accelerometer and rate gyroscopes), and attitude and heading reference system (AHRS), and one downward-looking camera with a resolution of 720 × 480 pixels and a focal length of 49 degrees of vertical field of view.

To simulate a moving target, we will use a Husky [30] Unmanned Ground Vehicle (UGV), a field robotics platform that support ROS and can be loaded with a variety of sensors. In our case, a 2 × 2 m square platform will be placed on top of the Husky UGV with an AprilTag of the same dimensions, in order for the quadrotor to see the moving platform. Simulations were performed on a Ubuntu 16.04 computer with an AMD Ryzen 3 1200 CPU with 8 GB of RAM and a Nvidia GTX 1050Ti GPU. The Gazebo simulation can be seen in Figure 5.

Figure 5. Pelican UAV on simulated environment with AprilTag on a platform on top of a Husky UGV.

7.2. Autonomous Tau-Based Control Simulation

Simulations are performed on Gazebo, the simulation environment, using ROS and RotorS with an AscTec Pelican. To accurately use the optic flow during the visual motion parameters calculation, the ground is covered with a print of randomly assembled lunar images taken by the personal telescope of Wes Higgins [31] (See Figure 5). The main source of light on the simulation has Gazebo default position and values. The quadrotor is flown manually to a 4 m height in the simulated environment, while the Husky is set in different location within camera frame. Four simulations were performed with different k_g, k_{gx} and k_{xy} constants and start from a hover position. The value of t_d is set to 4 across all simulations and x_0 is set to 10.

In Simulation 1, $k_g = 0.4$ while $k_{gx} = k_{xy} = 1.0$. The position of the Pelican and the Husky can be seen in the graphs on Figure 6. It can be observed that the Pelican is capable of tracking the objective and land on it, with a soft landing, when it is above it.

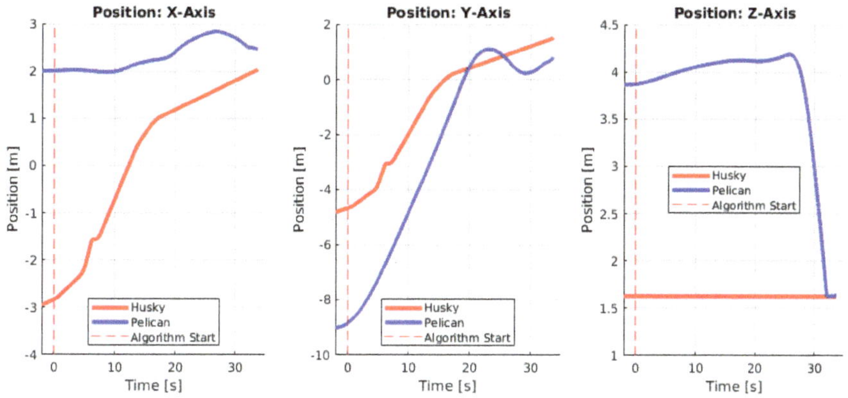

Figure 6. Position, in Gazebo's reference frame, of the Pelican UAV and Husky UGV over time during simulation, with k values of $k_g = 0.4$ and $k_{gx} = k_{xy} = 1.0$.

On Simulation 2 the values are set to: $k_g = 1.0$ and $k_{gx} = k_{xy} = 1.0$. The position of the Pelican and the Husky can be seen in the graphs on Figure 7. This manoeuvre is similar to the previously described, but due to the choice of k_g, landing is achieved with a higher vertical velocity. As previously discussed, this kind of movement can be useful during perching operations, just like some birds do to catch preys.

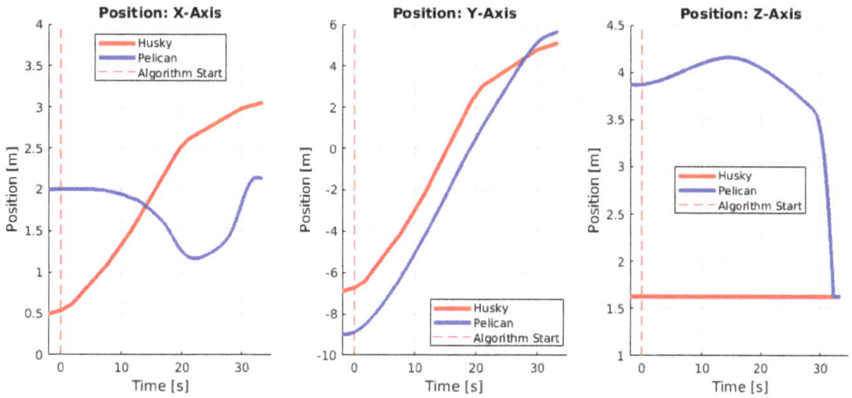

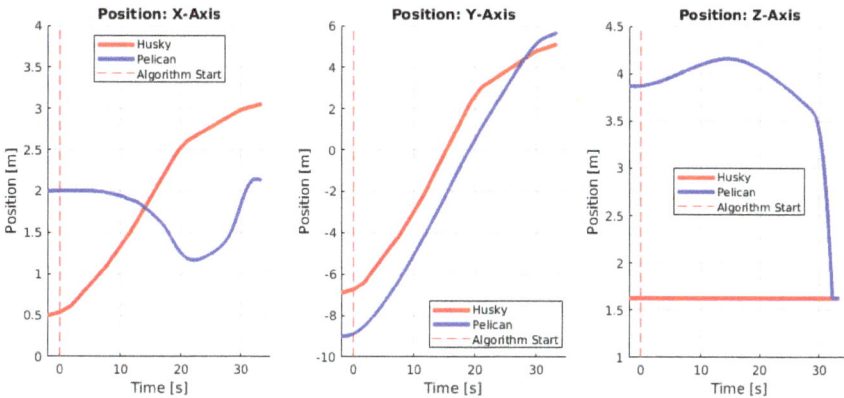

Figure 7. Position, in Gazebo's reference frame, of the Pelican UAV and Husky UGV over time during simulation, with k values of $k_g = 1.0$ and $k_{gx} = k_{xy} = 1.0$.

Note that the difference in distance in axes x and y at touchdown is due to the location of the IMU inside the Husky. The sensor is located on the vehicle's centre while the platform is a 2×2 m square. This explains why in the previously mentioned axes is not uncommon to end the manoeuvre with a difference of up to 1 m.

Velocities during the landing manoeuvres in Simulations 1 and 2 can be compared in Figure 8. They have a mean downward velocity of -0.0154 and -0.0035 m/s, and an execution time of 5.52 and 3.28 s, respectively. This confirms that the values of constant k_g modifies the vehicle dynamics during landing.

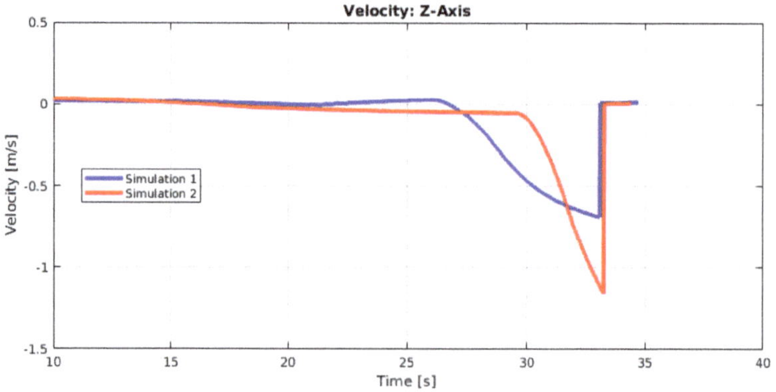

Figure 8. Velocity in the Z-Axis during Simulations 1 and 2.

Simulation 3 is performed with values of $k_g = 0.4$ and $k_{gx} = k_{xy} = 0.5$. The position of the Pelican and the Husky can be seen in the graphs on Figure 9. In this simulation, the Pelican will follow the Husky while keeping its distance, this is achieved due to the value of the constant k_{xy} and k_{gx}. This manoeuvre showcases the flexibility that Tau can achieve during near ground navigation. Just like observed in birds of prey, the quadrotor is capable of give chase to a target. During the simulation the Pelican and the Husky had a mean difference in distance of 0.7, 0.9 and 2.5 m in the x, y and z axes, respectively.

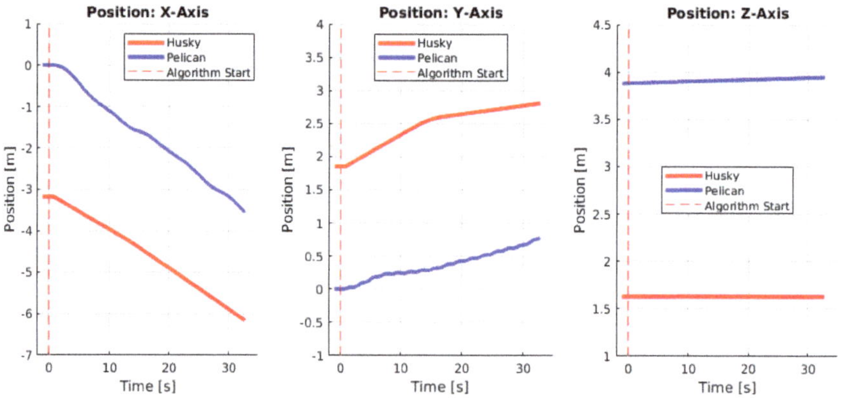

Figure 9. Position, in Gazebo's reference frame, of the Pelican UAV and Husky UGV over time during simulation, with k values of $k_g = 1.0$ and $k_{gx} = k_{xy} = 0.5$.

Finally, Simulation 4 is performed with the same values as Simulation 3 ($k_g = 0.4$ and $k_{gx} = k_{xy} = 0.5$). The position of the Pelican and the Husky can be seen in the graphs on Figure 10. In this simulation, just as in the previous one, the Pelican will follow the Husky while keeping its distance. During simulations, the Pelican and the Husky had a mean difference in distance of 2.48, 1.58 and 2.3 m in the x, y and z axes, respectively.

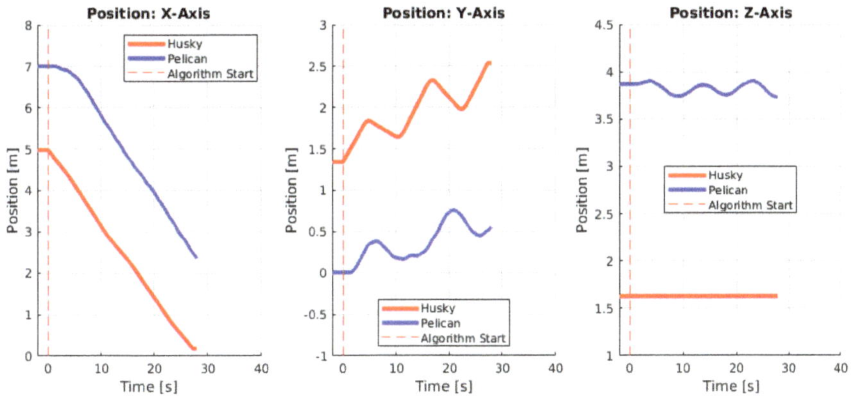

Figure 10. Position, in Gazebo's reference frame, of the Pelican UAV and Husky UGV over time during simulation, with k values of $k_g = 1.0$ and $k_{gx} = k_{xy} = 0.5$.

8. Discussion

From the simulations, it is clear that the proposed Tau theory based strategy for the control scheme is flexible enough to achieve different types of near-ground manoeuvres. In simulations 1 and 2, the quadrotor successfully performed a the detection, tracking, and, eventual, landing on a moving platform with different touchdown speeds. Simulations 3 and 4 showcase the flexibility of Tau, where the quadrotor is capable of follow the platform and keep itself at a viewing distance from the target without initializing landing. All this experiments start form hover, a new addition that, to the knowledge of the authors, has never been used during Tau theory based visual autonomous landing.

9. Conclusions

This paper shows a bio-inspired controller using Tau theory to achieve flexible visual autonomous vertical and horizontal control of a multi-rotor vehicle. The simulations confirm that near-ground manoeuvres, such as landing, and tracking of an objective can be performed visually without knowledge of the vehicle's height or objective's velocity. Practical applications of this method include target approach to perform inspection, tracking or landing; followed by perching or fly away based on the chosen constant values. Practical applications of the preposed method can be expanded into other VTOL vehicles, UGV and even spacecraft. Further work is required to automate the choice of manoeuvre parameters based on the vehicle's objective and context awareness.

Author Contributions: S.A.: analysis and design of control methods, programming and simulations, writing up. V.B.: conceptualization, supervision, proof reading, technical corrections; N.B.: supervision, proof reading, technical corrections.

Funding: The author Saul Armendariz is grateful for the funding from the "Consejo Nacional de Ciencia y Tecnologia" (CONACYT, Mexico). This research was funded by CONACYT, grant number 440239.

Conflicts of Interest: The authors declare no conflict of interest.

Abbreviations

The following abbreviations are used in this manuscript:

EKF	Extended Kalman Filter
UKF	Unscented Kalman Filter
CKF	Cubature Kalman Filter
RMS	Root Mean Square
FOE	Focus of Expansion
GPS	Global Positioning System
VMP	Visual Motion Parameters
ROS	Robot Operating System
UGV	Unmanned Ground Vehicle
FoV	Field of View
TTC	Time-to-Contact
UAV	Unmanned Aerial Vehicle

Appendix A

Low-level controller state feedback and reference feed-forward gains for Equation (31).

$$K = \begin{bmatrix} 0 & 613.77 & -472.22 & 0 & 63.01 & -51.003 \\ -613.77 & 0 & 472.22 & -63.01 & 0 & 51.003 \\ 0 & -613.77 & -472.22 & 0 & -63.01 & -51.003 \\ 613.77 & 0 & 472.22 & 63.01 & 0 & 51.003 \end{bmatrix} \tag{A1}$$

$$F = \begin{bmatrix} -211.61 & -54.09 & 0 & -21.52 \\ 211.61 & 0 & -54.05 & -21.52 \\ -211.61 & 54.09 & 0 & -21.52 \\ 211.61 & 0 & 54.05 & -21.52 \end{bmatrix} \tag{A2}$$

High-Level PI Controller Parameters in Equation (37)

$$K_P = 2.788 \tag{A3}$$

$$K_I = 0.067 \tag{A4}$$

References

1. Gupte, S.; Mohandas, P.I.T.; Conrad, J.M. A survey of quadrotor unmanned aerial vehicles. In Proceedings of the 2012 Proceedings of IEEE Southeastcon, Orlando, FL, USA, 15–18 March 2012; pp. 1–6.
2. Kendoul, F. Survey of Advances in Guidance, Navigation, and Control of Unmanned Rotorcraft Systems. *J. Field Robot.* **2011**, *23*, 245–267. [CrossRef]
3. Zhang, Z.; Xie, P.; Ma, O. Bio-inspired trajectory generation for UAV perching. In Proceedings of the 2013 IEEE/ASME International Conference on Advanced Intelligent Mechatronics: Mechatronics for Human Wellbeing (AIM 2013), Wollongong, Australia, 9–12 July 2013; pp. 997–1002.
4. Falanga, D.; Zanchettin, A.; Simovic, A.; Delmerico, J.; Scaramuzza, D. Vision-based autonomous quadrotor landing on a moving platform. In Proceedings of the 15th IEEE International Symposium on Safety, Security and Rescue Robotics, Conference (SSRR 2017), Shanghai, China, 11–13 October 2017; pp. 200–207.
5. Horridge, G.A. The evolution of visual processing and the construction of seeing systems. *Proc. R. Soc. Lond. Ser. B* **1987**, *230*, 279–292. [CrossRef]
6. Srinivasan, M.; Zhang, S.; Lehrer, M.; Collett, T. Honeybee navigation en route to the goal: Visual flight control and odometry. *J. Exp. Biol.* **1996**, *199*, 237–244. [CrossRef] [PubMed]
7. Lee, D.N. A Theory of Visual Control of Braking Based on Information about Time-to-Collision. *Perception* **1976**, *5*, 437–459. [CrossRef] [PubMed]

8. Lee, D.N.; Young, D.S.; Rewt, D. How do somersaulters land on their feet? *J. Exp. Psychol.* **1992**, *18*, 1195. [CrossRef]
9. Lee, D.N. Guiding movement by coupling taus. *Ecol. Psychol.* **1998**, *10*, 221–250. [CrossRef]
10. Herisse, B.; Russotto, F.X.; Hamel, T.; Mahony, R. Hovering flight and vertical landing control of a VTOL Unmanned Aerial Vehicle using optical flow. In Proceedings of the 2008 IEEE/RSJ International Conference on Intelligent Robots and Systems (IROS), Nice, France, 22–26 September 2008.
11. Bresciani, T. Modelling, Identification and Control of a Quadrotor Helicopter. Master's Theses, Lund University, Lund, Sweden, 2008.
12. Alkowatly, M.T.; Becerra, V.M.; Holderbaum, W. Body-centric Modelling, Identification, and Acceleration Tracking Control of a Quadrotor UAV. *Int. J. Model. Ident. Control* **2014**. [CrossRef]
13. Stevens, B.L.; Lewis, F.L.; Johnson, E.N. *Aircraft Control and Simulation: Dynamics, Controls Design, and Autonomous Systems*; Wiley: Hoboken, NJ, USA, 2016; p. 768.
14. Wang, J.H.; Sheng, J.; Tsai, H.; Chen, Y.C.; Guo, S.M.; Shieh, L.S. An Active Low-Order Fault-Tolerant State Space Self-Tuner for the Unknown Sample-Data Linear Regular System with an Input-Output Direct Feedthrough Term. *Appl. Math. Sci.* **2012**, *6*, 4813–4855.
15. Lee, D.N. General Tau Theory: Evolution to date. *Perception* **2009**, *38*, 837. [CrossRef] [PubMed]
16. Alkowatly, M.T.; Becerra, V.M.; Holderbaum, W. Estimation of Visual Motion Parameters Used for Bio-inspired Navigation. *J. Image Graphics* **2013**, *1*, 120–124. [CrossRef]
17. Menold, P.H.; Pearson, R.K.; Allgower, F. Online outlier detection and removal. In Proceedings of the 7th IEEE Mediterranean Conference on Control and Automation (MED '99), Haifa, Israel, 28–30 June 1999; pp. 1110–1133.
18. Harris, L.R.; Jenkin, M.R.; Zikovitz, D.; Redlick, F.; Jaekl, P.; Jasiobedzka, U.T.; Jenkin, H.L.; Allison, R.S. Simulating Self-Motion I: Cues for the Perception of Motion. *Virtual Real.* **2002**, *6*, 75–85. [CrossRef]
19. Hain, T.C.; Helminski, J.O. Anatomy and physiology of the normal vestibular system. *Vestibular Rehabil.* **2007**, *11*, 2–18.
20. Arasaratnam, I.; Haykin, S.; Hurd, T.R. Cubature Kalman filtering for continuous-discrete systems: Theory and simulations. *IEEE Trans. Signal Process.* **2010**, *58*, 4977–4993. [CrossRef]
21. Wan, E.A.; Van Der Merwe, R. The unscented Kalman filter for nonlinear estimation. *Technology* **2000**, 153–158. [CrossRef]
22. Dai, H.-D.; Dai, S.-W.; Cong, Y.-C.; Wu, G.-B. Performance Comparison of EKF/UKF/CKF for the Tracking of Ballistic Target. *Telecommun. Comput. Electron. Control.* **2012**, *10*, 1537–1542.
23. Chandra, K.; Gu, D. Cubature Kalman Filter based Localization and Mapping. In Proceedings of the 18th IFAC World Congress, Milano, Italy, 28 August–2 September 2011; pp. 2121–2125.
24. Wang, J.; Olson, E. AprilTag 2: Efficient and robust fiducial detection. *IEEE Int. Conf. Intell. Robots Syst.* **2016**, 4193–4198. [CrossRef]
25. Available online: http://www.ros.org/ (accessed on 29 December 2018).
26. Available online: http://gazebosim.org/ (accessed on 29 December 2018).
27. Furrer, F.; Burri, M.; Achtelik, M.; Siegwart, R. RotorS—A Modular Gazebo MAV Simulator Framework. In *Robot Operating System (ROS): The Complete Reference (Volume 1)*; Koubaa, A., Ed.; Springer International Publishing: Cham, Switzerland, 2016; pp. 595–625.
28. Available online: https://opencv.org/ (accessed on 29 December 2018).
29. Available online: http://eigen.tuxfamily.org/ (accessed on 29 December 2018).
30. Available online: http://www.clearpathrobotics.com/assets/guides/husky/ (accessed on 29 December 2018).
31. Available online: http://higginsandsons.com/astro/ (accessed on 29 December 2018).

electronics

Article

Multiple UAV Systems for Agricultural Applications: Control, Implementation, and Evaluation

Chanyoung Ju [1] **and Hyoung Il Son** [1,2,*]

[1] Department of Rural and Biosystems Engineering, Chonnam National University, 77 Yongbong-ro, Buk-gu, Gwangju 61186, Korea; cksdud15@gmail.com
[2] Hybrid Robotics Inc., 77 Yongbong-ro, Buk-gu, Gwangju 61186, Korea
* Correspondence: hison@jnu.ac.kr; Tel.: +82-62-530-2152

Received: 24 July 2018; Accepted: 22 August 2018; Published: 24 August 2018

Abstract: The introduction of multiple unmanned aerial vehicle (UAV) systems into agriculture causes an increase in work efficiency and a decrease in operator fatigue. However, systems that are commonly used in agriculture perform tasks using a single UAV with a centralized controller. In this study, we develop a multi-UAV system for agriculture using the distributed swarm control algorithm and evaluate the performance of the system. The performance of the proposed agricultural multi-UAV system is quantitatively evaluated and analyzed through four experimental cases: single UAV with autonomous control, multiple UAVs with autonomous control, single UAV with remote control, and multiple UAVs with remote control. Moreover, the performance of each system was analyzed through seven performance metrics: total time, setup time, flight time, battery consumption, inaccuracy of land, haptic control effort, and coverage ratio. Experimental results indicate that the performance of the multi-UAV system is significantly superior to the single-UAV system.

Keywords: agricultural UAV; multi-UAV system; distributed swarm control; performance evaluation; remote sensing

1. Introduction

Owing to the development of unmanned aerial vehicle (UAV) technology, there have been diverse studies on their applications in the agriculture field, which has the greatest potential for UAVs. According to the Association for Unmanned Vehicle Systems International (AUVSI), 80% of the commercial market for UAVs is expected to be occupied by agricultural UAVs in the future [1]. The reason why agricultural UAVs are popular is because they are expected to play an important role in overcoming some of the challenges of modern agriculture. In particular, an innovative agricultural UAV system is inevitable to ensuring the sustainability of agricultural productivity, which has become difficult to maintain because of climate change, and to meet the growing demand for agricultural products as the world's population increases. Currently, agricultural UAVs are operated mainly for pest control and monitoring numerous crops such as soybean, corn, vegetables, and rice. However, agricultural UAVs are expected to be used for soil and field survey, sowing, spraying, monitoring, irrigation, growth evaluation, mapping, remote sensing, reconnaissance and transportation [2].

By introducing a UAV into traditional agriculture, working hours and labor requirements have been significantly reduced, and the efficiency of agricultural works has improved significantly [3]. However, because a UAV uses a limited battery as its main power source, it is more efficient to use a multi-UAV system, than the current system of a single UAV, to perform agricultural works [4–6]. For example, a single UAV is used for agricultural works such as spraying or monitoring a large farmland; however, it is very inefficient because it requires considerable time and energy. In contrast, when using a multi-UAV, it is possible to carry out cooperative works at the same time (collaboration) or individual agricultural tasks on the assigned farmland (division of labor). As a result, it is possible to complete the agricultural tasks quickly on a large farmland. In others, when using multiple UAVs to find diseased crops, the accuracy of the agricultural tasks is also being increased or equal because there are overlapping areas between the mission areas of each UAV. Although the accuracy of the agriculture task may be superior for a single UAV with a well-planned path, it is greatly influenced by the path planning algorithm. Therefore, the multi-UAV system is more efficient in many ways than the single-UAV system currently in use.

However, when analyzing the existing application of UAV system for agriculture (see, for instance, [7–22]), most studies execute agricultural tasks using a single UAV with an autonomous control. There are few studies on the use of the multi-UAV system in performing agricultural works; thus, it is only at the advanced stage of research [19,21,23]. In [19], an autonomous system for use in inspections for precision agriculture based on the use of single and multiple UAVs was developed. In addition, in [21], precision agricultural technology based on the deployment of a team of UAVs that are able to take georeferenced pictures in order to create a full map by applying mosaicking procedures for post-processing was studied. Although [19,21] used the multi-UAV system for agricultural tasks, they used the centralized controllers through commercial software or a number of computers and did not perform a quantitative evaluation as the number of UAVs increased; thus, they overlooked the ease of the swarm controllers used.

Even if a multi-UAV system is used in agriculture, the most important factor is that the ease of control must be met such that a single operator can easily control multiple UAVs similar to controlling a single-UAV system. In our previous study [24], we developed a distributed swarm control algorithm and implemented a multi-UAV system into the simulator such that a single operator can easily control the multiple agricultural UAVs. Additionally, we argued that the agricultural task with a swarm control algorithm that efficiently and safely controls the multiple UAVs allows the operator to control the multiple UAVs more easily and intuitively and maximize the efficiency of agricultural works. To achieve this, this paper extends the previous study [24,25] by quantitatively evaluating the performance of multi-UAV systems with the proposed algorithm in agricultural scenarios.

For the agricultural scenarios, the remote sensing that represents the task of the agricultural UAV has been set as a benchmark test in this study, and the reason why remote sensing is a representative task is explained in detail in Section 2. In the evaluation, we focused on the ease with which the operator can control the multiple UAVs and improve the efficiency of agricultural works when performing remote sensing tasks using the developed agricultural multi-UAV system. Therefore, the experimental cases are divided into the use of a single-UAV system and the use of a multi-UAV system from the viewpoint of the number of UAVs. Furthermore, we compare the experimental cases by applying an automatic control method and remote-control method from the viewpoint of control. In other words, we perform a total of four experimental cases (single-UAV system using automatic control, hereafter, referred to as *Auto-Single-UAV*; multi-UAV system using automatic control, hereafter, referred to as *Auto-Multi-UAV*; single-UAV system using remote control, hereafter, referred to as *Tele-Single-UAV*; and multi-UAV system using remote control, hereafter, referred as to *Tele-Multi-UAV*) for remote sensing tasks. Finally, a total of seven performance metrics (total time, setup time, flight time, battery consumption, inaccuracy of land, haptic control effort, coverage ratio) were defined to describe and predict the performance of an agricultural UAV system.

2. Review about the Application of UAV in Agriculture

In order to apply the multi-UAV system with distributed swarm control algorithm for agriculture, it is necessary to confirm the type of agricultural UAV to be used and the type of agricultural task to be carried out. Therefore, in this section, the studies that utilized the existing agricultural UAV system are investigated and analyzed in Table 1.

Table 1 reveals an increasing interest in UAVs in the field of agriculture in recent years, and most agricultural UAVs currently in use are single-UAV systems except for [19,21]. The main research areas are remote sensing [7,8,10,13,17,18,20,22], mapping [7,8,11,15], and monitoring [9,12,14,19,26], and it is not yet used in various areas such as sowing and harvesting. Furthermore, the research for irrigation and pest control is on the rise nowadays [16,27,28]. In particular, the remote sensing task is the most widely used task of research for agricultural UAVs and is a basic task achieved by attaching additional hardware or controllers at any time. For this remote sensing, A. Barrientos et al. developed a path planning algorithm and performed the area coverage task [21]. As a result, in this study, the remote sensing task was set as a benchmark test because it is the basis for all agricultural tasks.

In sensors, RGB cameras [13–15,19], thermal cameras [7,8], and multi-spectral cameras [7–12,17,18,20,22] are used. Recent studies focused on agricultural UAVs through image processing, including preprocessing, onboard-processing and post-processing; thus, it is widely used in camera sensors. In addition, a spraying system was installed in the UAV for control, or a related sensor and controller was used in [13,16]. In particular, almost all UAVs are equipped with inertial measurement unit (IMU), pressure sensor and global positioning system (GPS) in common, and it is expected that agricultural UAVs for fully autonomous navigation using IMU and image processing will be developed in the future.

Recently, agricultural UAVs are mainly multi-copter type UAVs, and the fixed-wing type [7,18] or helicopter type [16,20] UAV that was used in the past is gradually disappearing. The reason for the increase in multi-copter type UAV is that the structure is simple, the noise and vibration are small, and it is easy to move and store by folding the frame. It also has the advantage of not requiring a large space for takeoff and landing; however, it also has a problem of low payload and flight time. One of the ways to solve this problem is to use multiple UAVs [29,30].

However, most agricultural UAV systems do not have a multi-UAV system and are still being developed to address the limitations of a single-UAV system [31]. In the case of research using multi-UAV, the completion time of the mission is remarkably shortened, and the efficiency of the work is greatly improved [21]. Taking this advantage into consideration, the agricultural multi-UAV system is essential for automation and unmanned technology of future agriculture, and it is considered as one way to solve the food shortage problem. In the case of Swarm Robotics for Agricultural Applications (SAGA) projects in Europe, for more details, see [32], agricultural swarm robotics is studying to prepare for the fourth industrial revolution and to build precision agriculture and smart farm [33]. Another project, Mobile Agricultural Robot Swarms (MARS), aimed to develop small and stream-lined mobile agricultural robot units to fuel a paradigm shift in farming practices. Recent research trends are focusing considerable attention on multi-robots and swarm robotics; furthermore, multiple agricultural UAVs are expected to become the core of future agricultural technology.

Therefore, the proposed agricultural multi-UAV system based on the distributed swarm control algorithm is a necessary study for the future agricultural technology, and quantitative evaluation of developed system contributes to the performance evaluation of the agricultural UAV system which has not been examined previously.

Table 1. Datasheet explaining reference, objective, agricultural task, UAV type, control method, sensors, and target crop for a recent study using an agricultural UAV.

Reference	Objective	Task	UAV	Control	Sensors	Crop
B.Allred et al. [7]	Evaluation of VIS, NIR, and TIR imagery for drainage pipe mapping	Remote Sensing and Mapping	Single Fixed-wing type UAV	Ground Control Station (Auto)	Multi-spectral camera, thermal camera	Corn, Soybeans
L. G. Santesteban et al. [8]	To estimate the instantaneous and seasonal variability of plat water status	Remote Sensing and Mapping	Single X8 type UAV	Ground Control Station (Auto)	Multi-spectral camera, thermal camera	Vineyard
F. A. Vega et al. [9]	To determine the capability of an UAV system to acquire multi-temporal images	Monitoring	Single Quadcopter type UAV	Ground Control Station (Auto)	Multi-spectral camera	Sunflower
P. Tokekar et al. [10]	To study the problem of maximizing the number of points visited by the UAV	Remote Sensing	Single Octocopter type UAV + Single UGV	Ground Control Station (Auto)	Multi-spectral camera	Field
J. Torres-Sánchez et al. [11]	To report an innovative procedure for a high-throughput and detailed 3D monitoring of agricultural tree plantations	Mapping	Single Quadcopter type UAV	Remote Control (Teleoperation)	Visible-light camera, Multi-spectral camera	Olive plantation
A. Noriega et al. [12]	Development of a path planning method to minimize the time required to scan a field	Monitoring	Single Octocopter type UAV	Ground Control Station (Auto)	Multi-spectral camera	Field
B. H. Alsala et al. [13]	To describe a modular and generic system that is able to control the UAV using computer vision	Remote Sensing	Single Quadcopter type UAV	Ground Control Station (Auto)	RGB camera, Ultrasonic, Spraying system	Weed
R. Jannoura et al. [14]	Evaluation of crop biomass using true colour aerial photographs	Monitoring	Single Hexacopter type UAV	Remote Control (Teleoperation)	RGB camera	Pea, Oat
M.P. Christiansen et al. [15]	Designing and testing a UAV mapping system for agricultural field surveying	Mapping	Single Quadcopter type UAV	Ground Control Station (Auto)	RGB camera, LiDAR	Wheat
B. S. Faiçal et al. [16]	To propose a computer-based system that able to adapt the UAV control rules	Spraying	Single Helicopter type UAV	Ground Control Station (Auto)	Spraying control system	Field
J. Torres-Sánchez et al. [17]	To describes the specifications and configurations of a UAV for site-specific weed management	Remote Sensing	Single Quadcopter type UAV	Ground Control Station (Auto)	Point-and-shoot camera, Multi-spectral camera	Sunflower
P. J. Zarco-Tejada et al. [18]	Development of methods for leaf carotenoid content estimation, using an UAV	Remote Sensing	Single Fixed-wing type UAV	Ground Control Station (Auto)	Multi-spectral/Hyper-spectral camera	Vineyard
D. Doering et al. [19]	Development of an autonomous system to perform inspections for agriculture based on the use of multiple UAVs	Monitoring	Multiple Quadcopter type UAV	Ground Control Station (Auto)	RGB camara	Field
H. Xiang et al. [20]	Development of an automatic aerial image georeferencing method for an UAV platform	Remote Sensing	Single Helicopter type UAV	Ground Control Station (Auto)	Multi-spectral camera	Field
A. Barrientos et al. [21]	Practical experimentation with an integrated tool to create a full map using multiple UAVs	Area Coverage and Path Planning	Multiple Quadcopter type UAV	Ground Control Station (Auto)	IMU, Pressure sensor, GPS	Vineyard
J. A. Arroyo et al. [22]	To propose a model to estimate Nitrogen nutrition level in crops using agricultural UAV	Remote Sensing	Single Quadcopter type UAV	Ground Control Station (Auto)	Multi-spectral camera	Corn

3. The Control of Multiple UAV System

3.1. UAV Dynamics

We consider N quadrotor-type UAVs with 3-DOF Cartesian positions that are denoted by $p_i \in \mathbb{R}^3$, $i = 1,2,...,N$. Flight control of UAVs is derived from the following under-actuated Lagrangian dynamics equation in $SE(3)$ [34]

$$m_i \ddot{p}_i = -\lambda_i R_i e_3 + m_i g e_3 + \delta_i \tag{1}$$

$$J_i \dot{w}_i + S(w_i) J_i w_i = \gamma_i + \zeta_i \tag{2}$$

with the following attitude kinematic equation

$$\dot{R}_i = R_i S(w_i) \tag{3}$$

where $m_i > 0$ denotes mass, $p_i := [p_1; p_2; \ldots, p_N] \in \mathbb{R}^{3N}$ denotes the Cartesian center-of-mass position represented in the north-east-down (NED) inertial frame $\{O\} := \{N^O, E^O, D^O\}$, $\lambda_i \in \Re$ denotes thrust control input, $R_i \in SO(3)$ denotes the rotational matrix describing the body frame $B := \{N^B, E^B, D^B\}$ of UAV w.r.t. to the inertial frame $\{O\}$, g is the gravitation constant, $e_3 = [0,0,1]^T$ denotes the basis vector representing the down direction and representing that thrust and gravity act in the D direction, $J_i \in \Re^{3\times3}$ denotes the UAV's inertia matrix with respect to the body frame $\{B\}$, $w_i \in \mathbb{R}^3$ denotes the angular velocity of the UAV relative to the inertial frame $\{O\}$ represented in the body frame $\{B\}$, $\gamma_i \in \mathbb{R}^3$ denotes the attitude torque control input, $\delta_i, \zeta_i \in \mathbb{R}^3$ denote the aerodynamic perturbations, and $S(w_i) : \mathbb{R}^3 \to so(3)$ denotes the skew-symmetric operator defined such that for $\alpha, \beta \in \mathbb{R}^3, S(\alpha)\beta = \alpha \times \beta$. For typical UAV flying, $\delta_i, \zeta_i \approx 0$.

3.2. Distributed Swarm Control

In the previous study [24], we developed the following distributed swarm control on each UAV, for the ith UAV,

$$\dot{p}_i(t) := u_i^u + u_i^f + u_i^o \tag{4}$$

where the meaning of the three control inputs $u_i^u \in \mathbb{R}^3, u_i^f \in \mathbb{R}^3$ and $u_i^o \in \mathbb{R}^3$ represents the velocity terms of the UAV.

3.2.1. UAV Control

The first velocity term, $u_i^u := \{u_i^a, u_i^n, u_i^t\} \in \mathbb{R}^3$ denotes a control input that directly controls the UAV and represents a velocity control input according to the control method. Normally, the UAV control method mainly uses the following three methods: the method of fully autonomous driving (u_i^a); the method of driving on a certain path specified by the operator (u_i^n); the method of teleoperation by the operator in real time (u_i^t). In the case of u_i^a, the position of the UAV $\hat{\mathbf{x}}_t$ at time t, given the previous k positions $\mathbf{x}_{t-k:t-1}$ and the corresponding laser measurements $b_{t-k:t}$, is as follows:

$$\hat{\mathbf{x}}_t = \text{argmax}\; p(\mathbf{x}_t \mid \mathbf{x}_{t-k:t-1}, \mathbf{b}_{t-k:t}), \quad u_i^u = \dot{\mathbf{x}}_t \tag{5}$$

where $\mathbf{x}_t = (x_t, y_t, z_t)$. We briefly review the control of autonomous UAVs and refer the reader to [35] for further details. In this study, because there are many limitations to apply to farming in the case of u_i^a, u_i^n was set as an automatic control method and u_i^t was set as a remote control method. Additionally, u_i^n and u_i^t are discussed in detail in Sections 3.3 and 3.4.

3.2.2. Formation Control

The second velocity term, $u_i^f \in \mathbb{R}^3$ denotes a control input to avoid a collision among UAVs, preserves connectivity, and achieves a certain desired formation as specified by the desired distances $d_{ij}^c \in \mathbb{R}^+ \quad \forall i = 1, \ldots, N,$ and $\forall j \in \mathcal{N}_i$, as defined by

$$u_i^f := -\sum_{j \in \mathcal{N}_i} \frac{\partial \varphi_{ij}^f(\|p_i - p_j\|^2)^T}{\partial p_i} \tag{6}$$

where φ_{ij}^c denotes a certain artificial potential function to create an attractive action if $\|p_i - p_j\| > d_{ij}^c$, a repulsive action if $\|p_i - p_j\| < d_{ij}^c$, and a null action if $\|p_i - p_j\| = d_{ij}^c$.

3.2.3. Obstacle Avoidance Control

The final velocity term, $u_i^o \in \mathbb{R}^3$, is expressed by the following equation as a control input based on a potential field that allows multiple UAVs to avoid obstacles through a certain distance threshold: $\mathcal{D}_o \in \mathbb{R}^+$

$$u_i^o := -\sum_{r \in \mathcal{O}_i} \frac{\partial \varphi_{ir}^o(\|p_i - p_r^o\|)^T}{\partial p_i} \tag{7}$$

where $\mathcal{O}_i$ denotes the set of obstacles of the ith UAV with an obstacle point p_r^o that corresponds to the position of the rth obstacle in the environment, and φ_{ir}^o denotes a certain artificial potential function that produces a repulsive action if $\|p_i - p_r^o\| < \mathcal{D}_o$, and a null action if $\|p_i - p_r^o\| \geq \mathcal{D}_o$. When the distance between the UAVs and the obstacles becomes closer to D_o, then the repulsive potential function increases to infinity.

Here, we briefly reviewed the distributed swarm control architecture and refer the reader to [24] for further details.

3.3. *Autonomous Control*

Automatic control of UAV through a ground station uses the navigation control based on GPS waypoint. The navigation control uses PID controller when UAV is in GUIDED mode, as defined by

$$u_i^n(t) = K_P(t)e_i(t) + K_I \int e_i(t)dt + K_D \frac{d}{dt} e_i(t) \tag{8}$$

where $r_i \in \mathbb{R}^3$ denotes the target position, $e_i(t) = r_i - p_i$ denotes the position error between target point and UAV, and K_P, K_I, and K_D are the gain values of the navigation controller, respectively.

In (8), the UAV follows the target point preset by the operator, and the position error decreases gradually. Here, the velocity of the UAV changes according to the position error. However, because the performance of the navigation controller changes depending on the gain value, appropriate values must be set through tuning.

3.4. *Teleoperation*

The teleoepration uses the haptic device to control the UAV. Therefore, we consider a 3-DOF haptic device for master as modeled by the following nonlinear Lagrangian dynamics equation [36]

$$M(q)\ddot{q} + C(q, \dot{q})\dot{q} = \tau + f_h \tag{9}$$

where $q \in \mathbb{R}^3$ denotes the configuration of the haptic device (e.g., the position of end effector), $M(q) \in \Re^{3\times3}$ denotes the positive-definite/symmetric inertia matrix, $C(q, \dot{q}) \in \Re^{3\times3}$ denotes the Coriolis matrix, and $\tau \in \mathbb{R}^3$, $f_h \in \mathbb{R}^3$ denote the control input and human forces, respectively.

The velocity term, $u_i^t \in \mathbb{R}^3$, represents the teleoperation command for the desired velocity input of the UAV that is directly controlled by the operator by using the configuration of the haptic device q

$$u_i^t = \Lambda q \qquad \forall i \tag{10}$$

where $\Lambda \in \mathbb{R}^+$ denotes a constant scale factor used to match different scales between q and the UAV desired velocity u_i^t, and $q \in \mathbb{R}^3$ denotes the position of end effector. In (10), multiple UAVs with an unbounded workspace can fly without the limitations of workspace by controlling the desired velocity by using the configuration of the haptic device with a bounded workspace.

4. Experimental Design

4.1. Remote Sensing Task

In this experiment, we set the remote sensing for the agricultural task as shown in Figure 1, and the reason for setting this task is explained in Section 2. The experiment is the operation of sensing using UAV with mounted sensors for a predetermined test area, and the experimental procedure includes the whole process from setup time before takeoff to landing after a flight time of mission. The starting point of the remote sensing task is the position where the UAV was originally located at the base station, and this point is also set as the ending point.

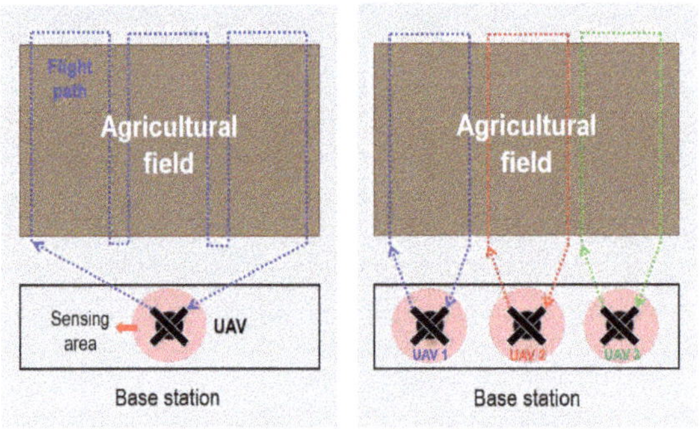

(**a**) (**b**)

Figure 1. The concept of the remote sensing tasks including sensing area, which the area covered by the camera mounted on the unmanned aerial vehicle (UAV). There is no reference path, and the point where the UAV is located in base station is set as the starting point and the ending point, and the UAV is controlled using the automatic controller and the remote controller according to the operator's judgment. (**a**) Case of single UAV (**b**) Case of multiple UAVs.

Experimental progress is required for the operator to control the agricultural UAV system based on the distributed swarm control algorithm while performing the remote sensing through the sensor attached to the UAV. In addition, the operator was required to look at the formation of the UAV from the remote site or to control it by looking at the camera screen mounted on the UAV. At this time, there is no reference path for the remote sensing tasks, and the UAV is remotely controlled by the intuitive judgment of the operator or is automatically controlled by setting a suitable waypoint. The time at which the UAV was landing properly was set as the criterion for the end of the experiment and the success of the experiment. Here, the operator decided to terminate the experiment by judging the moment when the UAV landed successfully.

Experiments consisted of four cases consisting of *Auto-Single-UAV*, *Auto-Multi-UAV*, *Tele-Single-UAV* and *Tele-Multi-UAV*. In the case of multi-UAV cases, a total of three quadcopters was used for remote sensing. When automatic control is used, the UAV is automatically controlled by specifying the GPS-based waypoint using ground control station (GCS). However, in the case of teleoperation, the operator controls the UAV by controlling the haptic device. In other words, the experimental cases are defined by

- *Auto-Single-UAV* : $i = 1$, $\dot{p}_i(t) := u_i^n$
- *Auto-Multi-UAV* : $i = 3$, $\dot{p}_i(t) := u_i^n$ where the target position $r_i \in \{r_1, r_2, r_3\}$
- *Tele-Single-UAV* : $i = 1$, $\dot{p}_i(t) := u_i^t$
- *Tele-Multi-UAV* : $i = 3$, $\dot{p}_i(t) := u_i^t + u_i^f + u_i^o$

In the case of *Tele-Multi-UAV*, it is a multi-UAV system applying our proposed distributed swarm control algorithm. A total of three trials were performed for each case and a total of 12 trials were performed in agricultural experiments.

4.2. Performance Metric

We used a total of seven performance metrics to evaluate the performance of agricultural UAV systems. The performance metrics are mainly focused on the control effort of the operator and the performance of the system for the agricultural task, and total time, setup time, flight time, battery consumption, inaccuracy of land, haptic control effort, and coverage ratio were used as the metrics.

Definition 1. ***Total time*** *is the completion time during the agricultural task as defined by*

$$P_{TT} := \int_{t_0}^{t_c} dt \tag{11}$$

where t_0 is the start time, t_c is the completion time of the agricultural task.

Definition 2. ***Setup time*** *is defined as the time that the operator prepares before the UAV executes the agricultural task,*

$$P_{ST} := \int_{t_0}^{t_s} dt \tag{12}$$

where t_s is the time that UAV takes off to perform the agricultural task.

Definition 3. *The metric for the* ***Flight time*** *is*

$$P_{FT} := P_{TT} - P_{ST} \tag{13}$$

Definition 4. ***Battery consumption*** *is defined as*

$$P_{BC} := \frac{\int_{t_0}^{t_c} B_{consumed}(t)dt}{B_{total}} \times 100 \tag{14}$$

where B_{total} is the total amount of batter and $B_{consumed}$ is the consumption of the battery.

Definition 5. *The metric for the* ***Inaccuracy of land*** *is*

$$P_{IL} :=\parallel p_i(t_0) - p_i(t_c) \parallel \tag{15}$$

Definition 6. ***Haptic control effort*** *is defined as the total distance of haptic device moved by operator shown in below,*

$$P_{HC} := \sum_{t=0}^{t_c-1} \parallel q(t+1) - q(t) \parallel \tag{16}$$

where $q(t)$ is the configuration of the haptic device.

Definition 7. ***Coverage ratio*** *is defined as*

$$P_{CR} := \frac{A_{covered}(t)}{t \times A_{unit}} \times 100 \tag{17}$$

where $A_{covered}$ is the area covered by the sensor mounted on UAV, and A_{unit} is the area covered by sensor per time.

P_{TT}, P_{ST}, and P_{FT} are basically the most important time factors for the UAV to perform agricultural tasks. As the value of these metrics increases, it implies that energy and costs for agricultural task increase. Therefore, the smaller the value of P_{TT}, P_{ST}, and P_{FT}, the better the performance of the system. Similarly, the lower the value of P_{BC}, P_{IL} and P_{HC}, the lower the energy consumption of the UAV, the lower the error of the landing, and the lower the control effort of the operator. However, the values of P_{CR} indicate the performance of the remote sensing tasks; therefore, the higher the value, the better the performance.

4.3. Experimental Setup

The experimental environment was built to allow the UAV to control and communicate with ROS on the notebook of the 16.04 LTS version Ubuntu. In the experiment, a remote sensing task is performed while recording a real-time image by attaching an RGB camera to the UAV. The experimental environment is shown in Figure 2 and the experiment was carried out on a clear day with low geomagnetic coefficient. The UAV used in the developed system was a quadcopter type UAV (3DR SOLO), which is suitable for remote sensing because of low vibrations. As shown in Figure 3, the UAV is basically composed of a frame and battery, a GPS receiver and a flight controller (FC), a camera, an IMU consisting of an accelerometer, a gyroscope, and a magnetic field, supplementary battery that supplies power to the onboard computer, onboard computer for controller, and a printed circuit board (PCB) for connection between the UAV and onboard computer. The payload of this UAV is 450 g, and it flies without problems when all the components are connected; in this state, it can fly up to 20 min.

For the distributed system, we constructed a multi-UAV system using the above UAV, and the developed system consists of a number of UAVs and a base station. As shown in Figure 4, the base station consists of a PC with a ROS-based controller and a haptic device, which is used as the master device for teleoperation, a wireless adapter, and a router for the user datagram protocol (UDP) communication. Here, each PC and the onboard computer mounted on the UAV communicate with each other through a router and exchange data, thereby constructing a distributed system. It is also possible to construct a centralized system easily using this system configuration.

Communication basically used UDP communication and changed the default port of each UAV to avoid interference between UAVs. After changing the default port of the UAV, we set up the onboard computer to automatically connect to the router's network used in this experiment. Therefore, when the UAV is booted, it is automatically located on the same network with a computer without any configuration and recognizes and communicates with each other through different IP address and ports. The channel used 2.4 GHz frequency, and the optimum channel was set to receive the data out of interference.

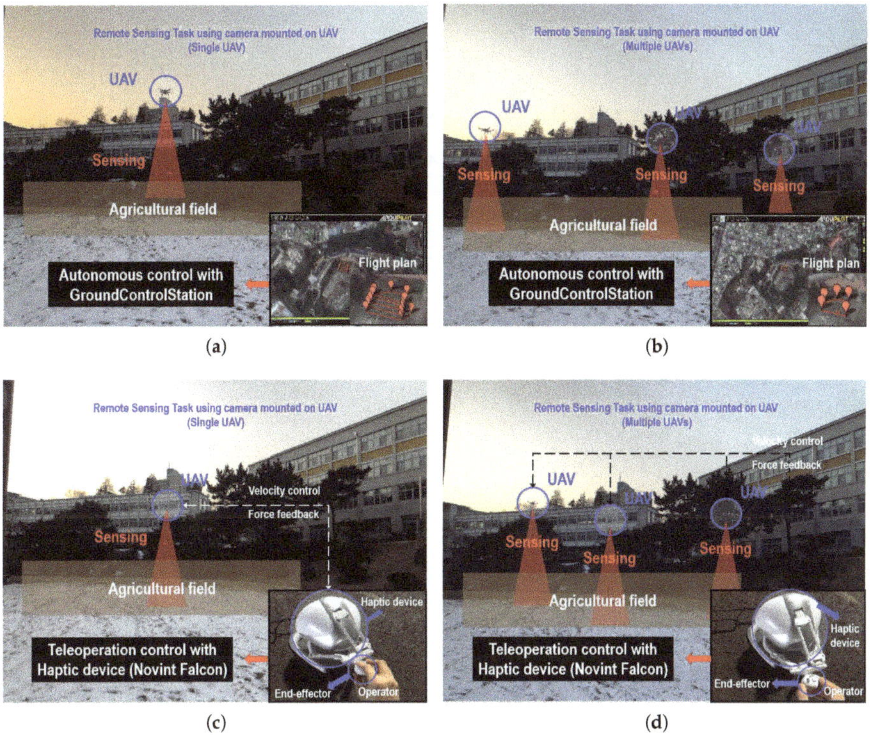

Figure 2. Experimental setup for experiments: Unmanned aerial vehicle (UAV) performs remote sensing task in the test area (**a**) case of *Auto-Single-UAV* (**b**) case of *Auto-Multi-UAV* (**c**) case of *Tele-Single-UAV* (**d**) case of *Tele-Multi-UAV*.

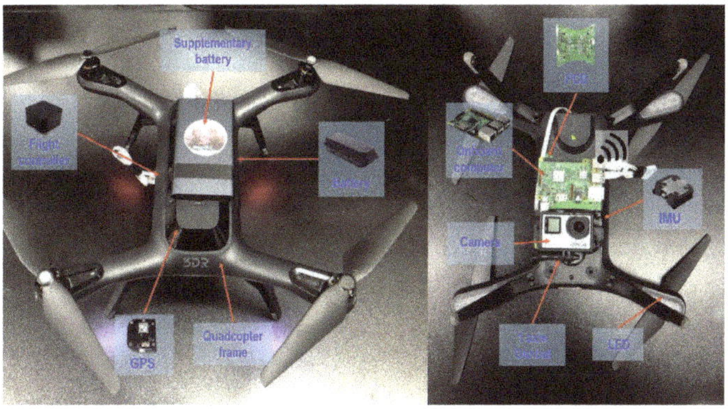

Figure 3. Quadcopter type unmanned aerial vehicle: 3DR SOLO. We attached additional hardware to the 3DR SOLO and performed a remote sensing task. The left picture is from the top view and the right is the bottom view of the 3DR SOLO.

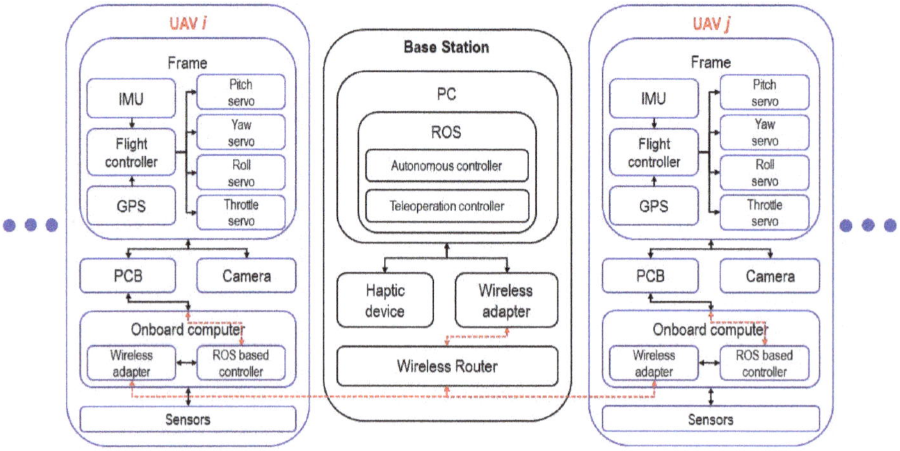

Figure 4. Scheme of multiple unmanned aerial vehicle (UAV) system: Robot operating system (ROS) based distributed system. For this, additional onboard computers were mounted on the UAV and wireless router was used for communications. In addition, the ROS based controller is mounted not only on the computer but also on the UAV.

4.4. Data Acquisition and Analysis

During the experiment, we recorded the local position, global position, linear velocity, angular velocity, battery state, and heading value of the UAV, as well as the experiment time, position and force of the haptic device, and the raw date of sensors at 1000 Hz in the ground station. All data were transferred from UAV to the ground station through Micro Air Vehicle Communication Protocol (MAVLink), and we monitored the data via *rostopic*, which is command-line tool for displaying debug information about ROS topics, including publishers, subscribers, publishing rate, and ROS Messages, and stored it using *rosbag*, which is a set of tools for recording from and playing back to ROS topics.

5. Experimental Results

Figure 5 shows the results of one flight trial. We performed a statistical analysis of performance metrics after all experiments (3 trials per case, 12 trials in total), which are summarized in Table 2.

Table 2. Experimental results for each case and performance metric.

Metric	*Auto-Single-UAV*	*Auto-Multi-UAV*	*Tele-Single-UAV*	*Tele-Multi-UAV*
P_{TT} [s]	96.2	78.8	65.1	32.6
P_{ST} [s]	48.7	64.5	13.5	18.9
P_{FT} [s]	47.5	14.3	51.6	13.7
P_{BC} [%]	3.9	1.6	4.2	1.2
P_{IL} [cm]	18.0	19.3	8.2	13.8
P_{HC} [cm]	0.0	0.0	31.1	15.3
P_{CR} [%]	100.0	300.0	100.0	300.0

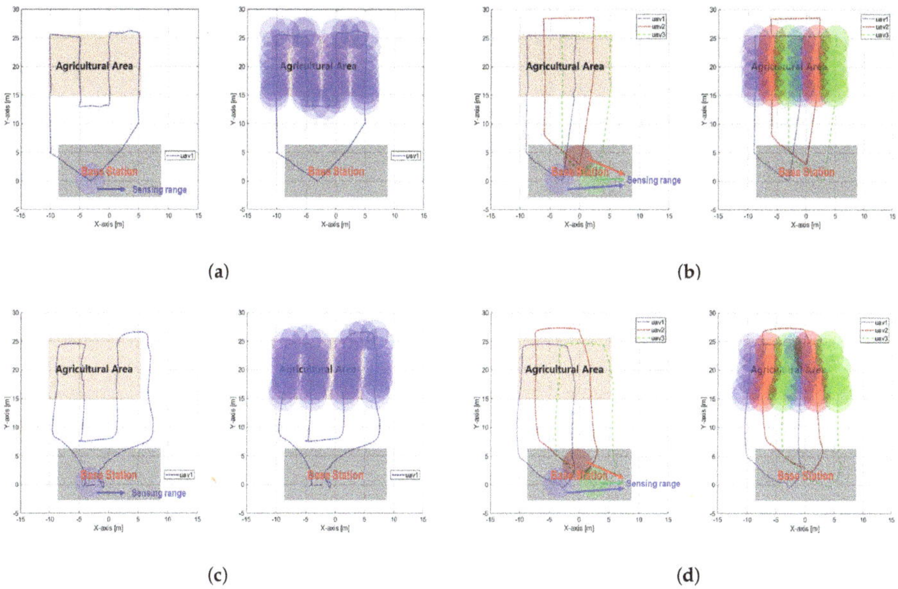

Figure 5. Experimental results of remote sensing for each case: Flight trajectory for one trial. (**a**) case of *Auto-Single-UAV* (**b**) case of *Auto-Multi-UAV* (**c**) case of *Tele-Single-UAV* (**d**) case of *Tele-Multi-UAV*.

5.1. Total Time

Because P_{TT} is the sum of P_{ST} and P_{FT}, a detailed evaluation should identify two metrics. However, P_{TT} is one of the most important factors when evaluating the system, because it shows intuitively the completion time of the remote sensing task. In addition, a smaller P_{TT} reduces the overall energy consumption and saves the cost of operating the system. P_{TT} is the highest for *Auto-Single-UAV* and lowest for *Tele-Multi-UAV* in experiment results. Additionally, P_{TT} is less in the teleoperation method than in the automatic control method, and the multi-UAV system is less P_{TT} than the single-UAV system. In detail, P_{TT} decreased by 31.1 s from 96.2 s (*Auto-Single-UAV*) to 65.1 s (*Tele-Single-UAV*) and decreased by 46.2 s from 78.8 s (*Auto-Multi-UAV*) to 32.6 s (*Tele-Multi-UAV*).

When using the multi-UAV system, the decrease was 17.4 s from 96.2 s (*Auto-Single-UAV*) to 78.8 s (*Auto-Multi-UAV*) and 32.5 s from 65.1 s (*Tele-Single-UAV*) to 32.6 s (*Tele-Multi-UAV*). Moreover, when comparing the proposed *Tele-Multi-UAV* and *Auto-Single-UAV*, experimental results show that T_T for *Tele-Multi-UAV* is approximately 66.1% (from 96.2 s to 32.6 s) lower than *Auto-Single-UAV*.

5.2. Setup Time

P_{ST} is what an operator does before the UAV performs an agricultural task, which is related to the operator's control effort aspect. No matter how good a system is, it is not good if the control effort of the operator is significant. Therefore, P_{ST} is a very important metric and should be considered when developing a system. In experiments, P_{ST} is the highest at 64.5 s (*Auto-Multi-UAV*) and P_{ST} is the lowest at 13.5 s (*Tele-Single-UAV*). The tendency is that P_{ST} is less when using teleoperation method compared to automatic control method; however, when the multi-UAV system is used, P_{ST} increases more than the single-UAV system. Quantitatively, P_{ST} decreased by 35.2 s from 48.7 s (*Auto-Single-UAV*) to 13.5 s (*Tele-Single-UAV*) and decreased by 45.6 s from 64.5 s (*Auto-Multi-UAV*) to 18.9 s (*Tele-Multi-UAV*). However, P_{ST} increased from 48.7 s (*Auto-Single-UAV*) to 64.5 s (*Auto-Multi-UAV*) in 15.8 s and from 13.5 s (*Tele-Single-UAV*) to 18.9 s (*Tele-Multi-UAV*) in 5.4 s.

Most importantly, P_{ST} of *Tele-Multi-UAV* compared to *Auto-Single-UAV* was reduced by 61.2% (from 48.7 s to 18.9 s). Additionally, P_{ST} of *Tele-Multi-UAV* compared to *Tele-Single-UAV* was increased by 39.9% (from 13.5 s to 18.9 s). P_{ST} for *Auto-Multi-UAV* increased by 32.4% (from 48.7 s to 64.5 s) compared to *Auto-Single-UAV*. This result means that the use of multiple UAVs unconditionally increases the work efficiency; however, the operator's control effort and fatigue increased even more. However, this result is heavily influenced by the user interface (UI), controller and feedback [37,38].

5.3. Flight Time

P_{FT} is the time that UAV travels for agricultural task and is directly related to the energy consumption of UAV. In other words, P_{FT} is the working time of UAV, and the smaller the P_{FT}, the shorter the working time and the less energy consumption. However, it can vary greatly depending on the gain value of the velocity control input. The experimental results show that P_{FT} is the lowest for *Tele-Multi-UAV* (13.7 s) and the highest for *Tele-Single-UAV* (51.6 s). However, there is no significant difference between *Tele-Multi-UAV* (13.7 s) and *Auto-Multi-UAV* (14.3 s). Considering *Auto-Single-UAV* (47.5 s) and *Tele-Single-UAV* (51.6 s), P_{FT} increase when the teleoperation is used rather than automatic control.

It is seen that P_{FT} is significantly reduced when using multiple UAVs rather than a single-UAV system. In the case of *Auto-Single-UAV* and *Auto-Multi-UAV*, the decrease was 33.2 s (from 47.5 s to 14.3 s). Additionally, in the case of *Tele-Single-UAV* and *Tele-Multi-UAV*, the decrease was 37.9 s (from 51.6 s to 13.7 s). Obviously, the case of *Tele-Multi-UAV* had a 71.2% (from 47.5 s to 13.7 s) decrease in P_{FT} compared to *Auto-Single-UAV* in the experiment. These results indicate that using a multi-UAV system can save the battery by reducing P_{FT} over a single-UAV system.

5.4. Battery Consumption

The UAV typically consumes considerable battery power when flying; P_{BC} is similar to the P_{FT}. This metric is very important as an intuitive indicator of the potential for solving the battery shortage problems facing current agricultural UAVs. Therefore, the smaller the P_{BC}, the better the performance of the agricultural UAV system.

In experiments, P_{BC} is the smallest at 1.2% for *Tele-Multi-UAV* and the largest at 4.2% for *Tele-Single-UAV*. The difference between *Tele-Multi-UAV* and *Tele-Single-UAV* is 3.0%; however, if P_{FT} is longer, the difference in P_{BC} increases even more. Additionally, P_{BC} decreased by 2.3% from 3.9% (*Auto-Single-UAV*) to 1.6% (*Auto-Multi-UAV*) when using the multi-UAV system. Furthermore, in the case of *Tele-Multi-UAV*, the results show that P_{BC} is 2.6% (from 1.2% to 3.9%) less than *Auto-Single-UAV*. As a result, it is more efficient to use multiple UAVs than to use a single UAV, because when nth UAV performs the agricultural task, the agricultural area is divided by n, and each UAV performs an agricultural task only on $1/n$ areas. However, if we proceed to the same accuracy of agricultural task for a given farmland, the teleoperation method consumes much more P_{BC} than the automatic control method. This is because the control is limited when the operator performs teleoperation on the remote site.

5.5. Inaccuracy of Land

P_{IL} is not an index related to the performance of agricultural task; however, it is an element that affects the performance of the system. This metric is set to determine the accuracy of landing and is a very important performance metric when the base station is a narrow or dangerous area or when the UAV lands on the unmanned ground vehicle (UGV). Therefore, this metric must be considered to build smart farming in the future.

P_{IL} is the highest for *Auto-Multi-UAV* (19.3 cm) and lowest for *Tele-Single-UAV* (8.3 cm) in experiment results. The reason for this is that the disturbance can not be ignored when performing the experiment in an outdoor environment, and error is particularly affected by GPS, which is considered to be inaccurate because of the performance of the device or the weather and wind. Generally,

P_{IL} tends to increase when using multiple UAVs. In detail, P_{IL} increased 1.3 cm from 18.0 cm (*Auto-Single-UAV*) to 19.3 cm (*Auto-Multi-UAV*) and increased 5.5 cm from 8.3 cm (*Tele-Single-UAV*) to 13.8 cm (*Tele-Multi-UAV*). The reason why P_IL increases when using multi-UAV is because signal disturbance occurs. Additionally, P_{IL} decreased by 23.3% (4.2 cm) from 18.0 cm (*Auto-Single-UAV*) to 13.8 cm (*Tele-Multi-UAV*). However, this result is reversed when using a more accurate and expensive GPS receiver.

5.6. *Haptic Control Effort*

P_{HC} numerically shows the control input of the operator when using the teleoperation. In order to more precisely measure the control effort of the operator, it is necessary to measure the input force; however, in this study, P_{HC} is regarded as a general control effort (e.g., see [39]). Experimental results show that P_{HC} is significantly reduced when using a multi-UAV system than when using a single-UAV system. Quantitatively, P_{HC} decreased by 15.8 cm from 31.1 cm at *Tele-Single-UAV* to 15.3 cm at *Tele-Multi-UAV*.

As a percentage, the control effort at *Tele-Multi-UAV* tended to decrease by 50.9% (from 31.1 cm to 15.3 cm) in the experiment compared to *Tele-Single-UAV*. The reason for this is that when using a single-UAV system, basically it is necessary to carry out multiple flying and agricultural tasks, and therefore, the effort of the operator to control the haptic device is inevitable. These results indicate that using the multi-UAV system rather than a single-UAV system, as opposed to P_{ST}, reduced the operator's control effort. P_{HC} can be regarded as a limitation of teleoperation rather than automatic control; however, if the proper haptic feedback adds to the operator, the UAV can be controlled almost without operator's control input, similar to automatic control [40].

5.7. *Coverage Ratio*

P_{CR} yields the performance of the agricultural task by calculating the covered area at the same time. This metric should be considered when developing a system as a very important indicator along with P_{TT} in performing agricultural works. No matter how fast P_{TT} is, if P_{CR} is low, the efficiency of the agricultural task will be low. Therefore, P_{CR} represents the simultaneous covered area of the agricultural UAV system. In the experiment, the recording was done for the test area through the RGB-camera mounted on UAV. As a result, P_{CR} of a single-UAV system is only one-third of the performance compared to a multi-UAV system. In particular, when multi-UAV system is used, P_{CR} is increased by as many as the number of UAVs; thus, it offers a much better performance.

6. Discussions

Table 3 summarizes the experimental results on the comparison between single and multiple systems and the comparison between automatic control and teleoperation. The results show the increase and decrease in teleoperation based on the single-UAV system when *Single* → *Multi* and automatic control when *Auto* → *Tele*.

Table 3. Experimental results: comparison between single-UAV system and multi-UAV system and comparison between automatic control method and teleoperation method. For example, *Auto-UAV* and *Single* → *Multi*, $result = \frac{(Auto-Multi-UAV)-(Auto-Single-UAV)}{Auto-Single-UAV} \times 100$.

Metric	*Auto-UAV* *Single* → *Multi*	*Tele-UAV* *Single* → *Multi*	*Single-UAV* *Auto* → *Tele*	*Multi-UAV* *Auto* → *Tele*
P_{TT} [s]	−18.1%	−50.0%	−32.3%	−58.7%
P_{ST} [s]	+32.4%	+39.9%	−72.2%	−70.7%
P_{FT} [s]	−69.8%	−73.5%	+8.6%	−4.7%
P_{BC} [%]	−59.3%	−70.5%	+9.1%	−21.0%
P_{IL} [cm]	+7.1%	+66.8%	−54.0%	−28.4%
P_{HC} [cm]	0.0%	−50.9%	+	+
P_{CR} [%]	+200.0%	+200.0%	0.0%	0.0%

6.1. *Single vs. Multiple*

Currently, a method for solving the problems of battery and payload shortage in an agricultural UAV system is to use a multi-UAV system. Using multiple UAVs requires more time to set up and extra initial cost; however, it brings about results such as improved accuracy of agricultural task, reduced working time, and reduced operator's control efforts. As a result, agricultural multi-UAV systems are regarded as better systems than single-UAV systems. However, it is necessary to thoroughly confirm that it has acceptable performance before introducing the agricultural multi-UAV system. Therefore, in this subsection we will quantitatively evaluate and analyze the single-UAV system and multi-UAV system.

First, if *Multi-UAV* is used, P_{TT} is reduced by 18.1% at *Auto-UAV* and reduced by 50.0% at *Tele-UAV*. These results show a clear reduction in P_{TT} for *Multi-UAV*, which improves the efficiency of agricultural works. Although three UAVs were used in this study, the agricultural multi-UAV system based on distributed swarm control showed better performance as the number of UAVs increased and the farmland became larger. However, experimental results show that P_{ST} increases with *Multi-UAV*. An 32.4% and a 39.9% increase in *Auto-UAV* and *Tele-UAV* were confirmed, respectively. These values are disadvantages of the multi-UAV system; however, it is a more efficient system because multiple UAVs are controlled with a few P_{ST}. Generally, to control three UAVs, a P_{ST} of three times is required. However, if the operator controls the multi-UAV with additional P_{ST} of only 30.0%$\sim$40.0%, the agricultural works are economically beneficial. First, P_{ST} is greatly influenced by UI; thus, P_{ST} is significantly reduced if human-centered GUI and PUI are developed.

Even though P_{ST} increases, multiple UAVs reduce P_{FT} of each UAV through collaboration. This is the main reason why P_{TT} decreases even if P_{FT} increases. In the experimental results, *Auto-UAV* and *Tele-UAV* decreased by 69.8% and 73.5%, respectively. Because three UAVs are used for *Multi-UAV*, theoretically it should be reduced by approximately 66.0%. However, in the experiment, it is confirmed that it is lower than the reference value (66.0%), which means that the energy of the UAV is further reduced. Furthermore, because P_{FT} decreases, P_{BC} is reduced, and the experimental results show that P_{BC} is reduced by 59.3% (*Auto-UAV*)$\sim$70.5% (*Tele-UAV*) when three UAVs are used. As a result, it is considered that the multi-UAV system overcomes the battery shortage problem of current agricultural UAV systems. Therefore, no matter how vast the area of farmland is, multiple UAVs collaborate to perform agricultural tasks without encountering battery shortage.

Even though P_{IL} tends to increase when using *Multi-UAV*, this metric is subject to a change by other factors. For example, in the case of *Auto-UAV*, P_{IL} is greatly influenced by GPS. However, GPS varies with device resolution, wind, weather, and geomagnetic factors. In the case of *Tele-UAV*, P_{IL} can be greatly influenced by UI because the operator directly watches the UAV or the camera mounted on the UAV for takeoff and landing. Interestingly, experiments show that P_{HC} decreases when using multiple UAVs. This metric is only for *Tele-UAV*, which decreased by 50.9% in the experiment. These results are related to P_{FT}, because the area allocated to each UAV is reduced; thus, it is natural that P_{HC} is reduced. Unlike P_{ST}, P_{HC} decreases as UAV increases; thus, it is advantageous to use agricultural multi-UAV systems based on the distributed swarm control.

Finally, P_{CR} is significantly improved. When multiple UAVs are used, P_{CR} increases (200.0%); thus, accuracy of remote sensing also increases, which lead to an increase in the efficiency of the farming. P_{CR} clearly shows that the accuracy of the agricultural works when using *Multi-UAV* is improved.

As a result, when using the multi UAV system, a little P_{ST} is required because it offers improved results in almost metrics ($P_{TT}, P_{FT}, P_{BC}, P_{HC}, P_{CR}$). In other words, *Multi-UAV* reduces the time, cost and operator's environment, including the control effort in agricultural works. In addition, the battery shortage problem and low payload are easily solved, which are the current challenges of agricultural UAVs.

6.2. *Autonomous vs. Teleoperation*

The use of automatic control when controlling an agricultural UAV saves much control effort on the operator side. However, there are many limitations to applying the automatic control to actual farming, and there are moments when the teleoperation command of the operator is needed. Additionally, when teleoperation is used, it offers a better performance in working duration than automatic control. Each control method has advantages and disadvantages, and it is necessary to quantitatively evaluate the performance of the system.

P_{TT} decreased by 32.3% (*Single-UAV*) and 58.7% (*Multi-UAV*) when *Tele-UAV* was used. Additionally, experimental results show that *Tele-UAV* has excellent performance in terms of P_{ST}. In particular, P_{ST} is reduced by 72.2% (*Single-UAV*) to 70.7% (*Multi-UAV*) compared to *Auto-UAV*, and the simulation is also reduced by 81.3% (*Single-UAV*) to 82.1% (*Multi-UAV*). These results mean that there is nothing to set in the case of *Tele-UAV*; however, in the case of *Auto-UAV*, a long P_{ST} is required because it is necessary to specify the path to each UAV. Unusually, P_{FT} increased for *Single-UAV* but decreased for *Multi-UAV* in the experiment results. However, the teleoperation method basically requires more P_{FT}. The reason for this result in the experiment was that when using *Tele-Multi-UAV*, the operator did not control the UAV carefully and this carelessness caused the low accuracy of the agricultural task by flying fast. However, *Auto-UAV* running on GPS based waypoints is accurate and faster.

For other metrics, P_{BC} is similar to P_{FT} as mentioned above. In P_{IL}, the results shows excellent performance when using *Tele-UAV* than using Auto-UAV. These results are due to the fact that GPS is interfered with the outdoor environment and is very variable. It means that performance is worse, and the UAV is dangerous when using *Auto-UAV* where GPS is not accurate. Particularly, it is a great advantage of *Auto-UAV* that the operator does not need P_{HC}. However, *Auto-UAV* has the disadvantage that while the UAV is in flight, it is comfortable because the operator has no control effort, but it takes a lot of P_{ST}. Additionally, there is no difference between *Auto-UAV* and *Tele-UAV*, because P_{CR} represents the simultaneous covered area.

Determining which control method is the better one depends on which performance metric is the priority; however, if time (P_{TT}, P_{ST}) is important, *Tele-UAV* is better than *Auto-UAV*. However, *Auto-UAV* is a good method, given the working time (P_{FT}), energy consumption (P_{BC}), and the fatigue of the operator (P_{HC}).

7. Conclusions

In this study, we developed an agricultural multi-UAV system using quadcopters based on the distributed swarm control algorithm. To evaluate the developed system and proposed control algorithm, in this experiment, the remote sensing was set as the benchmark test. Thereafter, using the agricultural multi-UAV system, the performance evaluation was performed through four experiment cases consisting of *Auto-Single-UAV*, *Auto-Multi-UAV*, *Tele-Single-UAV*, and *Tele-Multi-UAV*. A total of seven metrics were used to evaluate the performance, and the experimental results show that the multi-UAV system improved the performance obtained with a single-UAV system. As a result, the developed agricultural multi-UAV system with the distributed swarm control solves the problem of battery shortage and reduces working time and control effort. Most importantly, using the agricultural multi-UAV system improves the efficiency of agricultural work.

Author Contributions: C.J. developed the UAV systems, designed and implemented the experiments, measured and analyzed the data, and wrote the paper. H.I.S. provided some useful suggestions, performed overall revision and supervision for these experiments and paper, also performed project administration and funding acquisition.

Funding: This research was supported in part by the Basic Science Research Program through the National Research Foundation of Korea (NRF) funded by the Ministry of Science, ICT and Future Planning under Grant NRF- 2018R1D1A1B07046948, in part by grants (115062-2 and 316038-3) funded by the Ministry of Agriculture, Food, and Rural Affairs (MAFRA, Korea), and in part by a grant (100768) funded by the Ministry of Trade, Industry & Energy (MOTIE, Korea).

Acknowledgments: This research was supported in part by the Basic Science Research Program through the National Research Foundation of Korea (NRF) funded by the Ministry of Science, ICT and Future Planning under Grant NRF- 2018R1D1A1B07046948, and in part by a grant (115062-2) funded by the Ministry of Agriculture, Food, and Rural Affairs (MAFRA, Korea).

Conflicts of Interest: The authors declare no conflict of interest.

References

1. Valavanis, K.P.; Vachtsevanos, G.J. Future of unmanned aviation. In *Handbook of Unmanned Aerial Vehicles*; Springer: Dordrecht, The Nederland, 2015; pp. 2993–3009.
2. Zhang, C.; Kovacs, J.M. The application of small unmanned aerial systems for precision agriculture: A review. *Precis. Agric.* **2012**, *13*, 693–712. [CrossRef]
3. Kavvadias, A.; Psomiadis, E.; Chanioti, M.; Gala, E.; Michas, S. Precision agriculture-comparison and evaluation of innovative very high resolution (UAV) and landsat data. In Proceedings of the International Conference on Information & Communication Technologies in Agriculture, Food and Environment (HAICTA), Kavala, Greece, 17–20 September 2015; pp. 376–386.
4. Avellar, G.S.; Pereira, G.A.; Pimenta, L.C.; Iscold, P. Multi-uav routing for area coverage and remote sensing with minimum time. *Sensors* **2015**, *15*, 27783–27803. [CrossRef] [PubMed]
5. Franchi, A.; Giordano, P.R.; Secchi, C.; Son, H.I.; Bülthoff, H.H. A passivity-based decentralized approach for the bilateral teleoperation of a group of UAVs with switching topology. In Proceedings of the IEEE International Conference on Robotics and Automation (ICRA), Shanghai, China, 9–13 May 2011; pp. 898–905.
6. Lee, D.; Franchi, A.; Giordano, P.R.; Son, H.I.; Bülthoff, H.H. Haptic teleoperation of multiple unmanned aerial vehicles over the internet. In Proceedings of the IEEE International Conference on Robotics and Automation (ICRA), Shanghai, China, 9–13 May 2011; pp. 1341–1347.
7. Allred, B.; Eash, N.; Freeland, R.; Martinez, L.; Wishart, D. Effective and efficient agricultural drainage pipe mapping with uas thermal infrared imagery: A case study. *Agric. Water Manag.* **2018**, *197*, 132–137. [CrossRef]
8. Santesteban, L.; Di Gennaro, S.; Herrero-Langreo, A.; Miranda, C.; Royo, J.; Matese, A. High-resolution UAV-based thermal imaging to estimate the instantaneous and seasonal variability of plant water status within a vineyard. *Agric. Water Manag.* **2017**, *183*, 49–59. [CrossRef]
9. Vega, F.A.; Ramírez, F.C.; Saiz, M.P.; Rosúa, F.O. Multi-temporal imaging using an unmanned aerial vehicle for monitoring a sunflower crop. *Biosyst. Eng.* **2015**, *132*, 19–27. [CrossRef]
10. Tokekar, P.; Vander Hook, J.; Mulla, D.; Isler, V. Sensor planning for a symbiotic UAV and UGV system for precision agriculture. *IEEE Trans. Robot.* **2016**, *32*, 1498–1511. [CrossRef]
11. Torres-Sánchez, J.; López-Granados, F.; Serrano, N.; Arquero, O.; Peña, J.M. High-throughput 3-d monitoring of agricultural-tree plantations with unmanned aerial vehicle (UAV) technology. *PLoS ONE* **2015**, *10*, e0130479. [CrossRef] [PubMed]
12. Noriega, A.; Anderson, R. Linear-optimization-based path planning algorithm for an agricultural UAV. In Proceeding of the Infotech of American Institute of Aeronautics and Astronautics (AIAA), San Diego, CA, USA, 13–16 September 2016; p. 1003.
13. Alsalam, B.H.Y.; Morton, K.; Campbell, D.; Gonzalez, F. Autonomous UAV with vision based on-board decision making for remote sensing and precision agriculture. In Proceedings of the IEEE Aeropace Conference, Big Sky, MT, USA, 4–11 March 2017; pp. 1–12.
14. Jannoura, R.; Brinkmann, K.; Uteau, D.; Bruns, C.; Joergensen, R.G. Monitoring of crop biomass using true colour aerial photographs taken from a remote controlled hexacopter. *Biosyst. Eng.* **2015**, *129*, 341–351. [CrossRef]
15. Christiansen, M.P.; Laursen, M.S.; Jørgensen, R.N.; Skovsen, S.; Gislum, R. Designing and testing a UAV mapping system for agricultural field surveying. *Sensors* **2017**, *17*, 2703. [CrossRef] [PubMed]
16. Faiçal, B.S.; Freitas,H.; Gomes, P.H.; Mano, L.Y.; Pessin, G.; de Carvalho, A.C.; Krishnamachari, B.; Ueyama, J. An adaptive approach for UAV-based pesticide spraying in dynamic environments. *Comput. Electron. Agric.* **2017**, *138*, 210–223.

17. Torres-Sãnchez, J.; López-Granados, F.; De Castro, A.I.; Peña-Barragán, J.M. Configuration and specifications of an unmanned aerial vehicle (UAV) for early site specific weed management. *PLoS ONE* **2013**, *8*, e58210. [CrossRef] [PubMed]
18. Zarco-Tejada, P.J.; Guillén-Climent, M.; Hernández-Clemente, R.; Catalina, A.; González, M.; Martín, P. Estimating leaf carotenoid content in vineyards using high resolution hyperspectral imagery acquired from an unmanned aerial vehicle (UAV). *Agric. For. Meteorol.* **2013**, *171*, 281–294. [CrossRef]
19. Doering, D.; Benenmann, A.; Lerm, R.; de Freitas, E.P.; Muller, I.; Winter, J.M.; Pereira, C.E. Design and optimization of a heterogeneous platform for multiple uav use in precision agriculture applications. *IFAC Proc. Vol.* **2014**, *47*, 12272–12277. [CrossRef]
20. Xiang, H.; Tian, L. Method for automatic georeferencing aerial remote sensing (RS) images from an unmanned aerial vehicle (UAV) platform. *Biosyst. Eng.* **2011**, *108*, 104–113. [CrossRef]
21. Barrientos, A.; Colorado, J.; Cerro, J.D.; Martinez, A.; Rossi, C.; Sanz, D.; Valente, J. Aerial remote sensing in agriculture: A practical approach to area coverage and path planning for fleets of mini aerial robots. *J. Field Robot.* **2011**, *28*, 667–689. [CrossRef]
22. Arroyo, J.A.; Gomez-Castaneda, C.; Ruiz, E.; de Cote, E.M.; Gavi, F.; Sucar, L.E. Assessing nitrogen nutrition in corn crops with airborne multispectral sensors. In Proceedings of the International Conference on Industrial, Engineering and Other Applications of Applied Intelligent Systems (IEA/AIE), Arras, France, 27–30 June 2017; pp. 259–267.
23. Skobelev, P.; Budaev, D.; Gusev, N.; Voschuk, G. Disigning Multi-agent Swarm of UAV for Precise Agriculture. In Proceedings of the International Conference on Practical Applications of Agents and Multi-Agent Systems, Toledo, Spain, 20–22 June 2018; pp. 47–59.
24. Ju, C.; Park, S.; Park, S.; Son, H.I. A haptic teleoperation of agricultural multi-UAV. In Proceedings of the Workshop on Agricultural Robotics: Learning from Industry 4.0 and Moving into the Future at the IEEE/RSJ International Conference on Intelligent Robots and Systems (IROS), Vancouver, BC, Canada, 24–28 September 2017; pp. 1–6.
25. Ju, C.; Son, H.I. Performance Evaluation of Multiple UAV Systems for Remote Sensing in Agriculture. In Proceedings of the Workshop on Robotic Vision and Action in Agriculture at the IEEE International Conference on Robotics and Automation (ICRA), Brisbane, Australia, 21–26 May 2018; pp. 1–6.
26. Long, D.; MrCarthy, C.; Jensen, T. Row and water front detection from UAV thermal-infrared imagery for furrow irrigation monitoring. In Proceedings of the International Conference on Advanced Intelligent Mechatronics (AIM), Banff, AB, Canada, 12–15 July 2016; pp. 300–305.
27. Albornoz, C.; Giraldo, L.F. Trajectory design for efficient for crop irrigation with a UAV. In Proceedings of the Colombian Conference on Automatic Control (CCAC), Cartagena, Colombia, 18–20 October 2017; pp. 1–6.
28. Romero, M.; Luo, Y.; Su, B.; Fuentes, S. Vineyard water status estimation using multispectral imagery from an UAV platform and machine learning algorithms for irrigation scheduling management. *Comput. Electron. Agric.* **2018**, *147*, 109–117. [CrossRef]
29. Franchi, A.; Secchi, C.; Son, H.I.; Bülthoff, H.H.; Giordano, P.R. Bilateral teleoperation of groups of mobile robots with time-varying topology. *IEEE Trans. Robot.* **2012**, *28*, 1019–1033. [CrossRef]
30. Lee, D.; Franchi, A.; Son, H.I.; Ha, C.; Bülthoff, H.H.; Giordano, P.R. Semiautonomous haptic teleoperation control architecture of multiple unmanned aerial vehicles. *IEEE/ASME Trans. Mechatron.* **2013**, *18*, 1334–1345. [CrossRef]
31. Li, J.; Ye, D.H.; Chung, T.; Kolsch, M.; Wachs, J.; Bouman, C. Multi-target detection and tracking from a single camera in unmanned aerial vehicles (UAVs). In Proceedings of the IEEE/RSJ International Conference on Intelligent Robots and Systems (IROS), Deajeon, Korea, 9–14 October 2016; pp. 4992–4997.
32. Trianni, V.; IJsselmuiden, J.; Haken, R. The Saga Concept: Swarm Robotics for Agricultural Applications; Technical Report. 2016. Available online: http://laral.istc.cnr.it/saga/wp-content/uploads/2016/09/saga-dars2016.pdf (accessed on 23 August 2018).
33. Albani, D.; IJsselmuiden, J.; Haken, R.; Trianni, V. Monitoring and mapping with robot swarms for agricultural applications. In Proceedings of the IEEE International Conference on Advanced Video and Signal Based Surveillance (AVSS), Lecce, Italy, 31 August–1 September 2017; pp. 1–6.
34. Yang, H.; Lee, Y.; Jeon, S.; Lee, D. Multi-rotor drone tutorial: Systems, mechanics, control and state estimation. *Intell. Serv. Robot.* **2017**, *10*, 79–93. [CrossRef]

35. Grzonka, S.; Grisetti, G.; Burgard, W. A fully autonomous indoor quadrotor. *IEEE Trans. Robot.* **2012**, *28*, 90–100. [CrossRef]
36. Rodríguez-Seda, E.J.; Troy, J.J.; Erignac, C.A.; Murray, P.; Stipanovic, D.M.; Spong, M.W. Bilateral teleoperation of multiple mobile agents: Coordinated motion and collision avoidance. *IEEE Trans. Control Syst. Technol.* **2010**, *18*, 984–992. [CrossRef]
37. Hong, A.; Lee, D.G.; Büulthoff, H.H.; Son, H.I. Multimodal feedback for teleoperation of multiple mobile robots in an outdoor environment. *J. Multimodal User Interfaces* **2017**, *11*, 67–80. [CrossRef]
38. Son, H.I.; Cho, J.H.; Bhattacharjee, T.; Jung, H.; Lee, D.Y. Analytical and psychophysical comparison of bilateral teleoperators for enhanced perceptual performance. *IEEE Trans. Ind. Electron.* **2014**, *61*, 6202–6212. [CrossRef]
39. Son, H.I.; Franchi, A.; Chuang, L.L.; Kim, J.; Bülthoff, H.H.; Giordano, P.R. Human-centered design and evaluation of haptic cueing for teleoperation of multiple mobile robots. *IEEE Trans. Cybern.* **2013**, *43*, 597–609. [PubMed]
40. Son, H.I.; Kim, J.; Chuang, L.; Franchi, A.; Giordano, P.R.; Lee, D.; Bülthoff, H.H. An evaluation of haptic cues on the tele-operator's perceptual awareness of multiple UAVs' environments. In Proceedings of the World Haptics Conference (WHC), Istanbul, Turkey, 21–24 June 2011; pp. 149–154.

electronics

MDPI

Article

Longitudinal Attitude Control Decoupling Algorithm Based on the Fuzzy Sliding Mode of a Coaxial-Rotor UAV

Kewei Li [1], Yiran Wei [1], Chao Wang [2] and Hongbin Deng [1,*]

1 School of Mechatronical Engineering, Beijing Institute of Technology, Beijing 100081, China; likewei@bit.edu.cn (K.L.); 2120160257@bit.edu.cn (Y.W.)
2 China North Industries Corp., Beijing 100053, China; wch@norinco.cn
* Correspondence: denghongbin@bit.edu.cn; Tel.: +86-1861-050-1349

Received: 21 December 2018; Accepted: 14 January 2019; Published: 18 January 2019

Abstract: A longitudinal attitude decoupling algorithm based on the fuzzy sliding mode control for a small coaxial rotor unmanned aerial vehicle (UAV) is presented in this paper. The attitude system of a small coaxial rotor UAV is characterized by nonlinearity, strong coupling and uncertainty, which causes difficulties pertaining to its flight control. According to its six-degree-of-freedom model and structural characteristics, the dynamic model was established, and a longitudinal attitude decoupling algorithm was proposed. A fuzzy sliding mode control was used to design the controller to adapt to the underactuated system. Compared with the uncoupled fuzzy sliding mode control, simulation results indicated that the proposed method could improve the stability of the system, presented with a better adapting ability, and could effectively suppress the modeling error and external interference of the coaxial rotor aircraft attitude system. The proposed method also has the advantages of high accuracy, good stability, and the ease of implementation.

Keywords: coaxial-rotor; UAV; aircraft; longitudinal motion model; decoupling algorithm; sliding mode control

1. Introduction

In recent years, due to small unmanned aerial vehicles' (UAV) characteristics regarding maneuverability, flexibility and location difficulties, research on this type of UAV has drawn wide attention. With the unprecedented development of small aircrafts, the autonomous flight control of UAVs has become a research priority in the field of aviation [1]. Compared with fixed-wing aircrafts, the coaxial rotor uses a pair of coaxial reversing rotors which compensate for each other's torque, instead of balancing the yaw moment of the aircraft without the tail rotor [2]. Therefore, the aircraft has a compact structure, a small radial size, and a higher power efficiency. The data indicate that it is 35–40% smaller than the single rotor structure with a tail rotor, and in the same hovering conditions, the coaxial-rotor consumes 5% less energy than the single rotor [3]. In addition, with the reduction of the radial size of the aircraft along the rotor, the inertia of the aircraft decreases and its controllability and maneuverability are enhanced. The design without the tail rotor has also eliminated some hidden problems [4]. Research on coaxial-rotor helicopters has already had significant achievements, but the small coaxial-rotor UAV has received special attention in recent years. The small size of the aircraft and the different maneuvering modes brings about differences in control methods. The operation mode of the coaxial vehicle is different from that of an ordinary vehicle which is also a typical underactuated system [5]. For the small coaxial-rotor UAV, the six-degree-of-freedom non-linear coupling problem is prominent, and decoupling is important for stability and control of the vehicle.

The general attitude control system of the UAV coordinates and controls the longitudinal, lateral and heading channels. The design of the longitudinal channel controller is the most critical and complex of the three channels, and its control rate significantly affects the UAV's flight performance [6,7]. In the literature [8], the decoupling control method is used to design the aircraft control system. Both the adaptability and control effect of the system, however, need improvement. In Reference [9], by combining the advantages of feedback linearization and variable structure control, the attitude controller of the aircraft was designed. However, it was unable to effectively weaken the sliding mode chattering of the system, and the controller parameters could not be adjusted in real time according to the disturbance, which caused poor control performance. Reference [7] proposed a control law which was designed using the adaptive backstepping method and which did not require any knowledge of aircraft aerodynamics. Simulation results showed good performance of the feedback law, but the actual implementation was complicated and difficult to achieve. In Reference [10], a fuzzy logic control of the longitudinal motion of an aircraft based on the Takagi–Sugeno modeling approach was presented, and while the stability and tracking effect were good, the problem of system coupling had not been solved well and the control precision needed improvement. There are many studies of the decoupling controls of aircraft, but few are focused specifically on coaxial aircrafts [11].

The main role of this paper is to propose a decoupling algorithm that improves the reliability of the attitude control for the longitudinal motion stability of the coaxial rotor UAV. In order to satisfy the stability requirements of a coaxial-rotor UAV's longitudinal motion [12], a suitable control algorithm and controller needed to be designed. Before this, we required a dynamic model which featured a qualified and effective vehicle longitudinal motion [13]. In accordance to the lab-developed coaxial rotor UAV, a rigorous and effective non-linear mathematical model of longitudinal motion was established, and an under-actuated controller was designed using the fuzzy sliding mode. Simulation results showed that the position control performance of the aircraft was improved when the decoupling algorithm was applied to the coaxial rotor longitudinal motion control system. The position and attitude were significantly improved [14] and the method was simple and effective.

This paper is organized as follows: In Section 2, according to the self-developed coaxial vehicle, the modeling and derivation processes are given. In Section 3, the decoupling algorithm design is introduced. The controller design and stability analysis based on the fuzzy sliding mode control are described in Section 4. Finally, the simulation results and comparison with the decoupling algorithm are shown in Section 5.

2. Aircraft Longitudinal Flight Model

There is a large degree of coupling among the control inputs of the aircraft. The general method is to regard these coupling quantities as external disturbances, but this method introduces large errors. To solve this problem, we used the method of controlling the correlation coefficient of the input, by selecting the appropriate correlation coefficient so that coupling among the control inputs would be handled better.

2.1. Rotor System Modeling

In order to establish a simplified model that could both reflect the aerodynamic characteristics of a coaxial-rotor and be suitable for controller design, we first made the following assumptions: The blade was rigid, the blade root truncation effect was ignored, and the tip loss and the flapping hinge extension were assumed without considering the unsteady effect [15]. The structure design and force analysis of the coaxial-rotor are shown in Figure 1.

Based on the blade-element theory, the integral expression of the rotor pulling force and torque could be obtained:

$$\begin{aligned} T &= \frac{N_b}{2\pi}\int_0^{2\pi}\int_0^{R-e}(lc_v - ds_v)drd\psi_f \\ Q &= \frac{N_b}{2\pi}\int_0^{2\pi}\int_0^{R-e}(lc_v - ds_v)rdrd\psi_f \end{aligned} \tag{1}$$

where N_b is the number of blades, R is the radius of the rotor, e is the amount of hinge extension of blade swing, ψ_f is the local azimuth, r is the local radial coordinates, l and d respectively are local lift and resistance, and c_v and s_v are the correction terms related to the aircraft and the flying environment [16]. The approximate inflow ratio of a blade-element was:

$$\lambda = \lambda_{in} + \lambda_{fs} = \frac{v_{in}}{\Omega R} - \frac{v_b}{\Omega R} \tag{2}$$

where v_{in} is the induced velocity, Ω is the rotor speed, v_b is the rotor speed, is the body speed, and λ_{in} and λ_{fs} are the inflow ratio corresponding to the induced velocity and body velocity, respectively.

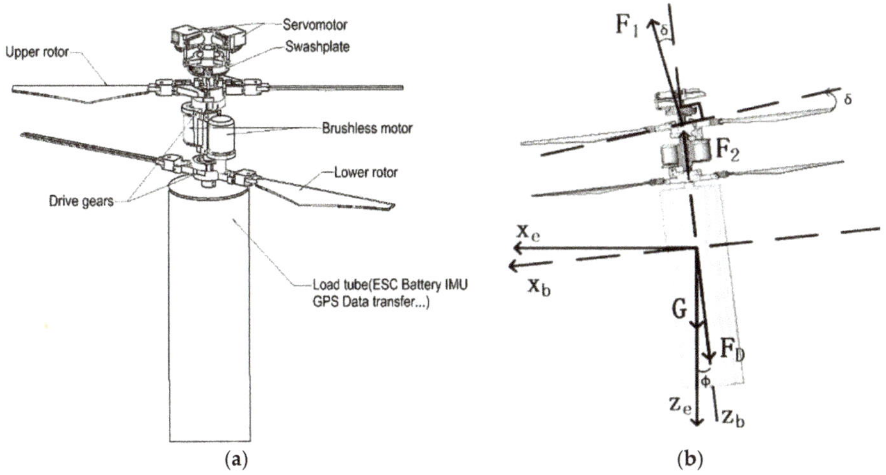

Figure 1. (**a**) The structure design and (**b**) force analysis of the coaxial-rotor Rotorcraft.

To solve the problem of interference between coaxial rotors, the Pitt–Peters dynamic inflow method was used to model the induced velocity. The specific method was to connect the dynamic variation of the induced velocity with the variation of aerodynamic parameters through a first-order linear differential equation. We could then clarify the relationship between the pull coefficient and the induced velocity through integral calculation. The method was simple and in good agreement with the experimental data. The proposed model could be well applied to the simulation and controller designs. According to the dynamic inflow model, the relationship between the induced velocity and the pull coefficient was as follows:

$$M\dot{\lambda}_{in} + VL^{-1}\lambda_{in} = C \tag{3}$$

where M, V and L are the parameter matrix of inflow dynamics, respectively, $\lambda_{in} = (\lambda_0 + \lambda_s + \lambda_c)^T$ represents the time-averaged, first-order horizontal, and vertical components of the induced inflow ratio. $C = (C_T + C_l + \lambda_m)^T$ represents the pull torque, roll torque and pitch torque coefficients of the rotors. The interaction of the induced velocity is expressed as:

$$\lambda_i = \lambda_{in,i} + K_{ji}\lambda_{in,j} + \lambda_{fs}e_1 \tag{4}$$

where i, j are the upper and lower rotors and K_{ji} is the parameter matrix, indicating the influence of the induced velocity between the upper and lower rotors related to the distance between the rotors, airfoil, and flight state, while $e_1 = \begin{pmatrix} 1 & 0 & 0 \end{pmatrix}^T$.

Considering that the interaction of the induced velocity mainly affected the channels of total distance and heading, the induced velocity in the plane of the propeller disk was almost unaffected.

Assuming that the induced velocity was uniformly distributed in the plane of the propeller disk, the average inflow ratio and the differential inflow ratio were defined respectively as:

$$\begin{aligned} \lambda_a &= \tfrac{1}{2}(\lambda_u + \lambda_l) \\ \lambda_r &= \tfrac{1}{2}(\lambda_u - \lambda_l) \end{aligned} \tag{5}$$

From the formulas above, the pull and torque coefficients could be obtained using integral calculation (τ is the rotor flapping time constant):

$$\begin{aligned} C_{Ti} &= \tfrac{\sigma a_0}{2}\left(\tfrac{1}{3}\theta_i - \tfrac{1}{2}\right)\lambda_i \\ C_{Qi} &= \lambda_i C_{Ti} + \tfrac{\sigma}{8} C_D \end{aligned} \tag{6}$$

So, the single rotor thrust and torque were presented:

$$\begin{aligned} T_i &= \rho A(\Omega_i R)^2 C_{Ti} \\ Q_i &= \rho A \Omega_i{}^2 R^3 C_{Qi} \end{aligned} \tag{7}$$

where subscript i represents the upper rotor (u) or the lower rotor (l), θ_i is the pitch, c is the chord length of the blade, A is the paddle area, a_0 is the slope of the lift line of the airfoil, C_D is the airfoil drag coefficient, ρ is the air density, and $\sigma = (N_b c)/(\pi R)$ is the real degree of the paddle.

2.2. External Force Modeling

In the steady state of hovering, the external force of the aircraft was determined by the lift F_1 and F_2, the gravity G, and the aerodynamic resistance, F_D:

$$F = F_1 + F_2 + G + F_D \tag{8}$$

2.2.1. The Lift of the Rotor System

The lift of the designed aircraft was provided by the upper and lower rotors. Since the swash plate related to the upper rotor, the upper rotor provided both lift and lateral force, while the lower rotor only provided lift. The following could be obtained from Formulas (2) and (7):

$$\begin{aligned} F_1 &= \rho A(\Omega_u R)^2 C_{Tu} \cos\delta \\ F_2 &= \rho A(\Omega_l R)^2 C_{Tl} \end{aligned} \tag{9}$$

2.2.2. The Gravity of Aircraft

The mass of aircraft is m, and the body gravity was expressed by the body coordinate system data:

$$G = mg\begin{bmatrix} -\sin\theta \\ \cos\theta\sin\phi \\ \cos\theta\cos\phi \end{bmatrix} \tag{10}$$

2.2.3. The Aerodynamic Resistance

According to the empirical formula of aerodynamics, the resistance of the fuselage in the hovering state could be expressed as:

$$F_D = \frac{1}{2}\rho V^2 A_{fus} C_{Dfus} \tag{11}$$

where V is the relative fly-forward speed of the aircraft, A_{fus} is the equivalent cross-sectional area of the body, and C_{Dfus} is the resistance coefficient of the whole body.

2.3. Establishing the Longitudinal Posture Model

$$\begin{cases} -m\ddot{x} = -(F_1\cos\delta + F_2 - F_D)\sin\phi + \varepsilon_0 F_1 d\sin\delta\cos\phi \\ -m\ddot{z} = (F_1\cos\delta + F_2 - F_D)\cos\phi + \varepsilon_0 F_1 d\sin\delta\sin\phi - mg \\ I_{xx}\ddot{\phi} = F_1 d\sin\delta \end{cases} \tag{12}$$

Take $x_1 = -x$, $x_2 = \dot{x}$, $z_1 = -z$, $z_2 = \dot{z}$, $u_1 = (F_1\cos\delta + F_2 - F_D)/m$, $u_2 = (F_1 d\sin\delta)/I_{xx}$, $\varepsilon = \frac{\varepsilon_0 I_{xx}}{m}$. Formula (12) [17,18] could be written as:

$$\begin{cases} \dot{x}_1 = x_2 \\ \dot{x}_2 = -u_1\sin\phi + \varepsilon u_2\cos\phi \\ \dot{z}_1 = z_2 \\ \dot{z}_2 = u_1\cos\phi + \varepsilon u_2\sin\phi - g \\ \dot{\phi} = \varphi \\ \dot{\varphi} = u_2 \end{cases} \tag{13}$$

where $x_1(t)$, $z_1(t)$ and $\phi(t)$ are the mass center and deflection angles of aircraft, u_1 and u_2 are the control inputs, g is the gravitational acceleration, and ε_0 represents the parasitic force of lateral displacement generated by rolling torque and $\varepsilon_0 \neq 0$ [19]. The system's output was:

$$\mathbf{y}(t) = \begin{bmatrix} x_1(t) & z_1(t) & \phi(t) \end{bmatrix}^T \tag{14}$$

3. Design of the Decoupling Algorithm

Formula (13) was a strong non-linear coupling model [20]. In order to transform the model into the under-driven standard form, the following method was used to decouple the model [21,22].

3.1. Eliminate the Control Coupling of $\dot{x}_2$ and $\dot{z}_2$

In order to eliminate the control coupling of $\dot{x}_2$ and $\dot{z}_2$, the following decoupling algorithm could be designed:

$$\begin{bmatrix} u_1 \\ u_2 \end{bmatrix} = \begin{bmatrix} -\sin\phi & \varepsilon\cos\phi \\ \cos\phi & \varepsilon\sin\phi \end{bmatrix}^{-1} \begin{bmatrix} u_{m1} \\ u_{m2} + g \end{bmatrix} \tag{15}$$

where u_{m1} and u_{m2} are the control items to be designed. Next, Formula (13) could be changed:

$$\begin{cases} \ddot{x}_1 = u_{m1} \\ \ddot{z}_1 = u_{m2} \\ \varepsilon\ddot{\phi} = u_{m1}\cos\phi + (u_{m2} + g)\sin\phi \end{cases} \tag{16}$$

3.2. Eliminate the Coupling of u_{m1} and u_{m2}

In order to eliminate the coupling of u_{m1} and u_{m2} in $\varepsilon\ddot{\phi}$, the following decoupling algorithm could be designed:

$$\begin{cases} x_m = x_1 - \varepsilon\sin\phi \\ z_m = z_1 + \varepsilon\cos\phi \\ u_{m1} = \left(u_{m3} - \varepsilon\dot{\phi}^2\right)\sin\phi + \varepsilon u_{m4}\cos\phi \\ u_{m2} = -\left(u_{m3} - \varepsilon\dot{\phi}^2\right)\cos\phi + \varepsilon u_{m4}\sin\phi - g \end{cases} \tag{17}$$

where u_{m3} and u_{m4} are the control items to be designed, then Formula (16) could be changed:

$$\begin{cases} \ddot{x}_m = u_{m3}\sin\phi \\ \ddot{z}_m = -u_{m3}\cos\phi - g \\ \ddot{\phi} = u_{m4} \end{cases} \tag{18}$$

3.3. *Eliminate the Coupling of* $\sin\phi$ *and* $\cos\phi$

In order to eliminate the coupling of $\sin\phi$ and $\cos\phi$, the following decoupling algorithm was designed:

We took $q_1 = x_m$, $q_2 = \dot{q}_1$, $q_3 = z_m$, $q_4 = \dot{q}_3$, $q_5 = \tan\phi$, and $q_6 = \dot{q}_5$, and when ϕ was very small, we took $\sin\phi \approx \phi$. Then we could obtain the following:

$$\ddot{q}_5 = \frac{u_{m4}\cos^2\phi + 2\dot{\phi}^2\cos^3\phi\sin\phi}{\cos^4\phi} \tag{19}$$

We took $h_1 = u_{m3}\cos\phi$, $h_2 = \tan''\phi$. Then we could get $u_{m4} = h_2\cos^2\phi - 2\dot{\phi}^2\tan\phi$. Now, h_1 and h_2 were the control items to be designed. Then, Formula (18) could be changed as:

$$\begin{cases} \dot{q}_1 = q_2 \\ \dot{q}_2 = q_5 h_1 \\ \dot{q}_3 = q_4 \end{cases} \quad \begin{cases} \dot{q}_4 = -h_1 - g \\ \dot{q}_5 = q_6 \\ \dot{q}_6 = h_2 \end{cases} \tag{20}$$

3.4. *Transform the Control Model into an Under-Driven Standard Form*

The under-driven standard form could be obtained:

$$\begin{cases} \dot{\mathbf{y}}_1 = \mathbf{y}_2 \\ \dot{\mathbf{y}}_2 = \mathbf{f}_1\begin{pmatrix} \mathbf{y}_1 & \mathbf{y}_2 & \mathbf{y}_3 \end{pmatrix} \\ \dot{\mathbf{y}}_3 = \mathbf{y}_4 \\ \dot{\mathbf{y}}_4 = \mathbf{f}_2\begin{pmatrix} \mathbf{y}_1 & \mathbf{y}_2 & \mathbf{y}_3 \end{pmatrix} + \mathbf{bh} + \mathbf{d} \end{cases} \tag{21}$$

where $\mathbf{f}_2\begin{pmatrix} \mathbf{y}_1 & \mathbf{y}_2 & \mathbf{y}_3 \end{pmatrix} = \begin{bmatrix} -g \\ 0 \end{bmatrix}$, $\mathbf{b} = \begin{bmatrix} -1 & 0 \\ 0 & 1 \end{bmatrix}$, $\mathbf{h} = \begin{bmatrix} h_1 \\ h_2 \end{bmatrix}$, and $\mathbf{d}$ is the control disturbance.

In order to make $\frac{\partial \mathbf{f}_1\begin{pmatrix} \mathbf{y}_1 & \mathbf{y}_2 & \mathbf{y}_3 \end{pmatrix}}{\partial \mathbf{y}_3}$ invertible, which would be helpful for the design of the control law, we took $\mathbf{y}_2 = \begin{bmatrix} q_2 + q_4 q_5 \\ \int_0^t q_3 dt \end{bmatrix}$, $\mathbf{y}_1 = \begin{bmatrix} q_1 + \int_0^t q_4 q_5 dt \\ \int_0^t \left(\int_0^t q_3 d_t\right) dt \end{bmatrix}$, $\mathbf{y}_3 = \begin{bmatrix} q_3 \\ q_5 \end{bmatrix}$, $\mathbf{y}_4 = \begin{bmatrix} q_4 \\ q_6 \end{bmatrix}$, and $\mathbf{f}_1\begin{pmatrix} \mathbf{y}_1 & \mathbf{y}_2 & \mathbf{y}_3 \end{pmatrix} = \begin{bmatrix} -g\mathbf{y}_3(2) \\ \mathbf{y}_3(1) \end{bmatrix}$.

4. Design of the Controller

The structure of the controller [16] is shown in Figure 2.

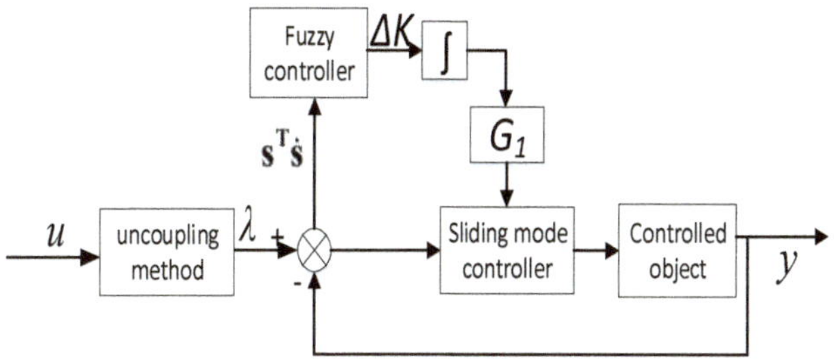

Figure 2. The structure of the controller.

4.1. Design of the Control Law

In order to design the control law for Formula (21), we took y_{1d} as the reference instruction of y_1, and the error variables were as follows [19]:

$$\begin{array}{ll} \mathbf{e}_1 = \mathbf{y}_1 - \mathbf{y}_{1\mathbf{d}} \quad , & \mathbf{e}_2 = \mathbf{y}_2 - \dot{\mathbf{y}}_{1\mathbf{d}} \\ \mathbf{e}_3 = \mathbf{f}_1 - \ddot{\mathbf{y}}_{1\mathbf{d}} \quad , & \mathbf{e}_4 = \frac{\partial \mathbf{f}_1}{\partial \mathbf{y}_3}\mathbf{y}_4 - \mathbf{y}_{1\mathbf{d}}{}^{(3)} \end{array} \tag{22}$$

Next, we designed the sliding surface:

$$\mathbf{s} = \mathrm{c}_1\mathbf{e}_1 + \mathrm{c}_2\mathbf{e}_2 + \mathrm{c}_3\mathbf{e}_3 + \mathbf{e}_4$$

where $c_i > 0$ and $i = 1,2,3$.

When $\dot{\mathbf{s}} = 0$, we could obtain $\mathbf{h} = \mathbf{u_{eq}}$ and $\mathrm{c}_1\dot{\mathbf{e}}_1 + \mathrm{c}_2\dot{\mathbf{e}}_2 + \mathrm{c}_3\dot{\mathbf{e}}_3 + \dot{\mathbf{e}}_4 = 0$.

The equivalent switch control items could thus be obtained [23]:

$$\begin{aligned} \mathbf{u_{eq}} = -\left(\frac{\partial \mathbf{f}_1}{\partial \mathbf{y}_3}\mathbf{b}\right)^{-1} & (\mathrm{c}_1\mathbf{y}_2 - \mathrm{c}_1\dot{\mathbf{y}}_{1\mathbf{d}} \\ & + \mathrm{c}_2\mathbf{f}_1 - \mathrm{c}_2\ddot{\mathbf{y}}_{1\mathbf{d}} + \mathrm{c}_3\frac{\partial \mathbf{f}_1}{\partial \mathbf{y}_3}\mathbf{y}_4 \\ & - \mathrm{c}_3\mathbf{y}_{1\mathbf{d}}{}^{(3)} + \frac{\partial \mathbf{f}_1}{\partial \mathbf{y}_3}\mathbf{f}_2 - \mathbf{y}_{1\mathbf{d}}{}^{(4)}) \end{aligned} \tag{23}$$

where $\dot{\mathbf{y}}_1 = \mathbf{y}_2, \ddot{\mathbf{y}}_1 = \mathbf{f}_1, \mathbf{y}_1{}^{(3)} = \dot{\mathbf{f}}_1 = \frac{\partial \mathbf{f}_1}{\partial \mathbf{y}_3}\mathbf{y}_4$ and $\mathbf{y}_1{}^{(4)} = \ddot{\mathbf{f}}_1 = \frac{\partial \mathbf{f}_1}{\partial \mathbf{y}_3}\dot{\mathbf{y}}_4$

The switching control item was then designed:

$$\mathbf{u_{sw1}} = -\left(\frac{\partial \mathbf{f}_1}{\partial \mathbf{y}_3}\mathbf{b}\right)^{-1}[\mathbf{K}(t)\mathrm{sgn}(\mathbf{s}) + \lambda\mathbf{s} + \mathbf{E}_1(t)] \tag{24}$$

where $\mathbf{E}_1(t)$ is unknown interference. Both $\mathbf{E}_1(t)$ and $\mathbf{K}(t)$ will be described in more detail in Section 4.2. The control law could be designed as follows:

$$\mathbf{h} = \mathbf{u_{eq}} + \mathbf{u_{sw}} \tag{25}$$

4.2. Stability Analysis of the Control System

Taking Formulas (23)–(25) into $\dot{\mathbf{s}}$, the following could be obtained:

$$\begin{aligned} \dot{\mathbf{s}} &= c_1\dot{e}_1 + c_2\dot{e}_2 + c_3\dot{e}_3 + \dot{e}_4 \\ &= c_1(\mathbf{y}_2 - \dot{\mathbf{y}}_{1\mathbf{d}}) + c_2(\mathbf{f}_1 - \ddot{\mathbf{y}}_{1\mathbf{d}}) + c_3\left(\frac{\partial \mathbf{f}_1}{\partial \mathbf{y}_3}\mathbf{y}_4 - \mathbf{y}_{1\mathbf{d}}{}^{(3)}\right) \\ &\quad + \frac{d}{dt}\left[\frac{\partial \mathbf{f}_1}{\partial \mathbf{y}_3}\right]\mathbf{y}_4 + \frac{\partial \mathbf{f}_1}{\partial \mathbf{y}_3}\mathbf{f}_2 - \mathbf{y}_{1\mathbf{d}}{}^{(4)} \\ &= -\mathbf{K}(t)\mathrm{sgn}(\mathbf{s}) - \lambda\mathbf{s} + \frac{\partial \mathbf{f}_1}{\partial \mathbf{y}_3}\mathbf{d} \end{aligned}$$

We took $\mathbf{K}(t) = \alpha\bar{\mathbf{d}} + \boldsymbol{\rho}$, where $\bar{\mathbf{d}}(1) \geq \left|\mathbf{d}(1)\right|$, $\bar{\mathbf{d}}(2) \geq \left|\mathbf{d}(2)\right|$, $\boldsymbol{\rho}(1) > 0$, $\boldsymbol{\rho}(2) > 0$, and $\alpha > 0$.

We took the Lyapunov function as $V = \frac{1}{2}\mathbf{s}^{\mathbf{T}}\mathbf{s}$, so:

$$\begin{aligned} \dot{V} = \mathbf{s}^{\mathbf{T}}\dot{\mathbf{s}} = \mathbf{s}^{\mathbf{T}}\left[-(\alpha\bar{\mathbf{d}} + \boldsymbol{\rho})\mathrm{sgn}(\mathbf{s}) - \lambda\mathbf{s} + \tfrac{\partial \mathrm{f}_1}{\partial \mathbf{y}_3}\mathbf{d}\right] \\ = -(\alpha\bar{\mathbf{d}} + \boldsymbol{\rho})\|\mathbf{s}\| - \lambda\|\mathbf{s}\|^2 + \mathbf{s}^{\mathbf{T}}\mathbf{E}(t) \\ \leq -\boldsymbol{\rho}\|\mathbf{s}\| - \lambda\|\mathbf{s}\|^2 \leq 0 \end{aligned} \tag{26}$$

where the gain of the switching $\mathbf{K}(t)$ was the cause of chattering, and the control disturbance can be expressed as $\mathbf{E}(t) = \frac{\partial \mathrm{f}_1}{\partial \mathbf{y}_3}\mathbf{d}$, which was used for ensuring that the necessary sliding mode presence conditions were met. When $\mathbf{s} = 0$, we could obtain $\mathbf{e}_4 = \mathrm{c}_1\mathbf{e}_1 + \mathrm{c}_2\mathbf{e}_2 + \mathrm{c}_3\mathbf{e}_3$. We took the following:

$$\mathbf{A} = \begin{bmatrix} 0 & 1 & 0 \\ 0 & 0 & 1 \\ -c_1 & -c_2 & -c_3 \end{bmatrix}$$

$\mathbf{A}$ is the Hurwitz function, and λ represents the Eigenvalues of $\mathbf{A}$, $\lambda > 0$.

Taking $\mathbf{E}_1 = \begin{bmatrix} \mathbf{e}_1 & \mathbf{e}_2 & \mathbf{e}_3 \end{bmatrix}^{\mathbf{T}}$, the error equation of the state could be written as $\dot{\mathbf{E}}_1 = \mathbf{A}\mathbf{E}_1$.

Taking $\mathbf{Q} = \mathbf{Q}^{\mathbf{T}} > 0$, we could get the Lyapunov equation $\mathbf{A}^{\mathbf{T}}\mathbf{P} + \mathbf{P}\mathbf{A} = -\mathbf{Q}$. The solution was $\mathbf{P} = \mathbf{P}^{\mathbf{T}} > 0$. We took the Lyapunov function as the following:

$$\begin{aligned} \dot{V}_1 = \dot{\mathbf{E}}_1^{\mathbf{T}}\mathbf{P}\mathbf{E}_1 + \mathbf{E}_1^{\mathbf{T}}\mathbf{P}\dot{\mathbf{E}}_1 = (\mathbf{A}\mathbf{E}_1)^{\mathbf{T}}\mathbf{P}\mathbf{E}_1 + \mathbf{E}_1^{\mathbf{T}}\mathbf{P}(\mathbf{A}\mathbf{E}_1) \\ = \mathbf{E}_1^{T}\mathbf{A}^{\mathbf{T}}\mathbf{P}\mathbf{E}_1 + \mathbf{E}_1^{\mathbf{T}}\mathbf{P}\mathbf{A}\mathbf{E}_1 = \mathbf{E}_1^{\mathbf{T}}\left(\mathbf{A}^{\mathbf{T}}\mathbf{P} + \mathbf{P}\mathbf{A}\right)\mathbf{E}_1 \\ = -\mathbf{E}_1^{\mathbf{T}}\mathbf{Q}\mathbf{E}_1 \leq -\lambda_{\min}(\mathbf{Q})||\mathbf{E}_1||_2^2 \leq 0 \end{aligned}$$

where $\lambda_{\min}(\mathbf{Q})$ is the minimum eigenvalue of positive definite matrix, $\mathbf{Q}$.

From $\dot{V}_1 \leq 0$, we could obtain: $\mathbf{e}_1 \to 0$, $\mathbf{e}_2 \to 0$, $\mathbf{e}_2 \to 0$, then $\mathbf{y}_1 \to \mathbf{y}_{1\mathbf{d}}$, $\mathbf{y}_2 \to \mathbf{y}_{2\mathbf{d}}$, and $\mathbf{y}_3 \to \mathbf{y}_{3\mathbf{d}}$. From the stability of the sliding mode, we could obtain $\mathbf{y}_4 \to \mathbf{y}_{4\mathbf{d}}$. In the end, $x \to x_{\mathrm{d}}$, $z \to z_{\mathrm{d}}$, and $\phi \to \phi_{\mathrm{d}}$.

4.3. Establish the Fuzzy System

The condition for the existence of sliding mode was $\mathbf{s}^{\mathbf{T}}\dot{\mathbf{s}} < 0$, and when the system reached the sliding surface, it would remain on the sliding surface [24]. From Formula (26), we could see that in order to ensure that the system movement reached the gain of the sliding surface, $\mathbf{K}(t)$ needed to be sufficient to eliminate the impact of uncertainty. Then we could ensure the existence of the sliding condition.

The idea of the fuzzy rules was represented as follows:

If $\mathbf{s}^{\mathbf{T}}\dot{\mathbf{s}} > 0$, then $\mathbf{K}(t)$ should increase;

if $\mathbf{s}^{\mathbf{T}}\dot{\mathbf{s}} < 0$, then $\mathbf{K}(t)$ should be reduced.

From the two types above, we could design the fuzzy system using $\mathbf{s}^{\mathbf{T}}\dot{\mathbf{s}}$ and $\Delta\mathbf{K}(t)$. In this system, $\mathbf{s}^{\mathbf{T}}\dot{\mathbf{s}}$ is the input, and $\Delta\mathbf{K}(t)$ is the output. The fuzzy sets of the system were defined as follows:

$$\begin{aligned} \mathbf{s}^{\mathbf{T}}\dot{\mathbf{s}} = \{ \; NB \quad NM \quad ZO \quad PM \quad PB\} \\ \Delta K = \{ \; NB \quad NM \quad ZO \quad PM \quad PB\} \end{aligned}$$

where NB represents the negative big, NM is the negative middle, ZO is the zero, PM is the positive middle, and PB is the positive big. The input and output membership functions of the fuzzy system are shown in Figures 3 and 4. The upper bound of $\hat{\mathbf{K}}(t)$ was estimated using the integral method:

$$\hat{\mathbf{K}}(t) = G_1 \int_0^t \Delta K dt \tag{27}$$

where G_1 is the proportion coefficient, determined according to experience. The control law was:

$$\mathbf{u_{sw2}} = -\left(\frac{\partial \mathbf{f}_1}{\partial \mathbf{y}_3}\mathbf{b}\right)^{-1}\left[\hat{\mathbf{K}}(t)\mathrm{sgn}(s) + \lambda \mathbf{s} + \mathbf{E}(t)\right] \tag{28}$$

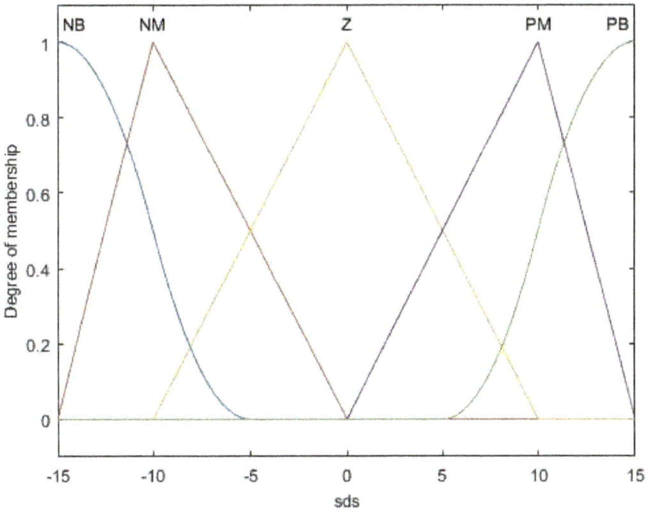

Figure 3. The input membership function.

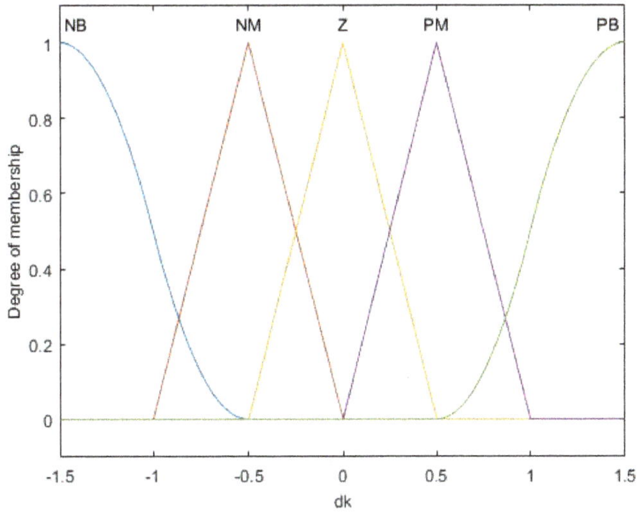

Figure 4. The output membership function.

5. Simulation Analysis

The physical parameters of the system model were obtained through the three-dimensional model established in the CAD software Inventor, and the model reference coefficients were calculated in accordance to the physical parameters and the modeling results. The parameters are shown in Table 1.

Table 1. The main symbols and parameters.

Parameter	Description	Value	Unit
ρ	Local air density	1.14	kg/cm^2
g	Local gravitational acceleration	9.804	m/s^2
R	Rotor radius	0.20	m
A	Area of rotor plane	0.126	m^2
m	Weight	1	kg
d	Distance between upper rotor and the gravity center	0.20	m
e	Amount of hinge extension of blade flapping	0.05	m
I_{xx}	Inertia about x-axis	9.16×10^{-4}	kg/m^2
C_{Tu}	Trust coefficient of upper rotor	9.42×10^{-3}	rad^{-1}
C_{Tl}	Trust coefficient of lower rotor	6.77×10^{-3}	rad^{-1}
C_{Qu}	Torque coefficient of upper rotor	6.14×10^{-4}	rad^{-1}
C_{Ql}	Torque coefficient of lower rotor	6.01×10^{-4}	rad^{-1}

For the controlled Formula (13), we took $\varepsilon = 10$ and $g = 9.8$, and set a predetermined track as $x_d = t,\ z_d = \sin t$ and $\phi_d = 0$.

In order to make **A** become the Hurwitz function, we took the control law parameters $c_1 = 27,\ c_2 = 27,\ c_3 = 9$ and $\lambda = 0.10$. The initial state of the controlled system was taken as $\begin{bmatrix} 5 & 0 & 0.5 & 0 & 0.1 & 0 \end{bmatrix}$. We used the control law (Formula (25)) and saturation function method, and took the thickness of the boundary layer Δ to be 0.10.

According to the structural characteristics of the coaxial-rotor UAV, a dynamic model of the longitudinal motion was established. The dynamic model of the aircraft was then decoupled, the fuzzy control and sliding mode controls were combined, and then a fuzzy sliding mode control based on the decoupling algorithm was designed for the coaxial-rotor. The control method was then simulated by MATLAB/Simulink. The results showed that the control method could track the command signal more quickly and efficiently compared to the method of the traditional sliding mode control. It could quickly reduce the yaw attitude angle deviation and the steady-state error could reach almost zero, and with a strong self-adaptive ability, it could achieve a better control effect. The response speed, tracking accuracy, and efficiency of the system were significantly improved.

The proposed control method could improve the stability of the system, which could effectively restrain the modeling errors and external disturbances of the aircraft's attitude system. This method had the advantages of high control precision, strong robustness, and ease of implementation in engineering. In future studies, we will focus on the design of the decoupling algorithm under the influence of more inputs and interferences, and will apply this algorithm to specific engineering practices.

Figures 5 and 6 show the performance of position tracking in the horizontal direction while the two control methods were used. The instruction given along x was a straight-line motion. The former figure indicates the position tracking with the decoupling algorithm and fuzzy control. The latter indicates the fuzzy control without the decoupling algorithm. From these figures, we could see that the performance of the control method with the decoupling algorithm and fuzzy control was faster, more accurate, and more stable than the general sliding mode control, ensuring that the aircraft would be more stable in actual movement and in improving the flight.

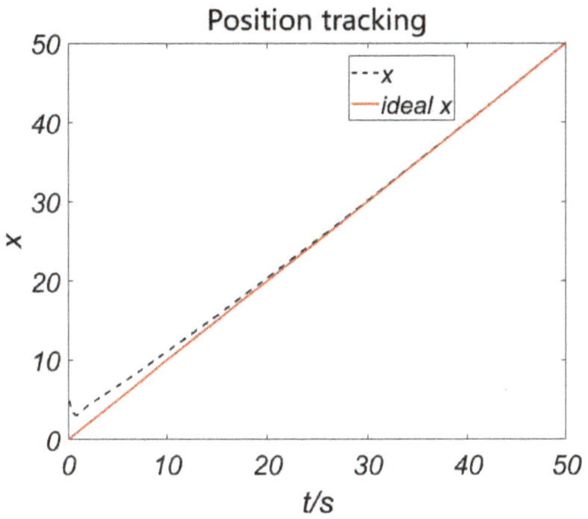

Figure 5. Position tracking with the decoupling algorithm of the x-axis.

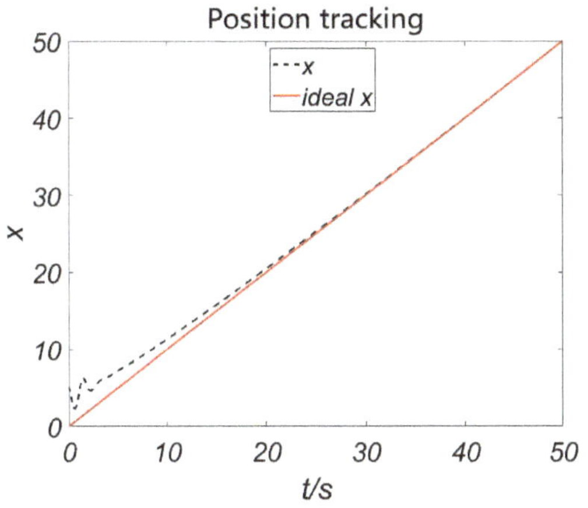

Figure 6. Position tracking without the decoupling algorithm of the x-axis.

Figures 7 and 8 show the performance of the position tracking in the vertical direction while the two control methods were used. The instruction given along z was a sinusoidal motion. The former figure indicates the position tracking with the decoupling algorithm and fuzzy control, and the latter indicates the general sliding mode control. From these figures we could see that the time required for the two methods to track from the initial position to the specified trajectory was almost the same, but the performance of the control method with the decoupling algorithm and fuzzy control was smoother, and the tracking error was also smaller. Thus, the flight of the aircraft would be more stable.

The angle tracking with the decoupling algorithm and fuzzy control are shown in Figure 9, and when compared with the angle tracking (Figure 10) without the decoupling algorithm and fuzzy control, it could be seen that in the former the tracking errors decreased while the response times were basically the same, and the movement accuracy and resistance to disturbances of the system improved. The system had better tracking performance. The impact of these disturbances, controller outputs chattering, external disturbances, and noise of the measurement were significantly reduced after the decoupling algorithm and fuzzy controller are added.

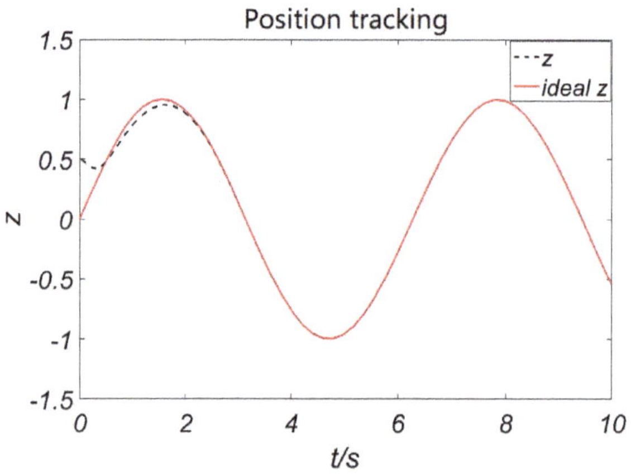

Figure 7. Position tracking with the decoupling algorithm of the z-axis.

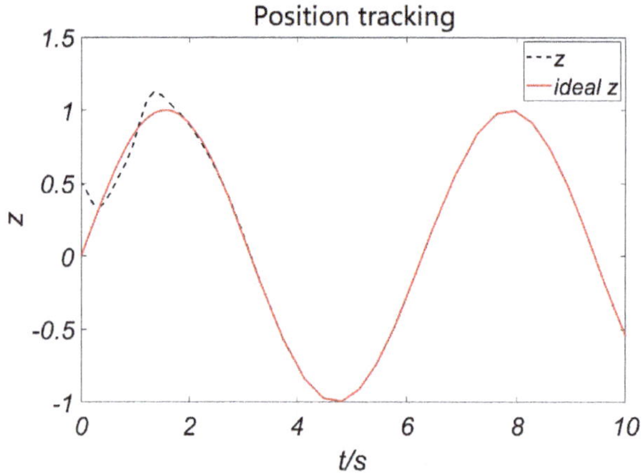

Figure 8. Position tracking without the decoupling algorithm of the z-axis.

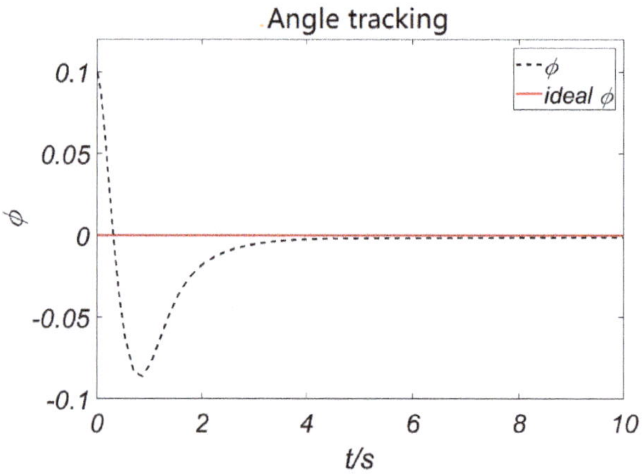

Figure 9. Angle tracking with the decoupling algorithm.

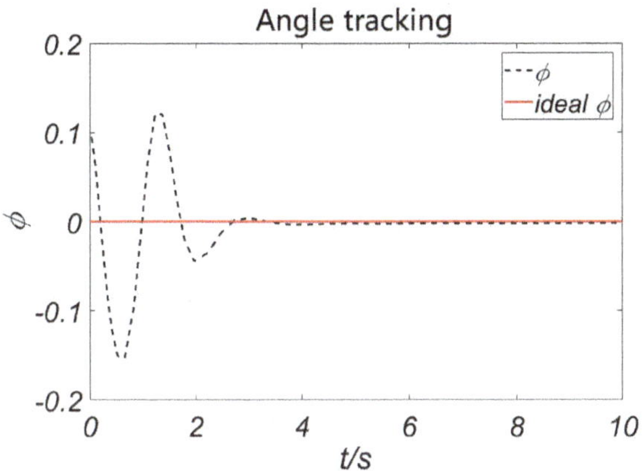

Figure 10. Angle tracking without the decoupling algorithm.

Figures 11–13 show the speed tracking and angular speed tracking with the decoupling algorithm and fuzzy control. It could be seen that the change trajectory was smooth, and the time required to reach a stable state from the initial state was very short. Figures 14 and 15 show the control input of the system. We could see that the control input curves were smooth and no chattering occurred. Theoretical analysis and experimental simulation results showed that the fuzzy sliding mode control based on the decoupling algorithm could improve the stability of the system, had a better self-adaptive ability, and effectively restrained the modeling errors and external disturbances of the co-rotating twin-rotor aircraft attitude system.

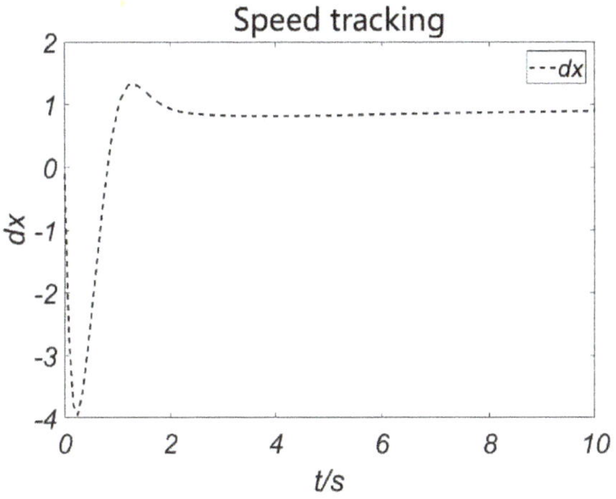

Figure 11. Speed tracking with the decoupling algorithm.

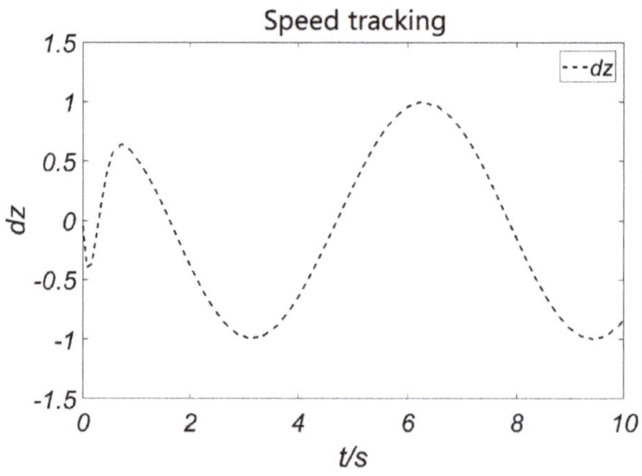

Figure 12. Speed tracking with the decoupling algorithm.

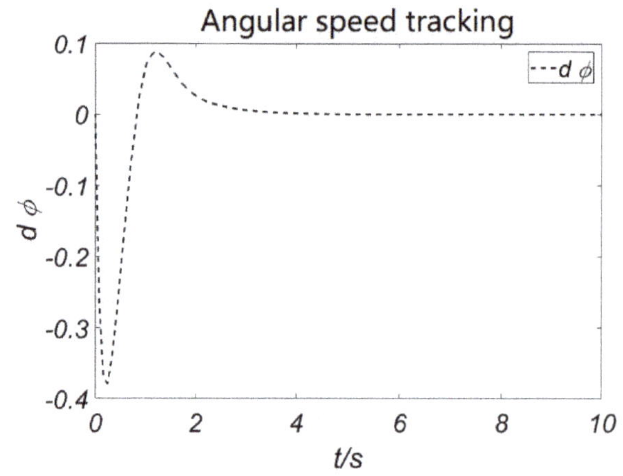

Figure 13. Position tracking with the decoupling algorithm.

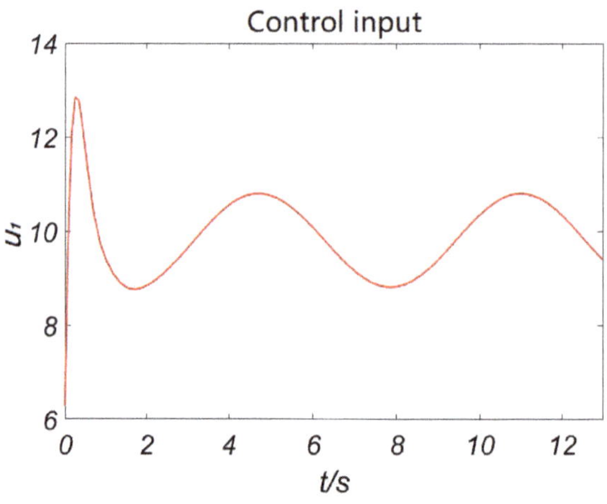

Figure 14. Control input u_1 with the decoupling algorithm.

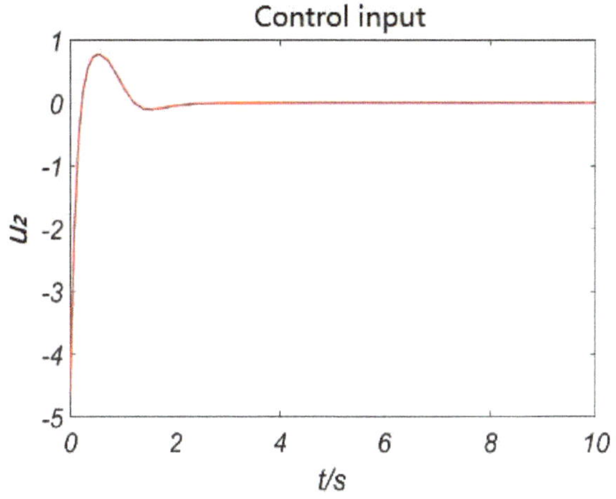

Figure 15. Control input u_2 with the decoupling algorithm.

6. Conclusions

According to the structural characteristics of the coaxial-rotor UAV, a dynamic model of longitudinal motion was established. The dynamic model of the aircraft was then decoupled, the fuzzy control and sliding mode controls were combined, and a fuzzy sliding mode control based on the decoupling algorithm was designed for the coaxial-rotor. The control method was then simulated by MATLAB/Simulink. The results showed that the control method could track the command signal more quickly and efficiently compared to the method of the traditional sliding mode control. It could quickly reduce the yaw attitude angle deviation and the steady-state error could reach almost zero. With a strong self-adaptive ability, it could achieve a better control effect. The response speed, tracking accuracy, and efficiency of the system had been significantly improved.

The proposed control method could improve the stability of the system, which could effectively restrain the modeling errors and external disturbances of the aircraft's attitude system. This method had the advantages of high control precision, strong robustness, and ease of implementation in engineering. In future studies, we will focus on the design of the decoupling algorithm under the influence of more inputs and interferences, and will apply this algorithm to specific engineering practices.

Author Contributions: K.L. designed the main parts of the method and was also responsible for writing the paper. Y.W. built the Simulation program. C.W. and H.D. provided some technical comments.

Funding: This research received no external funding.

Conflicts of Interest: The authors declare no conflicts of interest.

References

1. Kannan, N.; Bhat, M.S. Longitudinal H Infinity Stability Augmentation System for a Thrust Vectored Unmanned Aircraft. *J. Guid. Control Dyn.* **2005**, *28*, 1240–1250. [CrossRef]
2. Krämer, P.; Gimonet, B.; v Grünhagen, W. A systematic approach to nonlinear rotorcraft model identification. *Aerosp. Sci. Technol.* **2002**, *6*, 579–590. [CrossRef]
3. Wang, X.; Li, K.; Zhao, N.; Deng, H. Nonlinear dynamics modeling and simulation of cylindrical coaxial UAV. In Proceedings of the IEEE International Conference on Real-Time Computing & Robotics, Okinawa, Japan, 14–18 July 2017; pp. 505–510.
4. Niu, S.; Li, J.; Shen, Y. Design, modeling and disturbance rejection control of a bio-inspired coaxial helicopter MAV in Atmospheric Boundary Layer. In Proceedings of the 2015 IEEE International Conference on Robotics and Biomimetics (ROBIO), Zhuhai, China, 6–9 December 2016; pp. 1272–1277.

5. Mettler, B.; Tischler, M.B.; Kanade, T. System Identification Modeling of a Small-Scale Unmanned Rotorcraft for Flight Control Design. *J. Am. Helicopter Soc.* **2002**, *47*, 50–63. [CrossRef]
6. Khandani, K.; Majd, V.J.; Darestani, M.R.; Talebi, H. A stochastic sliding mode scheme for longitudinal control of an aircraft model. In Proceedings of the 25th Iranian Conference on Electrical Engineering, Tehran, Iran, 2–4 May 2017; pp. 795–800.
7. Gavilan, F.; Vazquez, R.; Acosta, J.Á. Adaptive control for aircraft longitudinal dynamics with thrust saturation. *J. Guid. Control Dyn.* **2015**, *38*, 651–661. [CrossRef]
8. Duan, F.H.; Han, C.Z. Application of Nonlinear System Decoupling Control Theory in Flight Control. *Electron. Opt. Control* **2001**, *84*, 30–34.
9. Ao, B.Q.; Li, J.L. Variable-structure control method of aircraft attitude based on feedback linearization. *Mod. Def. Technol.* **2003**, *31*, 41–44.
10. Petr, H.; Kashyapa, N. Aircraft longitudinal motion control based on Takagi–Sugeno fuzzy model. *Appl. Soft Comput.* **2016**, *49*, 269–278.
11. BC Campos, L.M. Nonlinear Longitudinal Stability of a Symmetric Aircraft. *J. Aircr.* **1997**, *34*, 360–369. [CrossRef]
12. McLean, D.; Zaludin, Z.A. Stabilization of longitudinal motion of a hypersonic transport aircraft. *Trans. Inst. Meas. Control* **1999**, *21*, 99–105. [CrossRef]
13. Jyothi, J.; Bindu, G.R.; Jayakumar, M. Robust Longitudinal Controller Design for an Unmanned Tailless Aircraft. In Proceedings of the 6th AIAA Aviation Technology, Integration and Operations Conference, Wichita, KS, USA, 25–27 September 2006; pp. 1–7.
14. Yue, T.; Wang, L.X.; Ai, J.Q. Longitudinal Linear Parameter Varying Modeling and Simulation of Morphing Aircraft. *J. Aircr.* **2013**, *50*, 1673–1681. [CrossRef]
15. Adam, M.W.; Ephrahim, G. Longitudinal Dynamics of a Perching Aircraft. *Proc. SPIE Int. Soc. Opt. Eng.* **2012**, *43*, 1386–1392.
16. Yuan, X.M.; Zhu, J.H.; Mao, M. Modeling and robust tracking control for coaxial unmanned helicopter. *Control Theory Appl.* **2014**, *31*, 1286–1294.
17. Yang, X.M.; Li, W.J. Four rotor aircraft control based on sliding mode controller. *J. Univ. Technol. Nat. Sci.* **2016**, *39*, 924–928.
18. Shiyang, R.; Dominick, A. Longitudinal Flying Qualities Prediction for Nonlinear Aircraft. *J. Guid. Control Dyn.* **2003**, *26*, 474–482.
19. Al-Hiddabi, S.A.; McClamroch, N.H. Aggressive Longitudinal Aircraft Path Tracking Using Nonlinear Control. *Asian J. Control* **2010**, *3*, 280–288. [CrossRef]
20. Speyer, J.; White, J.; Douglas, R.; Hull, D. MIMO controller design for longitudinal decoupled aircraft motion. In Proceedings of the Guidance and Control Conference, Boston, MA, USA, 19–22 August 2013; pp. 729–737.
21. Shan, S.Q.; Hou, Z.X.; Wang, W.K. Aircraft longitudinal decoupling based on a singular perturbation approach. *Adv. Mech. Eng.* **2017**, *9*, 1–8. [CrossRef]
22. Phillips, W.F.; Santana, B.W. Aircraft Small-Disturbance Theory with Longitudinal-Lateral Coupling. *J. Aircr.* **2002**, *39*, 973–980. [CrossRef]
23. Xu, R.; Ozguner, U. Sliding mode control of a class of underactuated systems. *Automatica* **2008**, *44*, 233–241. [CrossRef]
24. Wang, Y.C.; Sun, H. Sliding Mode Controller Design for VTOL Aircraft. *Comput. Meas. Control* **2016**, *6*, 102–105.

electronics

MDPI

Article

Motion Equations and Attitude Control in the Vertical Flight of a VTOL Bi-Rotor UAV

Sergio Garcia-Nieto [1,*], Jesus Velasco-Carrau [1], Federico Paredes-Valles [2], Jose Vicente Salcedo [1] and Raul Simarro [1]

1 Instituto Universitario de Automática e Informática Industrial, Universitat Politècnica de València, 46022 Valencia, Spain; jevecar@upv.es (J.V.-C.); jsalcedo@isa.upv.es (J.V.S.); rausifer@isa.upv.es (R.S.)

2 Department of Control and Simulation (Micro Air Vehicle Laboratory), Faculty of Aerospace Engineering, Delft University of Technology, Kluyverweg 1, 2629 HS Delft, The Netherlands; f.paredesvalles@tudelft.nl

* Correspondence: sgnieto@isa.upv.es; Tel.: +34-963877007 (ext. 85794)

Received: 1 January 2019; Accepted: 7 February 2019; Published: 12 February 2019

Abstract: This paper gathers the design and implementation of the control system that allows an unmanned Flying-wing to perform a Vertical Take-Off and Landing (VTOL) maneuver using two tilting rotors (Bi-Rotor). Unmanned Aerial Vehicles (UAVs) operating in this configuration are also categorized as Hybrid UAVs due to their ability of having a dual flight envelope: hovering like a multi-rotor and cruising like a traditional fixed-wing, providing the opportunity of facing complex missions in which these two different dynamics are required. This work exhibits the Bi-Rotor nonlinear dynamics, the attitude tracking controller design and also, the results obtained through Hardware-In-the-Loop (HIL) simulation and experimental studies that ensure the controller's efficiency in hovering operation.

Keywords: tilt rotors; nonlinear dynamics; simulation; hardware-in-the-loop; vertical take off

1. Introduction

In recent years, the continuous development in engineering-related fields, such as automatic systems, flight control and the aerospace industry as a whole, has contributed to the rapid growth of the area of Unmanned Aerial Vehicles (UAV), making it an appealing research topic in both military and civil applications. In terms of civil applications, it is important to mention those related with agricultural services, marine operations, natural disaster support, etc. Within the military field, UAVs are mostly used in missions in which there are high risks.

In order to increase the number and complexity, and hence, the performance efficiency of these applications, UAVs characterized by a dual flight envelop are currently needed. These unmanned vehicles, inheriting the advantages of both traditional fixed-wing aircraft and rotorcraft, have the ability to execute a VTOL maneuver and to aggressively inspect a certain area, as well as to perform a high-speed aerial surveillance over a wide region. For the aforementioned reasons, these vehicles are known as Hybrid UAVs.

According to [1], hybrid UAVs can be categorized into two main types: Convertiplanes and Tail-Sitters. Firt of all, Convertiplanes category regroup those aerial vehicles that take off, cruise, hover and land with the aircraft reference line remaining horizontal. Respect to this class, there exist several vehicles implementing the idea such as FireFLY6 [2] and TURAC [3]; and also projects researching in this direction [4,5]. Second, a Tail-Sitter is an aircraft that takes off and lands vertically on its tail and the whole aircraft tilts forward using differential thrust or control surfaces to achieve horizontal flight. This category, as it is considered as a complex challenge from the point of view of control systems engineering, has become an interesting research concept as shown by vehicles like Quadshot [6] or prototype [7].

This paper presents recent work concerning the first stage in the development of a hybrid UAV that can be categorized as Tail-Sitter with the exceptions that in this case the aircraft takes off and lands vertically on its nose (using an external ground-station) and that this platform changes the sense of the rotors in order to perform the transition phase between hovering and cruising. Figure 1 shows the manoeuvrability scheme of the proposed unmanned aerial vehicle. In addition, the prototype built based on this philosophy has been *nicknamed* V-Skye.

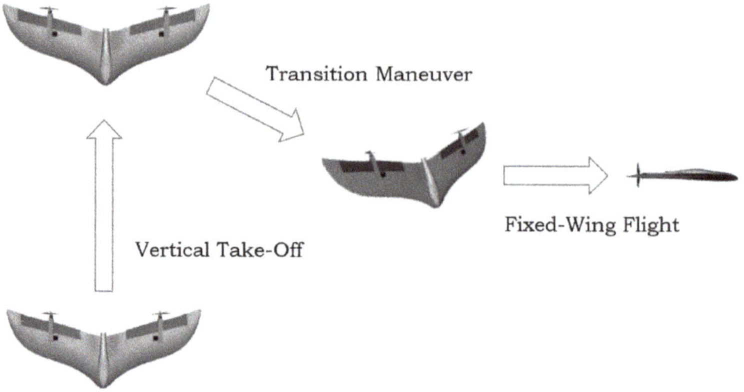

Figure 1. Scheme of the transition maneuver between flight modes.

In this article, the design of the control system is not only based on simulations, but also on an experimental procedure in which the controllers have to adequately stabilize the UAV allowing it to hover filtering external disturbances. In order to control the attitude, the vehicle is provided with two tilting rotors that allow alterations of its pitch angle and yaw rate and also, modifications in the motor throttles in order to handle roll and vertical speed variables. This is the first step of the development of the entire autonomous system, that will provide this UAV with the hybrid characteristics required by autonomous aviation market, as presented in [8].

Different types of controllers can be designed for UAVs. The simplest ones are linear PID based on linearized models of UAVs. In the literature it is possible to find several approaches which solve the problem of controlling non-linear UAVs: non-linear PID based solutions [9–11], non-linear robust approaches [12–14], back-stepping algorithms [15–17], sliding mode control [12,17], $H\infty$ control [16] or non-linear observer based [18,19].

As commented, the objective of this article is to adequately stabilize the designed hybrid UAV in an experimental procedure. This is performed by using 4 linear PIDs tuned by a genetic algorithm. The genetic algorithm searches for the PID parameters that minimize a performance index such as the integral squared error or the settling time.

Using 4 linear PIDs can be considered as a first approach to the design of the control system, and also the easiest way to implement a control system from an experimental point of view. In further researches authors will try to apply more complex techniques such as non-linear PID [9,10] and non-linear robust approaches [12–14].

The rest of this article is structured as follows. Section 2 covers the description of the airframe that has been used during this project, while Section 3 is focused on the explanation of the mathematical model that describes this aerial vehicle. In Section 4, the design of the attitude control system is presented. Section 5 gathers information related to the HIL simulation platform and finally, Section 6 presents the simulation and real-test results that ensure the controller's performance.

2. Airframe Description

A rigid body moving inside a three-dimensional (3D) space has a total of six-degrees-of-freedom (6DoF). In this way, a mechanical system formed by a single rigid body needs at least six independently manipulated interactions with the system (inputs) to drive it to an arbitrary orientation and position.

In a hovering maneuver, a flying vehicle is maintained motionless over a reference point at a constant altitude (constant reference position) and on a constant heading angle ψ. Hence, only four of the six degrees of freedom are forced to a reference value (controlled) when hovering. The other two, i.e., pitch θ and roll ϕ angles are dependent variables that evolve along time according to the system equations of motion.

The V-Skye is designed with two tilting-rotors moved by servo-mechanisms. The result is a vehicle with two motors for which thrust $\vec{T}_R$ and $\vec{T}_L$ can be independently modified, not only in magnitude, but also in one direction. The system is thus provided with the amount of independent inputs needed for the hovering manoeuvre.

Figure 2 shows an outline drawing of the V-Skye. In order to simplify the dynamics, all actuation parts (motors, motor frames, servomotors and their transmission parts) are allocated as symmetrically as possible about the fixed coordinate axis $\{\hat{X}_b, \hat{Y}_b, \hat{Z}_b\}$ of the aircraft reference frame. In particular, all elements are placed on the $\hat{Y}_b\hat{Z}_b$ plane and symmetrical to the $\hat{X}_b\hat{Z}_b$ plane.

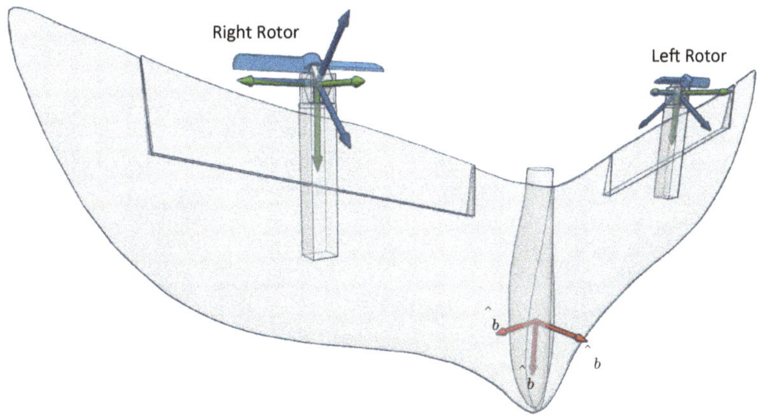

Figure 2. Local axis in the 3D graphical model of the V-Skye UAV.

For simplicity on the explanations, authors have divided the aircraft into three well-differentiated frames.

2.1. Main Body Frame

As depicted in Figure 2, the main body has the constructive shape of a Flying-Wing aircraft, such as the ones used in [20,21]. It is a rigid body housing all the electronics as well as the two servomotors that allow rotation of the motor frames (see Sections 2.2 and 2.3).

The reference system $\{\hat{X}_b, \hat{Y}_b, \hat{Z}_b\}$ has its origin at the aircraft centre of mass and it is fixed to the main body frame. For its vertical flight phase, the $\hat{X}_b$ direction points front, towards what would naturally be the upper part of the fuselage. The $\hat{Z}_b$ direction points down, towards the nose of the flying-wing. Finally, $\hat{Y}_b$ axis is perpendicular to the other two and points towards the right-side wing.

The earth coordinate axis $\{\hat{X}_e, \hat{Y}_e, \hat{Z}_e\}$ is a North-East-Down (NED) inertial frame of reference, also positioned at the aircraft centre of mass but fixed to the earth surface. Euler angles roll ϕ, pitch θ and yaw ψ define the main body orientation respect to the earth axis. Figure 3 shows those three independent rotations.

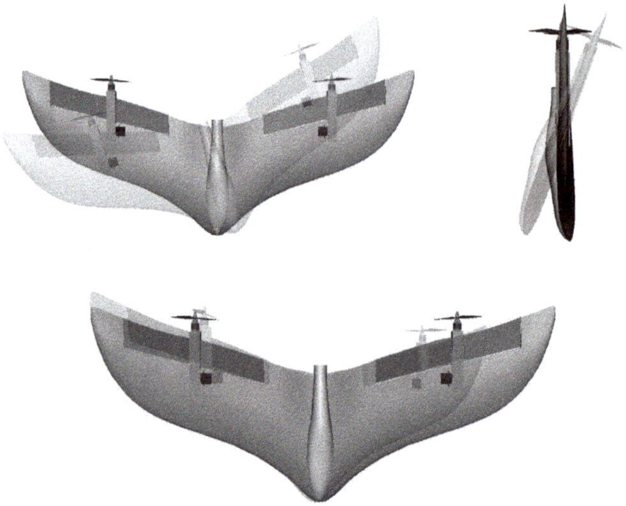

Figure 3. Roll, pitch and yaw motions.

2.2. Right Motor Frame

It is formed by the right motor, its right-handed propeller and a structure specially designed to hold it and stand any reaction force derived from the flight. It can be seen as a second rigid body attached to the aircraft by a rotatory joint, as Figure 4 describes in detail.

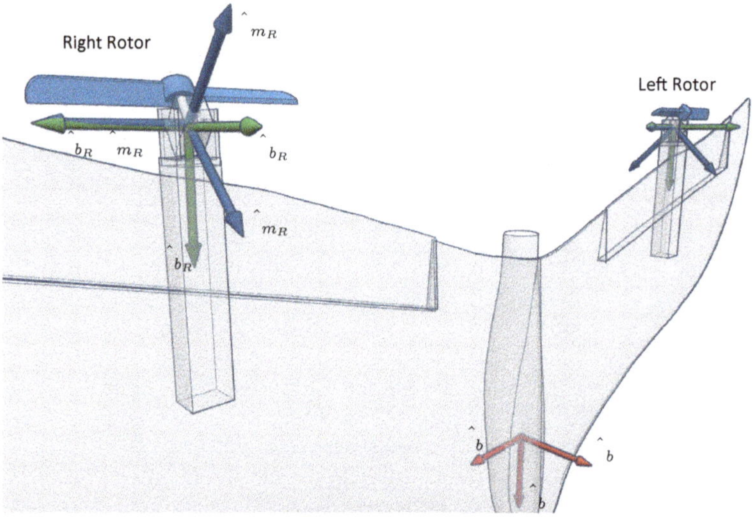

Figure 4. Right rotor coordinate reference system.

Two coordinate systems are defined to describe the motion of this frame with respect to the main body. $\{\hat{X}_{b_R}, \hat{Y}_{b_R}, \hat{Z}_{b_R}\}$ is fixed to the the main body frame and parallel to $\{\hat{X}_b, \hat{Y}_b, \hat{Z}_b\}$; its centre O_{b_R} is placed where the rotatory joint intersects the motor shaft axis. On the other hand, $\{\hat{X}_{m_R}, \hat{Y}_{m_R}, \hat{Z}_{m_R}\}$ have its origin O_{m_R} at O_{b_R}; the $\hat{Y}_{m_R}$ axis coincides with $\hat{Y}_{b_R}$ ($\hat{Y}_{b_R} \parallel \hat{Y}_{m_R}$) and $\hat{Z}_b$ axis coincides with the motor shaft axis.

Because the right motor frame is attached to the main body by a rotatory joint, it has one single DOF: the angle λ_R rotated about the axis $\hat{Y}_{b_R} \parallel \hat{Y}_{m_R}$. When $\lambda_R = 0$, both $\{\hat{X}_{m_R}, \hat{Y}_{m_R}, \hat{Z}_{m_R}\}$ and $\{\hat{X}_{b_R}, \hat{Y}_{b_R}, \hat{Z}_{b_R}\}$ have the exact same position and orientation. The direction of the right motor's thrust is changed by actuating on the λ_R value, since the thrust has always the $\hat{Z}_{m_R}$ direction. For this reason, a servomotor is used to manipulate λ_R.

2.3. Left Motor Frame

The left motor frame includes namely the left motor with a left-handed propeller and the structure to hold it. The right and left propellers are designed to be right and left handed respectively. This makes the motors to rotate in opposite senses, helping to compensate motors torques. Figure 5 illustrates the configuration and coordinate reference systems.

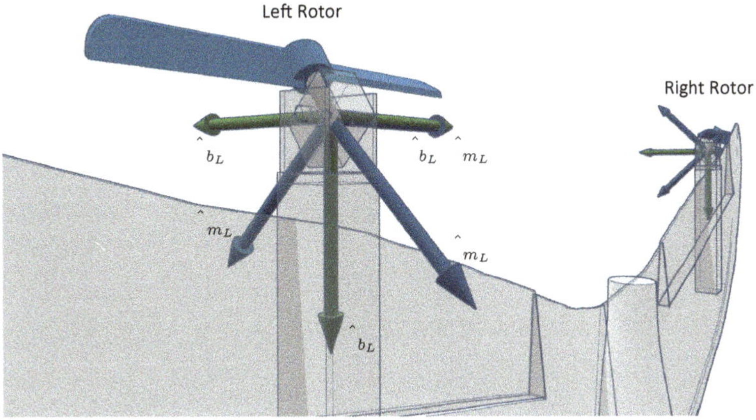

Figure 5. Left rotor coordinate reference system.

Similarly to the right motor frame, two coordinate systems are defined for the left motor frame: $\{\hat{X}_{b_L}, \hat{Y}_{b_L}, \hat{Z}_{b_L}\}$ and $\{\hat{X}_{m_L}, \hat{Y}_{m_L}, \hat{Z}_{m_L}\}$ with λ_L being the angle rotated by $\{\hat{X}_{m_L}, \hat{Y}_{m_L}, \hat{Z}_{m_L}\}$ with respect to $\{\hat{X}_{b_L}, \hat{Y}_{b_L}, \hat{Z}_{b_L}\}$ about the $\hat{Y}_{b_L} \parallel \hat{Y}_{m_L}$ axis.

3. Mathematical Model

The equations that conform the 6-DOF non-linear dynamical model are derived in this paper assuming the following hypothesis:

1. The whole aircraft is assumed to be rigid body; it means that the distance between any two points in the airframe remains constant. This is a fundamental condition because it allows to understand the movement of the vehicle as a translation and a rotation around the center of mass independently.
2. Derived from the previous item, the changes in λ_R and λ_L angles do not affect the mass distribution along the aircraft body.
3. The rotational movement of the Earth is negligible with respect to the accelerations on the vehicle. i.e., the Earth frame is an inertial frame of reference.
4. The atmosphere is assumed to be calm (no wind or turbulence)
5. The plane $\{Y_b = 0\}$ is a plane of symmetry. Hence, the inertia products about the Y_b axis $I_{y_b x_b} = I_{y_b z_b} = 0$

3.1. Translational Equations

Let $\vec{F}$ be the resultant force vector of all the external forces acting on the system, m the total mass of the aircraft and $\vec{V}$ the aircraft linear velocity with respect to the earth frame. Newton's second law can be written as:

$$\vec{F} = m \cdot \dot{\vec{V}} \quad (1)$$

$$\{F_{x_b}, F_{y_b}, F_{z_b}\} = m \cdot \frac{{}^{E}d}{dt}(\{u, v, w\}) \quad (2)$$

where F_{x_b}, F_{y_b}, F_{z_b}, u, v and w are the three components of the resultant force and the system velocity respectively, both magnitudes expressed in body axis. Velocity's time derivative with respect to the earth frame might be now rewritten as the summation of its time derivative with respect to the body frame and the cross product of angular and linear velocities as follows

$$\dot{\vec{V}} = \frac{{}^{B}d}{dt}(\{u, v, w\}) + {}^{E}\vec{\omega}^{B} \times \vec{V} \quad (3)$$

$$\dot{\vec{V}} = \{\dot{u}, \dot{v}, \dot{w}\} + \{p, q, r\} \times \{u, v, w\} \quad (4)$$

$$\dot{\vec{V}} = \{\dot{u} + q \cdot w - r \cdot v, \dot{v} + r \cdot u - p \cdot w, \dot{w} + p \cdot v - q \cdot u\} \quad (5)$$

where ${}^{E}\vec{\omega}^{B}$ is the angular velocity of the body frame with respect to the earth frame, and p, q and r its components expressed in body axis. Therefore, vector Equation (1) is separated into three independent equations as shown next

$$m \cdot (\dot{u} + q \cdot w - r \cdot v) = F_{x_b} \quad (6)$$

$$m \cdot (\dot{v} + r \cdot u - p \cdot w) = F_{y_b} \quad (7)$$

$$m \cdot (\dot{w} + p \cdot v - q \cdot u) = F_{z_b} \quad (8)$$

The external forces considered in this work are the rotors' thrust and the aircraft weight. Since the work is focused on the design of a control scheme for a hovering manoeuvre, the dynamics model does not consider aerodynamic effects on the aircraft body. In a VTOL procedure, the lift is totally generated through the thrust produced by the rotors. On the other hand, drag produced by the flying-wing airframe is taken as an external disturbance for the attitude tracking controller.

Denoting by T_R and T_L the thrust magnitudes of right and left rotors, respectively; the definition of the thrust forces in body axis is characterised by the angles λ_R and λ_L of the right and left motor frames with respect to the body frame.

$$\vec{T_R} = -T_R \cdot \{\sin \lambda_R, 0, \cos \lambda_R\} \quad (9)$$

$$\vec{T_L} = -T_L \cdot \{\sin \lambda_L, 0, \cos \lambda_L\} \quad (10)$$

In [22], a complete study of the performance of several types of propellers at different airflow conditions is presented. Along the paper, John B. Brandt and Michael S. Selig., explain how to model thrust and torque at low Reynolds numbers, gathering data of these values to calculate aerodynamic coefficients for a large number of commercial propellers. Now, taking the equations presented in [22] as a reference, and letting T and τ denote thrust and torque magnitudes of a propeller and C_T and C_τ its force and torque coefficients, then

$$T = C_T \rho n^2 D^4 \quad (11)$$

$$\tau = C_\tau \rho n^2 D^5 \quad (12)$$

Hence, the relation between torque and thrust on the propeller can be written as

$$T = \frac{C_T}{D \cdot C_\tau}\tau \quad (13)$$

If we denote by δ_R and δ_L the motors throttle, and given that the aircraft will be mounting two brushless DC motors managed by electronic speed controllers (ESC), the two motor torques are related to their throttle by the expressions (14) and (15).

$$\tau_R = k_\tau \delta_R \quad (14)$$
$$\tau_L = k_\tau \delta_L \quad (15)$$

and hence

$$T_R = \frac{C_T}{D \cdot C_Q} k_\tau \delta_R = k_T \delta_R \quad (16)$$
$$T_L = \frac{C_T}{D \cdot C_Q} k_\tau \delta_L = k_T \delta_L \quad (17)$$

k_τ is the torque's throttle coefficient and is assumed to be constant for a given motor and ESC combination whereas k_T is the thrust's throttle coefficient that depends on the airspeed and is constant for the static case. This leads to the following definition of thrust force:

$$F_{Tx} = -k_T\delta_R \cdot \sin\lambda_R - k_T\delta_L \cdot \sin\lambda_L \quad (18)$$
$$F_{Ty} = 0 \quad (19)$$
$$F_{Tz} = -k_T\delta_R \cdot \cos\lambda_R - k_T\delta_L \cdot \cos\lambda_L \quad (20)$$

The aforementioned force variables $\{F_{x_b}, F_{y_b}, F_{z_b}\}$ are now substituted by their corresponding terms of thrust and weight forces in the body axis.

$$m\left(\dot{u} + q \cdot w - r \cdot v\right) = -k_T\delta_R \cdot \sin\lambda_R - k_T\delta_L \cdot \sin\lambda_L - m \cdot g \cdot \sin\theta \quad (21)$$
$$m\left(\dot{v} + r \cdot u - p \cdot w\right) = m \cdot g \cdot \cos\theta \sin\phi \quad (22)$$
$$m\left(\dot{w} + p \cdot v - q \cdot u\right) = -k_T\delta_R \cdot \cos\lambda_R - k_T\delta_L \cdot \cos\lambda_L + m \cdot g \cdot \cos\theta\cos\phi \quad (23)$$

3.2. Rotational Equations

By definition of angular momentum about the mass centre $\vec{H}^{c.g.}$ and considering a rigid-body configuration

$$\vec{H}^{c.g.} = \begin{bmatrix} I_{x_bx_b} & -I_{x_by_b} & -I_{x_bz_b} \\ -I_{y_bx_b} & I_{y_by_b} & -I_{y_bz_b} \\ -I_{z_bx_b} & -I_{z_by_b} & I_{z_bz_b} \end{bmatrix}^{c.g.} \cdot^E \vec{\omega}^B \quad (24)$$

where $I_{ii}\forall i \in \{x_b, y_b, z_b\}$ are the moments of inertia of the aircraft body about its mass centre in body axis and $I_{ij}\forall ij \in \{x_b, y_b, z_b\}$with $j \neq i$ are the products of inertia.

Now, the total moment of forces about the aircraft's mass centre is equal to its angular momentum's time derivative with respect to the earth frame. Additionally, we can again split the time derivative of the angular momentum with respect to the earth frame into its time derivative with respect to the body frame and the cross product of the angular velocity and the angular momentum.

$$\vec{Q}^{c.g.} = \dot{\vec{H}}^{c.g.} \quad (25)$$
$$\vec{Q}^{c.g.} = \frac{^B d}{dt}\left(\vec{H}^{c.g.}\right) +^E \vec{\omega}^B \times \vec{H}^{c.g.} \quad (26)$$

Equation (26) gives the following three independent equations

$$L = I_{xx}\dot{p} - I_{xz}\dot{r} - I_{xz}pq + (I_{zz} - I_{yy})\, qr \quad (27)$$
$$M = I_{yy}\dot{q} - I_{xz}\left(p^2 - r^2\right) + (I_{xx} - I_{zz})\, pr \quad (28)$$
$$N = I_{zz}\dot{r} - I_{xz}\dot{p} + I_{xz}rq + (I_{yy} - I_{xx})\, pq \quad (29)$$

being L, M and N the three body axis components of the total external moments applied on the aircraft centre of mass.

The right rotor thrust applied at right motor frame generates an external moment about the aircraft centre of mass. Denoting $\vec{r}_{m_R} = \{x_{m_R}, y_{m_R}, z_{m_R}\}$ as the position vector of the right motor frame, the total moment of the right rotor thrust about the centre of mass is given by Equations (30) and (31). Note that $x_{m_R} = 0$ by design.

$$\vec{Q}_{m_R} = \vec{r}_{m_R} \times \vec{T}_R \quad (30)$$
$$\vec{Q}_{m_R} = k_T\delta_R \cdot \{-y_{m_R}\cos\lambda_R, -z_{m_R}\sin\lambda_R, y_{m_R}\sin\lambda_R\} \quad (31)$$

Additionally, the right motor torque applied to the propeller is translated to the right motor frame as a reaction torque (same direction but opposite sense). That torque is then translated to the body frame according to λ_R angle. From Equation (14)

$$\vec{\tau}_R = \tau_R \cdot \{\sin\lambda_R, 0, \cos\lambda_R\} \quad (32)$$
$$\vec{\tau}_R = k_\tau\delta_R \cdot \{\sin\lambda_R, 0, \cos\lambda_R\} \quad (33)$$

In accordance, thrust moment and motor torque for the left rotor are given by expressions (34) and (35)

$$\vec{Q}_{m_L} = k_T\delta_L \cdot \{-y_{m_L}\cos\lambda_L, -z_{m_L}\sin\lambda_L, y_{m_L}\sin\lambda_L\} \quad (34)$$
$$\vec{\tau}_L = -k_\tau\delta_L \cdot \{\sin\lambda_L, 0, \cos\lambda_L\} \quad (35)$$

Now, substituting the total external moments and torques into Equations (27)–(29)

$$I_{xx}\dot{p} - I_{xz}\dot{r} - I_{xz}pq + (I_{zz} - I_{yy})\, qr = \delta_R\left(k_\tau\sin\lambda_R - k_T y_{m_R}\cos\lambda_R\right) - \delta_L\left(k_\tau\sin\lambda_L + k_T y_{m_L}\cos\lambda_L\right) \quad (36)$$

$$I_{yy}\dot{q} - I_{xz}\left(p^2 - r^2\right) + (I_{xx} - I_{zz})\, pr = -\delta_L k_T z_{m_L}\sin\lambda_L - \delta_R k_T z_{m_R}\sin\lambda_R \quad (37)$$

$$I_{zz}\dot{r} - I_{xz}\dot{p} + I_{xz}rq + (I_{yy} - I_{xx})\, pq = \delta_R\left(k_\tau\cos\lambda_R + k_T\sin\lambda_R y_{m_R}\right) + \delta_L\left(k_T y_{m_L}\sin\lambda_L - k_\tau\cos\lambda_L\right) \quad (38)$$

3.3. Collection of Non-Linear Equations

In addition to the six dynamics equations derived above, six kinematic equations can be stated to express the transformation from the body to the earth system of reference. The aircraft behaviour is therefore, described by a total of twelve equations of motion.

As a summary, the aircraft model is divided into the following sets.

3.3.1. Translational Dynamics Equations

$$\dot{u} = rv - qw - g\sin\theta - \frac{k_T}{m}\left(\delta_R\sin\lambda_R + \delta_L\sin\lambda_L\right) \quad (39)$$
$$\dot{v} = pw - ru + g\cos\theta\sin\phi \quad (40)$$
$$\dot{w} = qu - pv + g\cos\theta\cos\phi - \frac{k_T}{m}\left(\delta_R\cos\lambda_R + \delta_L\cos\lambda_L\right) \quad (41)$$

3.3.2. Rotational Dynamics Equations

$$\begin{aligned} \dot{p} - \tfrac{I_{xz}}{I_{xx}}\dot{r} &= +\tfrac{I_{xz}}{I_{xx}}pq + \tfrac{I_{yy}-I_{zz}}{I_{xx}}qr + \tfrac{\delta_R}{I_{xx}}\left(k_\tau \sin\lambda_R - k_T y_{m_R}\cos\lambda_R\right) \\ &\quad -\tfrac{\delta_L}{I_{xx}}\left(k_\tau \sin\lambda_L + k_T y_{m_L}\cos\lambda_L\right) \end{aligned} \tag{42}$$

$$\dot{q} = \frac{I_{xz}}{I_{yy}}\left(p^2 - r^2\right) + \frac{I_{zz} - I_{xx}}{I_{yy}}pr - \frac{1}{I_{yy}}\delta_L k_T z_{m_L}\sin\lambda_L - \frac{1}{I_{yy}}\delta_R k_T z_{m_R}\sin\lambda_R \tag{43}$$

$$\begin{aligned} \dot{r} - \tfrac{I_{xz}}{I_{zz}}\dot{p} &= -\tfrac{I_{xz}}{I_{zz}}rq + \tfrac{I_{xx}-I_{yy}}{I_{zz}}pq + \tfrac{\delta_R}{I_{zz}}\left(k_\tau \cos\lambda_R + k_T \sin\lambda_R y_{m_R}\right) \\ &\quad +\tfrac{\delta_L}{I_{zz}}\left(k_T y_{m_L}\sin\lambda_L - k_\tau \cos\lambda_L\right) \end{aligned} \tag{44}$$

3.3.3. Kinematic Translational Equations

$$u = \dot{x}_e \cos\theta\cos\psi + \dot{y}_e\cos\theta\sin\psi - \dot{z}_e\sin\theta \tag{45}$$

$$\begin{aligned} v &= \dot{x}_e\left(\sin\phi\sin\theta\cos\psi - \cos\phi\sin\psi\right) + \dot{y}_e\left(\sin\phi\sin\theta\sin\psi + \cos\phi\cos\psi\right) \\ &\quad +\dot{z}_e\sin\phi\cos\theta \end{aligned} \tag{46}$$

$$\begin{aligned} w &= \dot{x}_e\left(\cos\phi\sin\theta\cos\psi + \sin\phi\sin\psi\right) + \dot{y}_e\left(\cos\phi\sin\theta\sin\psi - \sin\phi\cos\psi\right) \\ &\quad +\dot{z}_e\cos\phi\cos\theta \end{aligned} \tag{47}$$

3.3.4. Kinematic Rotational Equations (Euler Angles)

$$p = \dot{\phi} - \dot{\psi}\cdot\sin\theta \tag{48}$$

$$q = \dot{\theta}\cdot\cos\phi + \dot{\psi}\cdot\cos\theta\cdot\sin\phi \tag{49}$$

$$r = \dot{\psi}\cdot\cos\theta\cdot\cos\phi - \dot{\theta}\cdot\sin\phi \tag{50}$$

4. Control System

The control problem associated with the presented UAV stabilization is challenging for several reasons. The complexity of flight dynamics resides in the system non-linearity, unstable nature and high degree of coupling. Furthermore, in this case, the system is under-actuated because only four control inputs can be used during take off, landing and hover manoeuvres, while the whole system is six-degree-of-freedom (6DoF). Therefore, robust and reliable feedback control strategies are needed to regulate the attitude of the UAV within an operational range.

Due to their simplicity, ease of implementation and robust performance, a decentralised and linear control scheme based on four proportional-integral-derivative controllers (PID) has been chosen to design the attitude tracking controller. Figure 6 shows the block diagram of the proposed feedback control scheme. It is composed by a controller for each angle of orientation and one for the vertical velocity.

Attending to the nonlinear equations of motion presented in Section 3, a modification on one of the four system inputs δ_R, δ_L, λ_R or λ_L, excites more than one state variable at the same time. This means that the system dynamics are highly coupled. In an attempt to reduce the effect of coupled actuators, a set of four inputs u_1, u_2, u_3 and u_4 are defined as a combination of the real input variables. Equations (51) to (54) show the new system input variables and their relationship with motors throttle and thrust angles

$$u_1 = \frac{1}{2}\left(\delta_L - \delta_R\right) \tag{51}$$

$$u_2 = -\left(\lambda_R + \lambda_L\right) \tag{52}$$

$$u_3 = \lambda_R - \lambda_L \tag{53}$$

$$u_4 = \frac{1}{2}\left(\delta_R + \delta_L\right) \tag{54}$$

With this new definition of the system inputs, a change in u_1 has high effect on the $\dot{p}$ variable, u_2 on $\dot{q}$, u_3 on $\dot{r}$ and u_4 on $\dot{w}$. Now the different PID control schemes are defined accordingly.

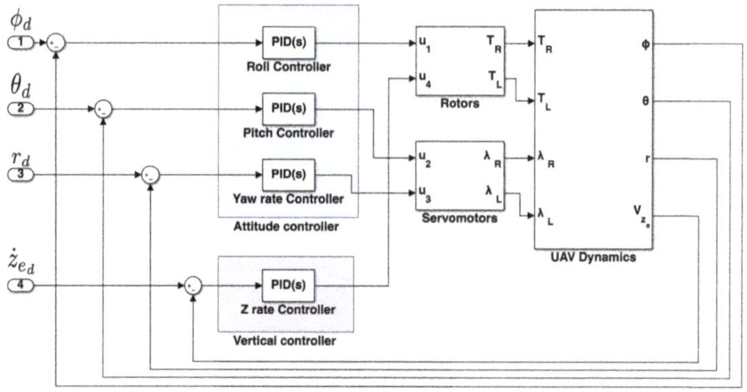

Figure 6. Attitude and vertical speed controller scheme of the UAV V-Skye.

4.1. Roll and Pitch Controllers

Controllers for roll and pitch angles are implemented similarly: both use standard PID controllers with feedback of estimated angles (roll or pitch) from a complementary filter, resulting in the following control law:

$$u_1 = K_{p,1}((\phi_d - \phi) + \frac{1}{T_{i,1}}\int(\phi_d - \phi)dt + T_{d,1}(\dot{\phi}_d - \dot{\phi})) \quad (55)$$

$$u_2 = K_{p,2}((\theta_d - \theta) + \frac{1}{T_{i,2}}\int(\theta_d - \theta)dt + T_{d,2}(\dot{\theta}_d - \dot{\theta})) \quad (56)$$

where $K_{p,i}$, $K_{i,i}$ and $K_{d,i}$ are the proportional, integral and derivative gains, respectively. ϕ_d and θ_d is associated with the reference angles or desired roll and pitch, while u_i is the controller action; on one hand, u_1 corresponds to the differential thrust between two rotors, while u_2 is associated with the tilt angle of both rotors.

4.2. Angular Velocity r Controller

Despite of the fact that the heading of the UAV is generally a less critical degree of freedom, external disturbances can generate undesired rotational movements around the Z axis during flying. In order to overcome this drift, a yaw rate PI controller has been implemented using the next expression:

$$u_3 = K_{p,3}((r_d - r) + \frac{1}{T_{i,3}}\int(r_d - r)dt) \quad (57)$$

In this case, the reference (r_d) is compared with the feedback yaw angular velocity (r) measured by a yaw rate gyroscope. The resulting difference is sent to the controller in order to generate opposite tilt angles by using a differential servo deflection (u_3).

4.3. Vertical Velocity Controller

Due to the high coupling, changes on angles λ_R and λ_L, from the roll and r controllers, induce variations on the vertical thrust, which leads to variations on vertical velocity. Therefore, a vertical velocity PID controller is implemented as shown in expression (58).

$$u_4 = K_{p,4}((\dot{z}_{e_d} - \dot{z}_e) + \frac{1}{T_{i,4}}\int(\dot{z}_{e_d} - \dot{z}_e)dt + K_{d,4}(\ddot{z}_{e_d} - \ddot{z}_e)) \quad (58)$$

where, $\dot{z}_{e_d}$ is the reference vertical velocity, $\dot{z}_e$ is the measured vertical velocity estimated from measures of the inertial magnetic unit (IMU) sensors. u_4 corresponds to the increase in the same amount of thrust in both rotors.

4.4. Local Stability

The global close loop stability has been considered from two aspects. Firstly, the control structures presented in this Section have been adjusted to ensure that the poles of the closed loop system have all of them negative real terms [23]. For this purpose, the particular transfer functions presented in Section 5 (Equation (59)) have been used. Obviously, this only guarantees stability close to the equilibrium point defined for the linearization and, therefore, it is a local stability constraint.

Secondly, in order to study the margin of the local stability, HIL simulations have been performed using the designed controllers against the full non-linear model of the UAV. In this way, the validity range of the controllers designed from the linear transfer function is tested by realistic simulations [24,25]. Obviously, this is not an overall guarantee of stability, but it is possible to define a range of operation with a high degree of confidence for the control design.

5. Test Prototype

The initial test platform presented in the work is a Flying-wing from Multiplex, model XENO UNI [26]. This RC plane has been modified to add two tilt rotors, following the main concept of VTOL UAV described previously. Figures 7 and 8 show the prototype assembled for real flight tests.

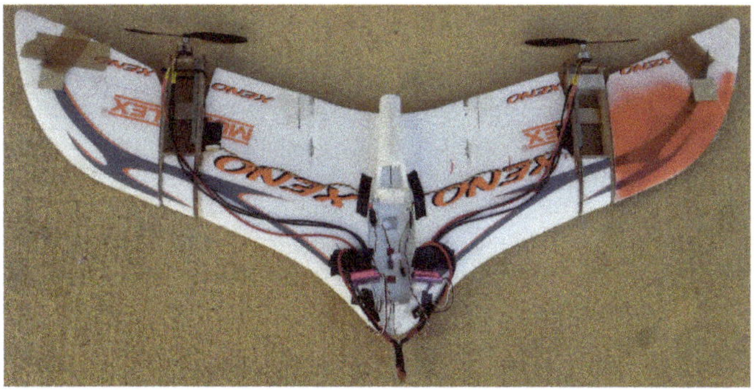

Figure 7. Xeno UNI form Multiplex with two customised tilt rotors.

Figure 8. Tilt rotor mechanical structure.

The set of aircraft parameters used for control design and simulation have been derived form the test platform described in the previous paragraph, and their particular values are as follows:

- Wingspan: 1.26 m
- Fuselage length: 0.526 m
- Empty weight: 0.1904 kg
- Operating weight: 0.7484 kg
- k_T = 15.7 and k_τ = 0.34
- Brushless motors: T-MOTOR Antimass MT2814 770 KV
- Propellers: T-MOTOR 12"x4" CF
- Electric Speed Controllers: HW-09-V2-OEM
- Tilt Rotor Servos: HITEC HS-5475HB
- Flight Controller: CC3D OpenPilot Revolution
- Battery: 2200 mAh 4 S 80/160 C

Apart from these geometrical characteristics, the engines provide a maximum thrust of 15.7 N and the tilt-mechanism admit a deflection up to 0.5235 rad. The combination of engines, propellers, Electronic Speed Controllers (ESC) and servos employed in this airframe provide enough power and maneuverability to successfully accomplish the hovering mission even with a larger payload, such as an action camera.

In order to complete the tilt-rotor model parameters, the corresponding moments of inertia are: $I_{xx} = 0.0015$ kg m^2, $I_{yy} = 0.0160$ kg m^2 y $I_{zz} = 0.0176$ kg m^2, and also a product of inertia of $I_{xz} = -1.4182 \times 10^{-5}$ m^4.

5.1. Prototype Model Linearization

In order to design the linear control approach described in Section 4, first of all, the attitude dynamic model described by expressions of Section 3 is decoupled into four linearized single-input single-output (SISO) models around the operational range of a hovering maneuver, which implies that $p \simeq q \simeq r \simeq 0$ and $u \simeq v \simeq w \simeq 0$.

The four linear models are obtained by replacing the nonlinear equations of motion with their Taylor series approximation truncated to the first order with respect to the controlled variables and inputs. This linearization approach is well know and, frequently, it is defined as *Small Perturbation Theory* [27–30]. Once the set of equations have a linear structure and after replacing the model parameters by their prototype value, the following transfer functions are obtained:

$$G_\phi(s) = \frac{\phi(s)}{u_1(s)} = \frac{554.78}{s^2(s+19.05)} \tag{59}$$

$$G_\theta(s) = \frac{\theta(s)}{u_2(s)} = \frac{-13.95(s-138.5)(s+138.5)}{s^2(s+153.2)(s+21.75)} \tag{60}$$

$$G_r(s) = \frac{r(s)}{u_3(s)} = \frac{1.1674(s-56160)}{s(s+153.2)(s+21.75)} \tag{61}$$

$$G_{\dot{z}_e}(s) = \frac{\dot{z}_e(s)}{u_4(s)} = \frac{-117.09}{s(s+19.05)} \tag{62}$$

5.2. Tuning PID Loops for Prototype Control

After calculating the transfer functions, the controllers parameters can be obtained either using classical techniques, such as the Root-Locus method, or more modern techniques based on optimisation with genetic algorithms.

The use of Root-Locus method for PID tuning proves to be quite easy and useful compared to others techniques, since it indicates the manner in which the open-loop poles and zeros should be modified so that the response meets system performance specifications [23]. However, a disadvantage of using this technique is that it is necessary to employ a linear model, which is only an approximation of the complex dynamics of the UAV.

On the other hand, PID tuning and optimisation using genetic algorithms is a modern technique that provides an adaptive searching mechanism inspired on Darwin's principle of reproduction and survival of the fittest. The individuals (solutions) in a population are represented by chromosomes that are associated to a fitness value (problem evaluation). The chromosomes are subjected to an evolutionary process which takes several cycles. Basic operations are selection, reproduction, crossover and mutation [31]. One of the main advantage of using genetic algorithms is that it is a global search technique of optimal and sub-optimal solutions of a problem and hence it can directly interact with the non-linear dynamics model.

Parameters of PID controllers obtained using both techniques are presented in Tables 1 and 2.

Table 1. Pid parameters using the Root-Locus method.

PID Controller	K_p	T_i	T_d
ϕ	0.25886	∞	0.952
θ	0.11936	∞	0.95
r	0.415	0.83	0
$\dot{z}_e$	0.85	∞	1.076

Table 2. Pid parameters using a genetic algorithm.

PID Controller	K_p	T_i	T_d
ϕ	0.2979	4.2997	0.2945
θ	0.2080	9.8801	0.368
r	0.2	4	0
$\dot{z}_e$	0.55	6.5	0.5

In order to compare the performance of the attitude and vertical speed control system tuned using both classical and modern strategies, a numerical simulation with the nonlinear model of the UAV is made using both set of parameters and the accuracy requirements for the system are formulated in terms of the settling time (t_s) and the integral squared error index (ISE), which is related to the time response of the system.

$$ISE = \int_0^t e^2 dt \tag{63}$$

The presented simulations consisted in transition with predefined dynamics from one steady state flight to another. Numerical results evaluating the proposed indexes are shown in Tables 3 and 4.

Table 3. Controller performance using the Root-Locus method.

PID Controller	t_s	ISE
ϕ	4.603	96,075
θ	3.618	128,310
r	3.986	95,095
$\dot{z}_e$	3.986	4922

Table 4. Controller performance using a genetic algorithm.

PID Controller	t_s	ISE	$Enhancement_{t_s}$ (%)	$Enhancement_{ISE}$ (%)
ϕ	2.931	95,385	36.32	0.72
θ	1.890	127,080	47.76	0.96
r	2.835	95,140	28.88	−0.05
$\dot{z}_e$	2.168	4419	45.61	10.22

Taking into consideration the proposed control effort indexes, PID tuning obtained by a genetic algorithm is the most comprehensive choice. For this reason, this set of parameters will be adopted during the simulations and flight tests (Section 6).

5.3. HIL Simulation Platform

The HIL simulation platform is shown in Figure 9. The main computation unit is a PXI laboratory computer from National Instruments. This equipment includes several boards to interface a great amount of external devices. As an example digital and analogue input/output boards, Ethernet and Serial ports or four USB ports. This hardware comes with a real-time operative system that can be configured to run Real-Time simulations [25,32,33].

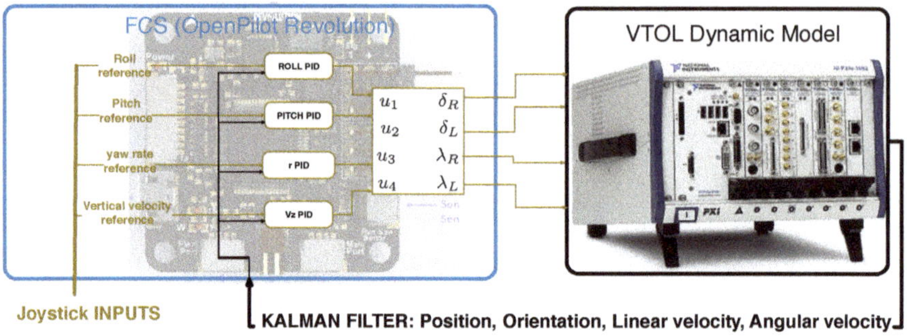

Figure 9. Hardware In the Loop platform diagram.

The set of 12 equations of motion derived in Section 3 have been particularised for the prototype model and implemented in Matlab/Simulink. Simulink allows to compile the model to be run on the PXI real-time target [34,35].

A CC3D OpenPilot Revolution Board [36] has been elected as the flight control unit (FCS). It is a digital board with a micro-controller that comes with the necessary onboard sensors and autopilot software already installed. The control algorithms have been modified and implemented on the FCS. Then the onboard sensors have been bypassed so that the values coming from the model are used to close the loop. Finally, the board has been connected to the PXI to send actuators values and receive state variables values. A joystick has also been connected to the board to send the reference value to the controllers.

6. Results

This section covers the analysis of the results obtained in simulation (using HIL simulation platform) and in real flight tests. The main goal is to understand the dynamical behaviour of the system while it is hovering at certain attitude configurations.

6.1. Simulation Results

The first simulation consists in coupled changes of roll and pitch that allow the reader to understand how the control actions (throttle and tilt-angle) have to be modified in order to follow the reference in attitude and vertical velocity.

Two conclusions can be drawn based on the previous Figure 10. First of all, as it was already mentioned, the system deflects the rotors in the same direction in order to achieve certain pitch angles. As it can be observed, until $t = 15$ seconds there have only been modifications in the pitch angle, and therefore, $\lambda_L = \lambda_R$.

However, the interesting aspect of this result is what happens with the tilt-angle when a roll manoeuvre is being carried out. As mentioned, this rotation needs a different thrust generated by each of the rotors until the air-frame reaches the desired configuration. Therefore, during this procedure the total torque is no longer null, meaning that there exists a tendency to modify the yaw of the system by increasing its angular velocity r. The result of the coupling of these dynamical movements is that the attitude tracking controller produces opposite tilt-angles in order to maintain $r = 0$.

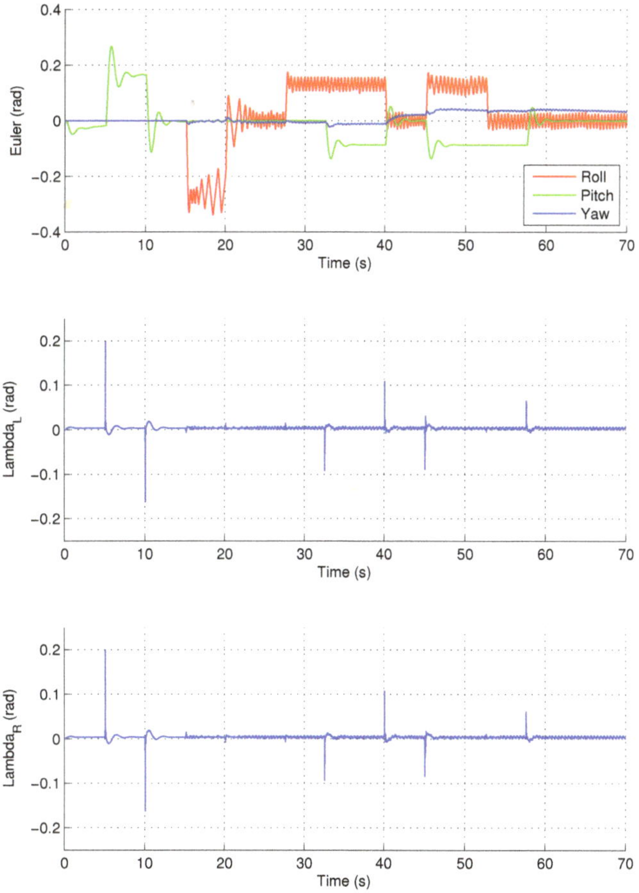

Figure 10. Rotor deflection against modifications in ϕ and θ.

This effect is better explained in Figure 11.

It is also important to remark that the system tends to modify its heading when the hovering takes place at configurations characterised by non-null pitch and roll angles. The reason of this azimuth variation is that the controller is based on the manipulation of the angular velocity r instead of the yaw angle ψ, and therefore, the system acts in order not to change the heading that it has at a specific time t but if this modification occurs, the new heading would be the new reference and the system would not try to recover the initial orientation.

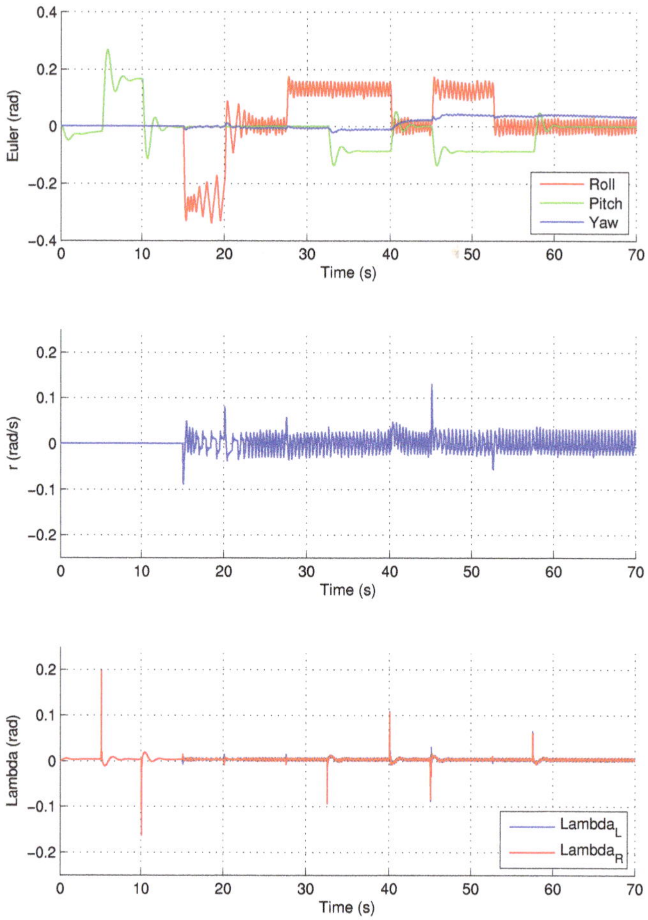

Figure 11. Evolution of r and λ against modifications in ϕ and θ.

Once the time-response of the tilt-angles has been analysed at different attitude configurations, the same study with the thrust generated by each of the rotors has been performed, as shown in Figure 12.

In this second study, it is important to remark two conclusions about the behaviour of the thrust when the system follows the attitude configurations shown in Figures 10 and 11. First of all, as it was established in Section 2, a change in roll implies the controller to modify the same absolute value of the rotational velocity in each of the rotor but with different sign. As a result, the thrust is not longer equal until the reference is reached.

The other conclusion refers to the thrust modification associated with changes in pitch from the equilibrium configuration. Although, this effect can not be graphically observed due to the resolution of the thrust signal, when a certain pitch angle is established as a reference, the engines have to create more thrust in order to compensate the fact that the entire system is tilted, and therefore, to maintain the equilibrium of external forces. This means that the higher the pitch angle is, the higher the increase in thrust has to be.

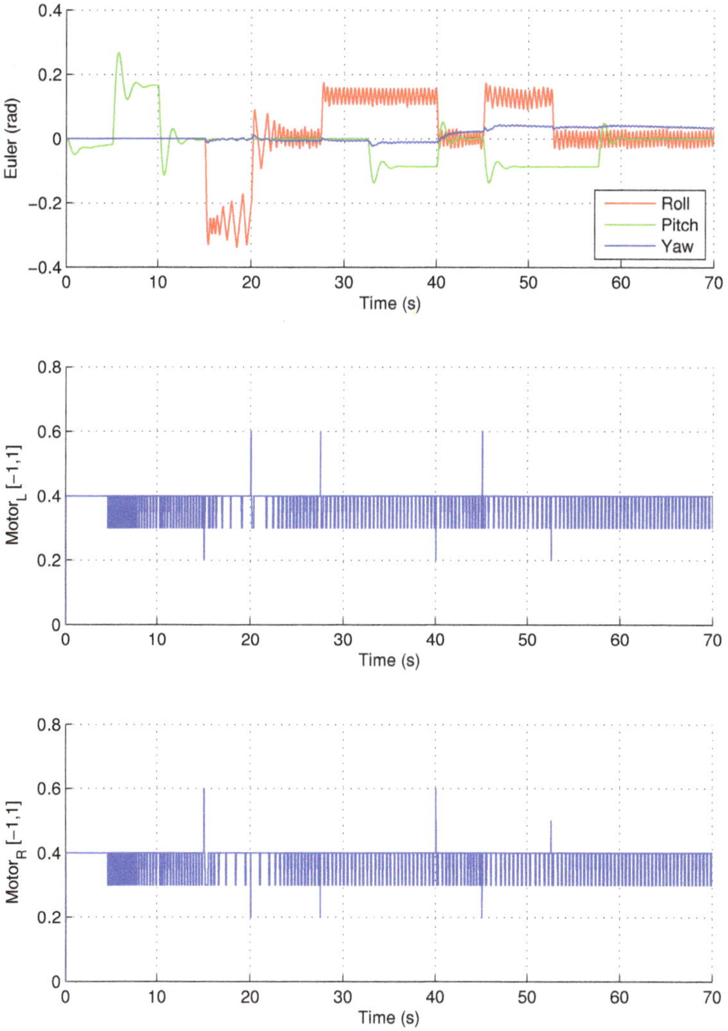

Figure 12. Thrust signal against modifications in ϕ and θ.

The last study linked to this changes in attitude is focused on the analysis of the time-response of the vertical velocity $\dot{z}_e$ shown in Figure 13. Due to the control scheme design, when the aircraft takes non-equilibrium attitude configurations, it has a tendency to descend, *i.e.*, to increase the positive value of its velocity in the Z_e axis of the earth reference frame. The vertical velocity controller tries to maintain a given reference value allowing to reduce variations in this velocity, but, because the altitude is not directly controlled, the system keeps changing its vertical position.

Figure 13 illustrates that each pitch or roll angle apart from the equilibrium ($\phi \neq 0$ or $\theta \neq 0$) implies an increase in $\dot{z}_e$, which is compensated by an increase in the thrust generated by the rotors. Again, since the controlled variable is the vertical velocity, each time a perturbation occurs there is a loss of altitude that is never recovered, but the vertical speed controller manages to stop falling and reaches again the $\dot{z}_e = 0$ set-point.

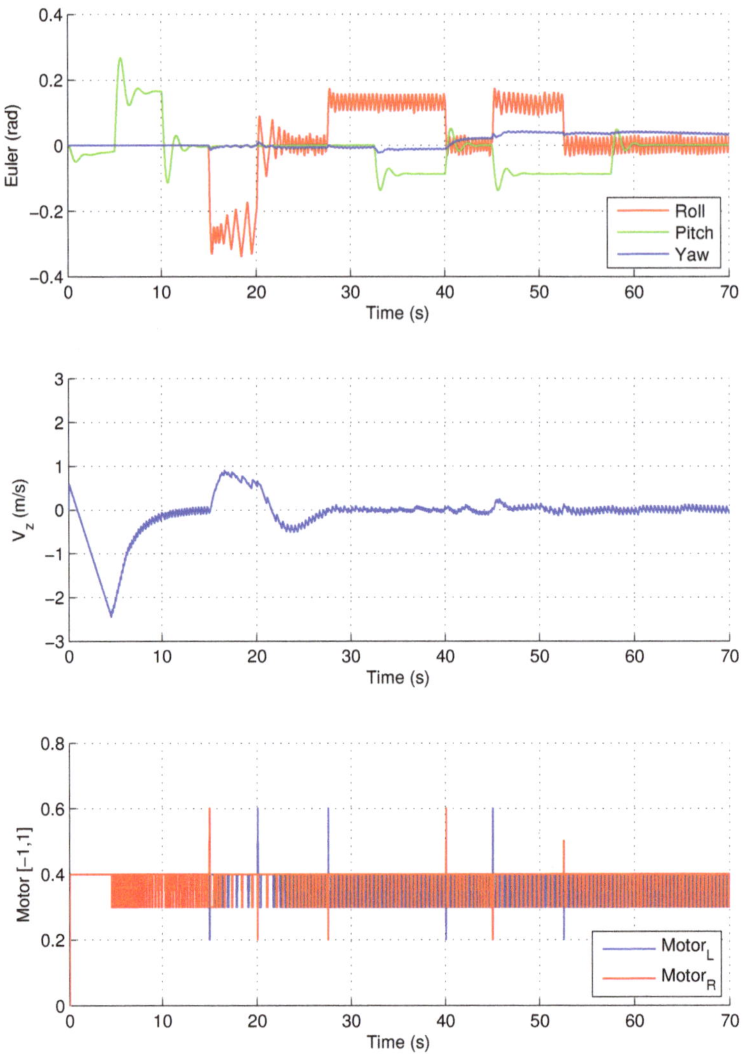

Figure 13. Evolution of V_z and thrust signal against modifications in ϕ and θ.

6.2. Real Flight Results

In order to verify the performance of attitude control system designed for this UAV, the proposed PID control laws have been implemented in the on-board hardware using the OpenPilot Revolution Board [36], which contains a full 10 DOF IMU with gyroscopes, accelerometers, magnetometers and barometric pressure sensors.

In the experimental test flight, the goal was the stabilisation of the UAV, compensating any external disturbance during vertical flight. Due to the lack of GPS, velocity and position can only be calculated by integration of accelerometers combined with the gyroscopes measures and with the barometric pressure sensor (sensor fusion [37,38]). As result, the measures of velocity and position are not reliable enough to close the loop. For this reason, in the experimental test presented, the vertical velocity controller has been annulled and the pilot directly acts on the u_4 system input. That means, direct modification of motors nominal thrust which implies a manual (piloted) control of the speed and vertical position.

The experimental results for the platform control are presented in Figure 14 as a time plot of all angles of the UAV and the controller actions.

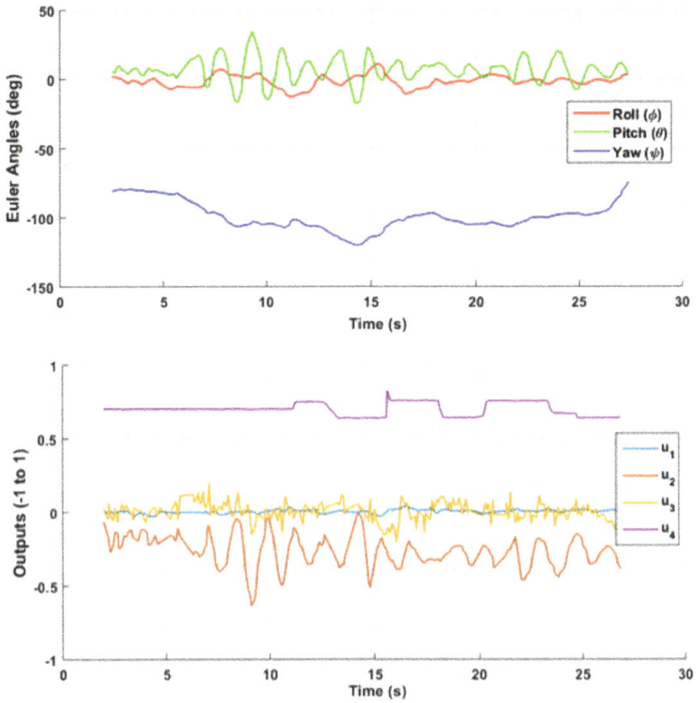

Figure 14. Time series of Euler angles and controller outputs.

As can be seen, the closed loop response of the UAV is stable because of the fact that Euler angles vary within a limited range, confirming the effectiveness of the proposed approach and theoretical results. Since vertical speed and altitude are manually piloted and the absence of GPS, it is impossible to define a 3D trajectory through waypoints. To cover this drawback, the real test flight demonstration video can be visualized on Youtube [39] (V-Skye Prototype: Test Flight—https://youtu.be/zS9oWur-Pss). Thanks to this video, readers can have an approximate idea of the 3D trajectory of the UAV during the test flight.

7. Conclusions and Future Work

The design and implementation of the attitude tracking controller for a Bi-Rotor VTOL UAV have been presented in this article. The simulations performed in a HIL environment showed the main aspects of the dynamical behaviour of this system, while the real test flight results presented verify the stability and viability for this UAV platform. Therefore, the UAV designed has the ability to hover despite its complex dynamics and allows future designs based on cascade control schemes focused on controlling global position and velocities. In addition, future works will explore the robust control design that allows the automatic transition between the two different flying modes: VTOL and cruise flight.

Author Contributions: Conceptualization, S.G.-N., J.V.; Funding acquisition, S.G.-N.; Investigation, S.G.-N., J.V., F.P.-V., J.V.S. and R.S.; Supervision, S.G.-N.

Funding: The authors would like to acknowledge the Spanish Ministry of Economy and Competitiveness for providing funding through the project DPI2015-71443-R and the local administration Generalitat Valenciana through the project GV/2017/029.

Acknowledgments: The authors would like to thank the editors and the reviewers for their valuable time and constructive 317 comments.

Conflicts of Interest: The authors declare no conflict of interest.

References

1. Saeed, A.S.; Younes, A.B.; Islam, S.; Dias, J.; Seneviratne, L.; Cai, G. A review on the platform design, dynamic modeling and control of hybrid UAVs. In Proceedings of the 2015 International Conference on Unmanned Aircraft Systems (ICUAS), Denver, CO, USA, 9–12 June 2015; pp. 806–815.
2. Vuruskan, A.; Yuksek, B.; Ozdemir, U.; Yukselen, A.; Inalhan, G. Dynamic modeling of a fixed-wing VTOL UAV. In Proceedings of the 2014 International Conference on Unmanned Aircraft Systems (ICUAS), Orlando, FL, USA, 27–30 May 2014; pp. 483–491.
3. Aktas, Y.O.; Ozdemir, U.; Dereli, Y.; Tarhan, A.F.; Cetin, A.; Vuruskan, A.; Yuksek, B.; Cengiz, H.; Basdemir, S.; Ucar, M.; et al. A low cost prototyping approach for design analysis and flight testing of the turac vtol uav. In Proceedings of the 2014 International Conference on Unmanned Aircraft Systems (ICUAS), Orlando, FL, USA, 27–30 May 2014; pp. 1029–1039.
4. Kendoul, F.; Fantoni, I.; Lozano, R. Modeling and control of a small autonomous aircraft having two tilting rotors. *IEEE Trans.Robot.* **2006**, *22*, 1297–1302. [CrossRef]
5. Papachristos, C.; Alexis, K.; Tzes, A. Design and experimental attitude control of an unmanned tilt-rotor aerial vehicle. In Proceedings of the 2011 15th International Conference on Advanced Robotics (ICAR), Tallinn, Estonia, 20–23 June 2011; pp. 465–470.
6. Sinha, P.; Esden-Tempski, P.; Forrette, C.A.; Gibboney, J.K.; Horn, G.M. Versatile, modular, extensible vtol aerial platform with autonomous flight mode transitions. In Proceedings of the 2012 IEEE Aerospace Conference, Big Sky, MT, USA, 3–10 March 2012; pp. 1–17.
7. Sanchez, A.; Escareno, J.; Garcia, O.; Lozano, R. Autonomous hovering of a noncyclic tiltrotor UAV: Modeling, control and implementation. In Proceedings of the 17th IFAC Wold Congress, Seoul, Korea, 6–11 July 2008; pp. 803–808.
8. Valavanis, K.P. *Advances in Unmanned Aerial Vehicles: State of the Art and the Road to Autonomy*; Springer Science & Business Media: Cham, Switzerland, 2008; Volume 33.
9. Moreno-Valenzuela, J.; Pérez-Alcocer, R.; Guerrero-Medina, M.; Dzul, A. Nonlinear PID-Type Controller for Quadrotor Trajectory Tracking. *IEEE/ASME Trans. Mechatron.* **2018**, *23*, 2436–2447. [CrossRef]
10. Gonzalez-Vazquez, S.; Moreno-Valenzuela, J. A new nonlinear pi/pid controller for quadrotor posture regulation. In Proceedings of the 2010 Electronics, Robotics and Automotive Mechanics Conference (CERMA), Cuernavaca, Morelos, Mexico, 28 September–1 October 2010; pp. 642–647.
11. Ortiz, J.P.; Minchala, L.I.; Reinoso, M.J. Nonlinear robust H-Infinity PID controller for the multivariable system quadrotor. *IEEE Lat. Am. Trans.* **2016**, *14*, 1176–1183. [CrossRef]
12. Zhao, B.; Xian, B.; Zhang, Y.; Zhang, X. Nonlinear robust sliding mode control of a quadrotor unmanned aerial vehicle based on immersion and invariance method. *Int. J. Robust Nonlinear Control* **2015**, *25*, 3714–3731. [CrossRef]
13. Pérez-Alcocer, R.; Moreno-Valenzuela, J.; Miranda-Colorado, R. A robust approach for trajectory tracking control of a quadrotor with experimental validation. *ISA Trans.* **2016**, *65*, 262–274. [CrossRef] [PubMed]
14. Xia, D.; Cheng, L.; Yao, Y. A Robust Inner and Outer Loop Control Method for Trajectory Tracking of a Quadrotor. *Sensors* **2017**, *17*, 2147. [CrossRef] [PubMed]
15. Ma, Z.; Zhang, Q.; Chen, L. Attitude control of quadrotor aircraft via adaptive back-stepping control. *CAAI Trans. Intell. Syst.* **2015**, *10*, 1–7.
16. Raffo, G.V.; Ortega, M.G.; Rubio, F.R. Backstepping/nonlinear H∞ control for path tracking of a quadrotor unmanned aerial vehicle. In Proceedings of the American Control Conference, Seattle, WA, USA, 11–13 June 2008; pp. 3356–3361.

17. Madani, T.; Benallegue, A. Sliding mode observer and backstepping control for a quadrotor unmanned aerial vehicles. In Proceedings of the American Control Conference, New York, NY, USA, 11–13 July 2007; pp. 5887–5892.
18. Mokhtari, M.R.; Braham, A.C.; Cherki, B. Extended state observer based control for coaxial-rotor UAV. *ISA Trans.* **2016**, *61*, 1–14. [CrossRef] [PubMed]
19. Ban, H.; Qi, Z.; Li, B.; Gong, W. Nonlinear Disturbance Observer based Dynamic Surface Control for Trajectory Tracking of a Quadrotor UAV. In Proceedings of the 2018 International Symposium in Sensing and Instrumentation in IoT Era (ISSI), Shanghai, China, 6–7 September 2018; pp. 1–6.
20. Wang, R.; Zhou, Z.; Shen, Y. Flying-Wing UAV Landing Control and Simulation Based on Mixed H 2/H. In Proceedings of the 2007 International Conference on Mechatronics and Automation, Harbin, China, 5–8 August 2007; pp. 1523–1528.
21. Quigley, M.; Barber, B.; Griffiths, S.; Goodrich, M.A. Towards real-world searching with fixed-wing mini-UAVs. In Proceedings of the 2005 IEEE/RSJ International Conference on Intelligent Robots and Systems, Edmonton, AB, Canada, 2–6 August 2005; pp. 3028–3033.
22. Brandt, J.; Selig, M. Propeller performance data at low reynolds numbers. In Proceedings of the 49th AIAA Aerospace Sciences Meeting including the New Horizons Forum and Aerospace Exposition, Orlando, FL, USA, 4–7 January 2011; p. 1255.
23. Ogata, K.; Yang, Y. *Modern Control Engineering*; Prentice Hall India: Delhi, India, 2002; Volume 4,.
24. Ayasun, S.; Fischl, R.; Vallieu, S.; Braun, J.; Cadırlı, D. Modeling and stability analysis of a simulation–stimulation interface for hardware-in-the-loop applications. *Simul. Model. Pract. Theory* **2007**, *15*, 734–746. [CrossRef]
25. Carrau, J.V.; Reynoso-Meza, G.; García-Nieto, S.; Blasco, X. Enhancing controller's tuning reliability with multi-objective optimisation: From Model in the loop to Hardware in the loop. *Eng. Appl. Artif. Intell.* **2017**, *64*, 52–66. [CrossRef]
26. Multiplex. XENO UNI. 2019. Available online: https://www.multiplex-rc.de/produkte/214241-bk-xeno-uni (accessed on 1 January 2019).
27. Stengel, R.F. *Flight Dynamics*; Princeton University Press: Princeton, NJ, USA, 2015.
28. Velasco-Carrau, J.; García-Nieto, S.; Salcedo, J.; Bishop, R.H. Multi-objective optimization for wind estimation and aircraft model identification. *J. Guid. Control Dyn.* **2015**, *39*, 372–389. [CrossRef]
29. Velasco, J.; García-Nieto, S. Unmanned aerial vehicles model identification using multi-objective optimization techniques. *IFAC Proc. Vol.* **2014**, *47*, 8837–8842. [CrossRef]
30. Schmidt, L.V. *Introduction to Aircraft Flight Dynamics*; American Institute of Aeronautics and Astronautics: Reston, VA, USA, 1998.
31. Griffin, I.; Bruton, J. *On-line PID Controller Tuning Using Genetic Algorithms*; Dublin City University: Dublin, Ireland, 2003.
32. Liao, Y.; Shi, X.; Fu, C.; Meng, J. Hardware in-the-loop simulation system based on NI-PXI for operation and control of microgrid. In Proceedings of the 2014 IEEE 9th Conference on Industrial Electronics and Applications (ICIEA), Hangzhou, China, 9–11 June 2014; pp. 1366–1370.
33. Fodor, D.; Enisz, K. Vehicle dynamics based ABS ECU verification on real-time hardware-in-the-loop simulator. In Proceedings of the 2014 16th International Power Electronics and Motion Control Conference and Exposition (PEMC), Antalya, Turkey, 21–24 September 2014; pp. 1247–1251.
34. Nemes, R.O.; Ruba, M.; Martis, C. Integration of Real-Time Electric Power Steering System Matlab/Simulink Model into National Instruments VeriStand Environment. In Proceedings of the 2018 IEEE 18th International Power Electronics and Motion Control Conference (PEMC), Budapest, Hungary, 26–30 August 2018; pp. 700–703.
35. Chang, H.; Wang, D.; Wei, H.; Zhang, Q.; Dong, G. Design of Tracked Model Vehicle Measurement and Control System Based on VeriStand and Simulink. In *MATEC Web of Conferences*; EDP Sciences: Les Ulis, France, 2018; p. 03047.
36. Community, L. Revolution Board Setup—LibrePilot/OpenPilot Wiki 0.1.4 Documentation. 2019. Available online: http://opwiki.readthedocs.io/en/latest/user_manual/revo/revo.html (accessed on 1 January 2019).
37. Marantos, P.; Koveos, Y.; Kyriakopoulos, K.J. UAV state estimation using adaptive complementary filters. *IEEE Trans. Control Syst. Technol.* **2016**, *24*, 1214–1226. [CrossRef]

38. Tailanian, M.; Paternain, S.; Rosa, R.; Canetti, R. Design and implementation of sensor data fusion for an autonomous quadrotor. In Proceedings of the 2014 IEEE International Instrumentation and Measurement Technology Conference (I2MTC), Montevideo, Uruguay, 12–15 May 2014; pp. 1431–1436.
39. CPOH. V-Skye Prototype: Test Flight. 2019. Available online: https://youtu.be/zS9oWur-Pss (accessed on 1 January 2019).

electronics

MDPI

Article

High-Order Sliding Mode-Based Fixed-Time Active Disturbance Rejection Control for Quadrotor Attitude System

Chunlin Song, Changzhu Wei *, Feng Yang and Naigang Cui

School of Astronautics, Harbin Institute of Technology, Harbin 150001, China; 14b918036@hit.edu.cn (C.S.); hitsyangfeng@gmail.com (F.Y.); cui_naigang@163.com (N.C.)

* Correspondence: weichangzhu@hit.edu.cn; Tel.: +86-0451-86415662

Received: 4 November 2018; Accepted: 21 November 2018; Published: 26 November 2018

Abstract: This article presents a fixed-time active disturbance rejection control approach for the attitude control problem of quadrotor unmanned aerial vehicle in the presence of dynamic wind, mass eccentricity and an actuator fault. The control scheme applies the feedback linearization technique and enhances the performance of the traditional active disturbance rejection control (ADRC) based on the fixed-time high-order sliding mode method. A switching-type uniformly convergent differentiator is used to improve the extended state observer for estimating and attenuating the lumped disturbance more accurately. A multivariable high-order sliding mode feedback law is derived to achieve fixed time convergence. The timely convergence of the designed extended state observer and the feedback law is proved theoretically. Mathematical simulations with detailed actuator models and real time experiments are performed to demonstrate the robustness and practicability of the proposed control scheme.

Keywords: quadrotor; ADRC; fixed-time extended state observer (FTESO); high-order sliding mode; wind disturbance; actuator fault; mass eccentricity

1. Introduction

Unmanned aerial vehicles (UAVs) are useful for tasks that are dangerous or unaccessible for human operation and popular in military and civilian applications such as investigation, inspection and surveillance with the advantage of moving in three-dimensional space flexibly. A quadrotor is a kind of vertical takeoff and landing (VTOL) aircraft lifted and controlled by four rotors [1]. Being simpler in structure, less sensitive to damage, easier to handle and more cost-effective, quadrotors have gained more attention in small unmanned aerial system (UAS) research than traditional helicopters. Therefore, plenty of remarkable research achievements related to quadrotors have been made in recent years [2,3].

In practical terms, the position control method is designed according to specific mission requirements and could be implemented onboard or remotely, while the attitude control is performed via an onboard processor to stabilize the attitude reliably [4]. Traditional efficient proportional–derivative (PD) control method has been used in the attitude control of quadrotor [5,6] and achieves good practical performance. Feedback linearization is a commonly used technique to control non-linear coupled systems and are adaptable to the controller design of quadrotor [7]. In [8], the attitude states are transformed into a new state space with a nonlinear transformation to obtain a linear system on which linear error-feedback control can be applied. The idea to design a controller on a transformed linear system instead of the original non-linear system is valuable, however the results in [8] show that the linear error-feedback controller is not robust in terms of uncertainties and measurement noises.

The small low-cost quadrotor is a non-redundant aircraft which suffers from external and internal disturbances such as dynamic wind and actuator faults. In order to design the attitude controller to be robust against unknown disturbances, the observer-based online disturbance estimation and attenuation strategies have been widely studied in recent years. A disturbance observer based on Q-filter is applied in [9,10] for robust hovering control. In [11], a neural network is used to learn the uncertain terms in UAV dynamics. Aboudonia integrated a disturbance observer with feedback linearization-based control in [12] for robustness, and combined a sliding mode controller with a disturbance observer in [13]. The control schemes are verified with the Dryden turbulence model and Von-Karman wind model, respectively. Disturbance observer-based trajectory tracking methods were studied in [14,15]. The sliding mode-based observer and high-order sliding mode-based observer are also extensively studied to enhance the robustness of feedback linearization-based controller [16], sliding mode controller [17] and back-stepping controller [18]. Shi designed extended state observers in [19,20] for the attitude control problem of quadrotors.

Active disturbance rejection control (ADRC) is proposed by Han in 2009 [21]. ADRC inherits from the classical proportional–integral–derivative (PID) controller and executes in an error-driven manner. By augmenting the system uncertainty into the state vector and constructing an extended state observer (ESO), the uncertainty is observed in real-time. Aiming at replacing PID in industrial applications, the ADRC scheme is developed experimentally in the first place. Recently, Guo [22] has proved the convergence of ADRC for non-linear systems mathematically. As a practical control scheme, ADRC has been applied to many engineering aspects, such as motor control, power plant control, ship control, etc [23] (pp. 6–9). ADRC has been adopted for attitude control of a quadrotor directly in [23–26].

In order to avoid the chattering effect in the sliding mode method, Levant [27,28] designed a general-form arbitrary-order exact robust differentiator based on high-order sliding mode (HOSM) algorithm. The proposed differentiator was proved to converge in finite time with bounded uncertainties. In [29], Levant's robust differentiator was improved via an extra exponential function with an exponent greater than 1 in order to achieve fixed-time convergence regardless of the initial deviation. In [30], a kind of fixed-time observer is derived and the constraints of the corrective term is formulated. In fact, the observer in [29] is an instance of the formula of the fixed-time observer in [30]. Angulo designed a fixed-time observer operated in a switching manner in [31]. The observer is switched from a uniform differentiator to Levant's finite-time differentiator. Compared with the methods used in [29,30], Angulo's switching-type observer performs only one exponential function at each step. Therefore, it is more practical on micro-processors with limited computing power. The switching-type observer has been applied to hypersonic vehicle flight control system [32] and the Brunovsky system [33]. An adaptive super-twisting-based controller is designed for a hypersonic vehicle in [34,35], in which a fixed-time observer is applied. In [36], a fixed-time observer and an integral terminal sliding mode method are studied for fault-tolerant control of a hypersonic vehicle.

In this paper, motivated by the ADRC structure and fixed-time observer, a fixed-time active disturbance rejection control (FTADRC) scheme based on the high-order sliding mode is proposed for the quadrotor attitude system in the presence of unknown disturbances including model uncertainty, dynamic wind, actuator fault and mass eccentricity. The proposed control scheme is analyzed theoretically and verified experimentally. The main contributions of this paper are as follows.

1. The ESO in ADRC is improved via robust uniform high-order sliding mode differentiator to achieve fixed time convergence given bounded differential of lumped disturbance.
2. A non-linear feedback control law combining a high-order sliding mode with feedback linearization is applied in the improved ADRC scheme. In this way, the attitude controller provides fixed-time stability.

The remainder of this paper is organized as follows. The mathematical models are presented in Section 2. The classical ADRC structure is described in Section 3. The proposed high-order sliding mode-based FTADRC is detailed in Section 4. Simulations with a Dryden wind model and experiments with simulated disturbance are carried out in Section 5. Finally, the discussions of the experiments and the conclusions are presented in Sections 6 and 7.

2. Mathematical Models

2.1. Rigid Body Dynamics

In this section, the quadrotor body is considered as a symmetrical rigid body attached with four sets of actuators. And the center of mass coincides exactly with the center of the body. Figure 1 shows a simplified model of an "X" type quadrotor and the coordinate frames used in this paper. The body coordinate frame $O_b x_b y_b z_b$ is fixed with the quadrotor body, and the Earth coordinate frame $O_e NED$ is fixed with the ground. As the flight time is short, the rotation of the earth is neglected such that the Earth coordinate frame is static in inertial space.

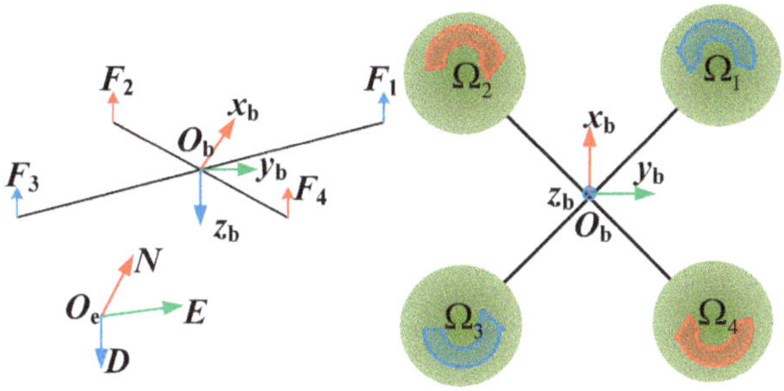

Figure 1. Definition of coordinate frames.

We use modified Rodrigues parameters (MRPs) to represent rigid rotation to avoid the singularity problem of the Euler angles and additional normalization requirement of the unit quaternions. MRPs are expressed in the form of three-dimensional vectors $\boldsymbol{\rho} = \begin{bmatrix} \rho_1 & \rho_2 & \rho_3 \end{bmatrix}^{\mathrm{T}}$. $\boldsymbol{\rho}$ can be expressed in the form of unit quaternion $\boldsymbol{q} = \begin{bmatrix} q_0 & \boldsymbol{q}_v^{\mathrm{T}} \end{bmatrix}^{\mathrm{T}}$ and angle ϕ and axis $\boldsymbol{n}$ of rotation as following.

$$\boldsymbol{\rho} = \frac{\boldsymbol{q}_v}{1+q_0} = \tan\frac{\phi}{4}\boldsymbol{n} \tag{1}$$

The rotation matrix represented in form of MRP is expressed as follows:

$$\boldsymbol{R}(\boldsymbol{\rho}) = \begin{bmatrix} 1-\frac{8(\rho_2^2+\rho_3^2)}{(|\rho|^2+1)^2} & -\frac{4(|\rho|^2\rho_3-2\rho_1\rho_2-\rho_3)}{(|\rho|^2+1)^2} & \frac{4(|\rho|^2\rho_2+2\rho_1\rho_3-\rho_2)}{(|\rho|^2+1)^2} \\ \frac{4(|\rho|^2\rho_3+2\rho_1\rho_2-\rho_3)}{(|\rho|^2+1)^2} & 1-\frac{8(\rho_1^2+\rho_3^2)}{(|\rho|^2+1)^2} & -\frac{4(|\rho|^2\rho_1-2\rho_2\rho_3-\rho_1)}{(|\rho|^2+1)^2} \\ -\frac{4(|\rho|^2\rho_2-2\rho_1\rho_3-\rho_2)}{(|\rho|^2+1)^2} & \frac{4(|\rho|^2\rho_1+2\rho_2\rho_3-\rho_1)}{(|\rho|^2+1)^2} & 1-\frac{8(\rho_1^2+\rho_2^2)}{(|\rho|^2+1)^2} \end{bmatrix} \tag{2}$$

The mathematical model of the quadrotor attitude described by MRPs is defined as follows:

$$\begin{cases} \dot{\rho}_1 = \frac{1}{4}\omega_x(\rho_1^2-\rho_2^2-\rho_3^2+1) + \frac{1}{2}\omega_y(\rho_1\rho_2-\rho_3) + \frac{1}{2}\omega_z(\rho_1\rho_3+\rho_2) \\ \dot{\rho}_2 = \frac{1}{4}\omega_y(\rho_2^2-\rho_1^2-\rho_3^2+1) + \frac{1}{2}\omega_x(\rho_1\rho_2+\rho_3) + \frac{1}{2}\omega_z(\rho_2\rho_3-\rho_1) \\ \dot{\rho}_3 = \frac{1}{4}\omega_z(\rho_3^2-\rho_1^2-\rho_2^2+1) + \frac{1}{2}\omega_x(\rho_1\rho_3-\rho_2) + \frac{1}{2}\omega_y(\rho_2\rho_3+\rho_1) \\ \dot{\omega}_x = \frac{I_y-I_z}{I_x}\omega_y\omega_z + \frac{\tau_x}{I_x} + \frac{J_r\Omega_r\omega_y}{I_x} \\ \dot{\omega}_y = \frac{I_z-I_x}{I_y}\omega_x\omega_z + \frac{\tau_y}{I_y} - \frac{J_r\Omega_r\omega_x}{I_y} \\ \dot{\omega}_z = \frac{I_x-I_y}{I_z}\omega_x\omega_y + \frac{\tau_z}{I_z} + \frac{J_r\dot{\Omega}_r}{I_z} \end{cases} \tag{3}$$

where $\omega = \begin{bmatrix} \omega_x & \omega_y & \omega_z \end{bmatrix}^T$ is the angular rate along body coordinate frame. I_x, I_y and I_z are the diagonal elements of inertial matrix. The products of inertia are assumed to be zero. $\boldsymbol{\tau} = \begin{bmatrix} \tau_x & \tau_y & \tau_z \end{bmatrix}^T$ is external torque. J_r is the rotational inertia of each propeller. Ω_r is linear combination of the rotor speeds $\Omega_i (i = 1, 2, 3, 4)$ and calculated as follows:

$$\Omega_r = \Omega_1 - \Omega_2 + \Omega_3 - \Omega_4 \tag{4}$$

Each MRP ρ has a shadow MRP ρ^S that represents the same orientation as ρ. Using the combination of MRPs and their shadow set, any rotation could be described without singularity. It is easy to be understand that rotating angle ϕ around the axis $\boldsymbol{n}$ is exactly the same as rotating angle $360° - \phi$ around the axis $-\boldsymbol{n}$, as shown in Figure 2. According to [37], the shadow point of MRP ρ is formulated as follows.

$$\rho^S = -\frac{\rho}{\|\rho\|^2} \tag{5}$$

It is worth noting that ρ is also the shadow point of ρ^S.

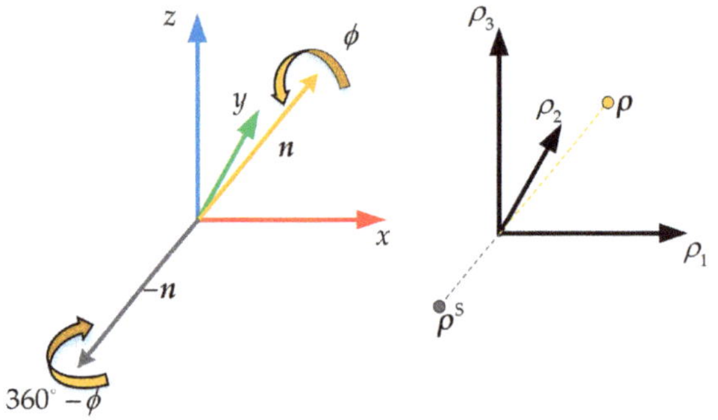

Figure 2. Graphic representation of original modified Rodrigues parameter (MRP) and shadow MRP.

2.2. Actuator Mode

Each of the four sets of actuators consists of an electron speed regulator, a brushless direct-current (DC) motor (BLDM) and a propeller. The electron speed regulator is responsive to input signals. Thus, it is modelled as a simple proportional component and neglected in following sections.

2.2.1. Motor Model

The circuit of the BLDM can be simplified into a series circuit composed of an equivalent resistance R_M, an equivalent inductance L_M and the induction power supply e_M produced by rotor rotation. The input voltage u satisfies the following voltage equation of the motor circuit:

$$u = R_M i + L_M \frac{di}{dt} + \frac{\Omega}{Kv} \tag{6}$$

where i is the current, Ω is the rotational speed of the rotor, K_v is the motor velocity constant. The motor torque is proportional to the current with coefficient K_T.

$$Q_M = K_T i \tag{7}$$

Because the propeller is mechanically attached with a motor rotor, we ignore the rotational inertia of the motor rotor and treat it as part of the rotational inertia of the propeller J_r. The rotation dynamics equation of the propeller is formulated as follows:

$$J_r\dot{\Omega} = Q_M - Q \tag{8}$$

where Q represents the aerodynamic drag torque, which will be discussed later in the propeller aerodynamic model. The equivalent inductance L_M of small BLDM is negligible. Therefore, the motor is modelled by the following first-order differential equation with input u and output Ω.

$$J_r\dot{\Omega} = K_T\left[\frac{1}{R_M}(u - \frac{\Omega}{K_v}) - i_0\right] - Q \tag{9}$$

2.2.2. Propeller Aerodynamic Model

We build the propeller aerodynamic model for quadrotors mainly by consulting the aerodynamics for a helicopter rotor in [38]. The model combines momentum theory and blade element analysis, and achieves the following equation of the inflow ratio λ:

$$4F\sqrt{\lambda^2 + \mu^2}(\lambda - \lambda_c)r - \frac{1}{4}\sigma C_{l\alpha}(\theta\mu^2 + 2\theta r^2 - 2\lambda r) = 0 \tag{10}$$

where μ is the rotor advance ratio; F is a function of λ named by Prandtl correction factor; λ_c is the climb inflow ratio; σ is the rotor solidity; $C_{l\alpha}$ is the lift-curve slope; θ is the blade pitch angle; r is the non-dimensional radial distance. The readers can refer to [38] for the detailed description and derivation of the above equation.

λ cannot be directly solved from the above equation as F is a non-linear function of λ. Instead, the equation could be solved in a nested iterative way as following.

1. Set $F = 1$.
2. Solve the equation through the Newton method with initial value in the case of $\mu = 0$:

$$\lambda = \lambda_0(r, \lambda_c) = \sqrt{(\frac{\sigma C_{l\alpha}}{16F} - \frac{\lambda_c}{2})^2 + \frac{\sigma C_{l\alpha}}{8F}\theta r} - (\frac{\sigma C_{l\alpha}}{16F} - \frac{\lambda_c}{2}) \tag{11}$$

3. Calculate $F(\lambda)$, go to step 2 and start the next iteration.

Generally, convergence is obtained after about 3 times of iteration (outer cycle).

Divide the blade into N elements with length of Δr. The rotor thrust coefficient C_T and torque coefficient C_Q are approximated with the sum of corresponding coefficients of each annulus with the width of Δr as shown in Figure 3.

$$\begin{array}{c} C_T = \sum_{n=1}^{N} \Delta C_{Tn} \\ C_Q = \sum_{n=1}^{N} \Delta C_{Qn} \\ \Delta C_{Tn} = 4F(r_n)\sqrt{\lambda(r_n)^2 + \mu^2}(\lambda(r_n) - \lambda_c)r_n\Delta r \\ \Delta C_{Qn} = \lambda(r_n)\Delta C_{Tn} + \frac{1}{4}\sigma(r_n)C_d(2r_n^2 + \mu^2)\Delta r \end{array} \tag{12}$$

where C_d is the drag coefficient, $n = 1,,N$. The trust and torque generated by the rotor could be calculated with the coefficients.

$$\begin{cases} T = C_T\rho\pi R^2(\Omega R)^2 \\ Q = C_Q\rho\pi R^3(\Omega R)^2 \end{cases} \tag{13}$$

where ρ is the air density; R is the radius of the rotor disk.

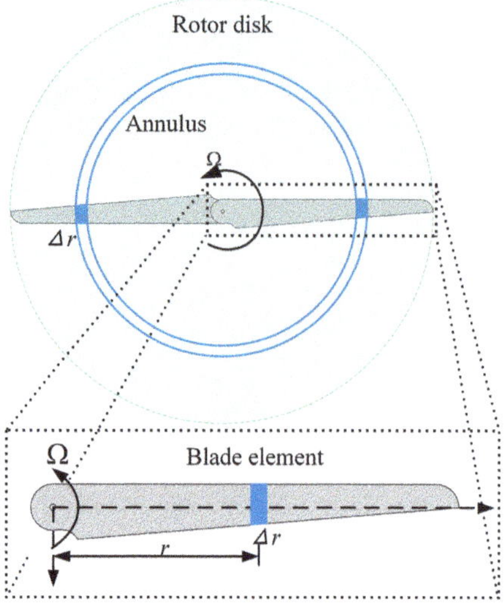

Figure 3. Blade element.

2.3. Example of Measuring and Calculating Propeller Aerodynamic Model Parameters

Multiple parameters need to be determined before applying the above propeller aerodynamic model in simulations. We summarize the parameters in Table 1, and present the method for determination with an example.

As Table 1 shows, there are four kinds of parameters required in propeller aerodynamic model. The atmospheric parameter is directly determined with the Earth's atmospheric parameters. Propeller parameters are manually measured. Velocity parameters are calculated online according to wind speed and quadrotor motion in simulations. Aerodynamic coefficients are usually determined with the aid of an accurate structural model, which is difficult to construct. We use an experimental method to estimate the coefficients through measuring rotor trust, torque and rotation speed.

Table 1. Parameters used in the aerodynamic model of propeller.

Parameters		Description	Determination method
Atmospheric parameter	ρ	Air density	1.2kg/m^3 in low altitude
Propeller parameters	R	Propeller radius	Measuring directly
	$c(r_n)$	Blade chord	Measuring directly
	$\theta(r_n)$	Blade pitch angle	Measuring directly
	N_b	Number of rotor blades	Measuring directly
Velocity parameters	λ_c	Climb inflow ratio	Calculated according to velocity of quadrotor and wind speed
	μ	Advance ratio	Calculated according to velocity of quadrotor and wind speed
Aerodynamic coefficients	$C_{l\alpha}$	Lift-curve slope	Estimated with trust-rotation speed curve
	C_d	Drag coefficient	Estimated with torque-rotation speed curve

We use 16″ carbon propellers as example for measuring and estimating. The propeller radius is 203.2 mm. The number of blades per propeller is 2. Blade chord and blade pitch angle vary with the radial position. We select 12 measurement points on the blade and measure the chord and pitch angle manually on the measurement points. The measured results are shown in Table 2.

Table 2. Chord length and pitch angle of the blade.

Radial Position (mm)	Chord (mm)	Pitch Angle (°)
32	29	35.2
48	40	28
64	48	21.4
80	52	17.6
96	55	14.2
112	56	12.7
128	55	11.3
144	51	10.6
160	47	6.6
176	38	6.0
192	28	6.0
203.2	0	6.0

We use a tachometer (Hobbywing, Shenzhen, China), a tensionmeter (Jnsensor, Bengbu, China) and a torquemeter (Jnsensor, Bengbu, China) to measure the thrust-speed and torque-speed curves of 16″ carbon propellers in static air. The three sensors are shown in Figure 4. In the measurement process, the tensionmeter and the torquemeter are fixed on a solid stable structure, while the motor with the propeller is mounted on the tensionmeter or the torquemeter with a flange. The tachometer is connected with arbitrary two phases of the three-phase motor and measures the rotation speed by counting the phase changes. The tensionmeter and the torquemeter output analog signals. The signals are converted into digital measurements with an analog-to-digital (A/D) converter and sent to the computer through a serial port.

Figure 4. Tensionmeter, torquemeter and tachometer.

According to Equation (13), the rotor thrust and torque increase with the square of the rotation speed with ratio of $C_T \rho \pi R^4$ and $C_Q \rho \pi R^5$ respectively. We take measurements in static air and fit the thrust-speed and torque-speed curves with 2-order polynomial function. The thrust coefficient and torque coefficient are recovered from the fitting parameters. We merge the measurements from four sets of motors and propellers to get the average coefficients. The resulting curves are shown in Figure 5. It is shown that the resulting 2-order polynomial functions fit the measurements well.

According to the fitting parameters, C_T and C_Q are 1.36×10^{-2} and 2.028×10^{-3} in the case of $\lambda_c = 0$ and $\mu = 0$. The whole calculation process of C_T described before is treated as a non-linear function of $C_{l\alpha}$, i.e.,

$$C_T = f_{CT}(C_{l\alpha}) \tag{14}$$

It can be easily solved by the classical dichotomy method. With the above results, we obtain $C_{l\alpha} = 23.91$. The solution of C_d is 0.0384 based on Equation (12) of C_Q.

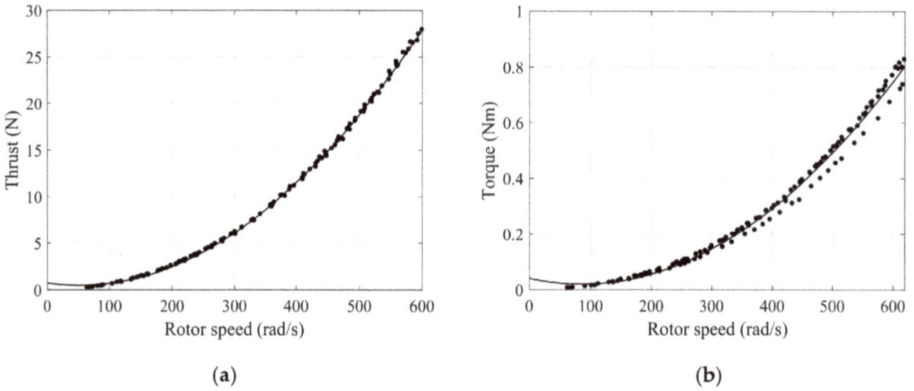

Figure 5. Measurements and fitting curves. (**a**) Trust curve; (**b**) torque curve.

3. Active Disturbance Rejection Control (ADRC) Method

The ADRC method applied to a second-order system is shown in Figure 6. The method consists of three parts: tracking differentiator (TD) is used to generate the desired states given the desired outputs; extended state observer (ESO) is used to estimate the system uncertainty; non-linear state error feedback (NLSEF) is used to feedback state errors effectively.

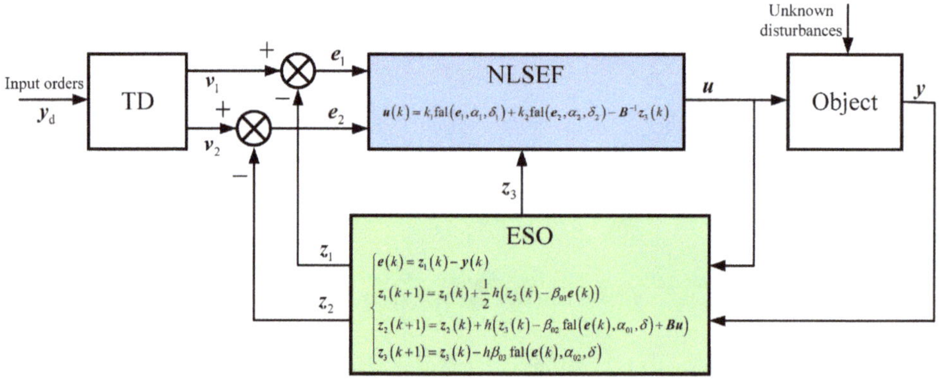

Figure 6. Structure of active disturbance rejection controller for second-order system.

The traditional ESO is a model-independent disturbance observer and can be employed universally in the non-linear system control problem. However, the convergence rate of the traditional ESO is not sufficient for a quadrotor attitude control problem. Therefore, this paper focuses on improving the ESO with a high-order slide mode algorithm for quadrotor attitude control problem. Additionally, the non-linear state error feedback control law is improved with a multivariable high-order slide mode algorithm.

4. High-Order Sliding Mode-Based Fixed-Time Active Disturbance Rejection Control (FTADRC)

The FTADRC scheme is designed within the traditional ADRC structure based on the feedback-linearization technique. A schematic block diagram of the proposed FTADRC scheme for quadrotor attitude control is shown in Figure 7. The attitude control scheme consists of five parts:

- Feedback linearization for regularizing the attitude dynamic model;
- Fixed-time extended state observer (FTESO) for observing the unknown disturbances accurately;

- MRP-TD for tracking the differential of input attitude described by MRP;
- Non-linear feedback control law for driving the orientation of quadrotor to track the desired attitude timely;
- Control allocation for generating pulse-width modulation (PWM) signals for motors.

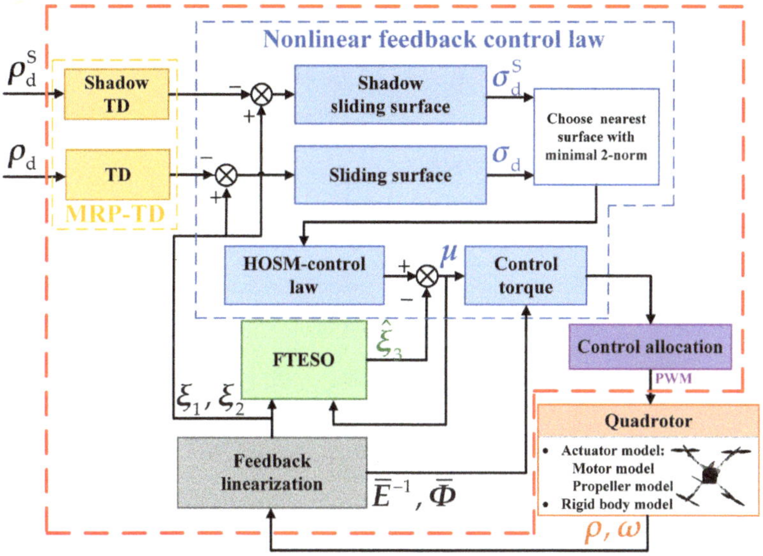

Figure 7. Block diagram of the proposed fixed-time active disturbance rejection control (FTADRC) attitude control scheme.

4.1. Feedback Linearization

We remove the gyroscopic effect and reactionary torque related to the rotor speed in the attitude dynamic equations and treat them as part of the unknown disturbance. A six-dimensional system is defined with state vector $\boldsymbol{x} = [\rho_1, \rho_2, \rho_3, \omega_x, \omega_y, \omega_z]^T$, input vector $\boldsymbol{u} = [\tau_x, \tau_y, \tau_z]^T$ and measurable output vector $\boldsymbol{y} = [y_1, y_2, y_3]^T$. A non-linear system is achieved according to the attitude model in Equation (3).

$$\underbrace{\begin{bmatrix} \dot{\rho}_1 \\ \dot{\rho}_2 \\ \dot{\rho}_3 \\ \dot{\omega}_x \\ \dot{\omega}_y \\ \dot{\omega}_z \end{bmatrix}}_{\dot{x}} = \underbrace{\begin{bmatrix} \frac{1}{4}\omega_x(\rho_1^2-\rho_2^2-\rho_3^2+1)+\frac{1}{2}\omega_y(\rho_1\rho_2-\rho_3)+\frac{1}{2}\omega_z(\rho_1\rho_3+\rho_2) \\ \frac{1}{4}\omega_y(\rho_2^2-\rho_1^2-\rho_3^2+1)+\frac{1}{2}\omega_x(\rho_1\rho_2+\rho_3)+\frac{1}{2}\omega_z(\rho_2\rho_3-\rho_1) \\ \frac{1}{4}\omega_z(\rho_3^2-\rho_1^2-\rho_2^2+1)+\frac{1}{2}\omega_x(\rho_1\rho_3-\rho_2)+\frac{1}{2}\omega_y(\rho_2\rho_3+\rho_1) \\ (I_y-I_z)\omega_y\omega_z/I_x \\ (I_z-I_x)\omega_x\omega_z/I_y \\ (I_x-I_y)\omega_x\omega_y/I_z \end{bmatrix}}_{f(x)} + \underbrace{\begin{bmatrix} 0&0&0 \\ 0&0&0 \\ 0&0&0 \\ \frac{1}{I_x}&0&0 \\ 0&\frac{1}{I_y}&0 \\ 0&0&\frac{1}{I_z} \end{bmatrix}}_{g} \underbrace{\begin{bmatrix} \tau_x \\ \tau_y \\ \tau_z \end{bmatrix}}_{u} \tag{15}$$

$$\underbrace{\begin{bmatrix} y_1 \\ y_2 \\ y_3 \end{bmatrix}}_{y} = \underbrace{\begin{bmatrix} \rho_1 \\ \rho_2 \\ \rho_3 \end{bmatrix}}_{h(x)}$$

The Lie derivative and k-th Lie derivative of function $h_i(\boldsymbol{x})$ with respect to a vector-valued function $f(\boldsymbol{x})$ are defined as follows:

$$\begin{aligned} L_f h_i(\boldsymbol{x}) &= \sum_{j=1}^{6} \frac{\partial h_i(\boldsymbol{x})}{\partial x_j} f_j(\boldsymbol{x}) \\ L_f^k h_i(\boldsymbol{x}) &= L_f(L_f^{k-1} h_i(\boldsymbol{x})) \end{aligned} \tag{16}$$

A 3×3 matrix $\boldsymbol{E}(\boldsymbol{x})$ is constructed with the Lie derivatives $L_{g_j} L_f h_i(\boldsymbol{x}), (i \in [1,3], j \in [1,3])$:

$$\boldsymbol{E}(\boldsymbol{x}) = \begin{bmatrix} L_{g_1} L_f h_1(\boldsymbol{x}) & L_{g_2} L_f h_1(\boldsymbol{x}) & L_{g_3} L_f h_1(\boldsymbol{x}) \\ L_{g_1} L_f h_2(\boldsymbol{x}) & L_{g_2} L_f h_2(\boldsymbol{x}) & L_{g_3} L_f h_2(\boldsymbol{x}) \\ L_{g_1} L_f h_3(\boldsymbol{x}) & L_{g_2} L_f h_3(\boldsymbol{x}) & L_{g_3} L_f h_3(\boldsymbol{x}) \end{bmatrix} \tag{17}$$

Substituting functions of (15) into $\boldsymbol{E}(\boldsymbol{x})$ yields:

$$\boldsymbol{E}(\boldsymbol{x}) = \begin{bmatrix} \frac{\rho_1^2-\rho_2^2-\rho_3^2+1}{4I_x} & \frac{\rho_1\rho_2-\rho_3}{2I_y} & \frac{\rho_1\rho_3+\rho_2}{2I_z} \\ \frac{\rho_1\rho_2+\rho_3}{2I_x} & \frac{\rho_2^2-\rho_1^2-\rho_3^2+1}{4I_y} & \frac{\rho_2\rho_3-\rho_1}{2I_z} \\ \frac{\rho_1\rho_3-\rho_2}{2I_x} & \frac{\rho_2\rho_3+\rho_1}{2I_y} & \frac{\rho_3^2-\rho_1^2-\rho_2^2+1}{4I_z} \end{bmatrix} \tag{18}$$

with a determinant of:

$$|\boldsymbol{E}(\boldsymbol{x})| = \frac{(\rho_1^2+\rho_2^2+\rho_3^2+1)^3}{64 I_x I_y I_z} \tag{19}$$

Obviously $|\boldsymbol{E}(\boldsymbol{x})| > 0$ holds for all $\boldsymbol{\rho} \in \mathbb{R}^3$. Therefore, $\boldsymbol{E}(\boldsymbol{x})$ is non-singular. The relative degree vector of system (15) is $\boldsymbol{r} = [2,2,2]^{\mathrm{T}}$. System (15) can be transformed into a regular form.

$$\begin{bmatrix} \ddot{y}_1 \\ \ddot{y}_2 \\ \ddot{y}_3 \end{bmatrix} = \underbrace{\begin{bmatrix} L_f^2 h_1(\boldsymbol{x}) \\ L_f^2 h_2(\boldsymbol{x}) \\ L_f^2 h_3(\boldsymbol{x}) \end{bmatrix}}_{\boldsymbol{\phi}(\boldsymbol{x})} + \boldsymbol{E}(\boldsymbol{x})\boldsymbol{u} \tag{20}$$

The formula of a three-dimensional vector $\boldsymbol{\phi}(\boldsymbol{x})$ is in Appendix A. Defining a new six-dimensional state vector $\boldsymbol{\xi} = [\boldsymbol{\xi}_1{}^{\mathrm{T}}, \boldsymbol{\xi}_2{}^{\mathrm{T}}]^{\mathrm{T}} = [\rho_1, \rho_2, \rho_3, \dot{\rho}_1, \dot{\rho}_2, \dot{\rho}_3]^{\mathrm{T}}$, a new state equation is derived from (20).

$$\begin{cases} \dot{\boldsymbol{\xi}}_1 = \boldsymbol{\xi}_2 \\ \dot{\boldsymbol{\xi}}_2 = \boldsymbol{\phi} + \boldsymbol{E}\boldsymbol{u} \end{cases} \tag{21}$$

where $\boldsymbol{\phi}$ and $\boldsymbol{E}$ are determined by the system model parameters. We divide them into the nominal parts $\overline{\boldsymbol{\phi}}, \overline{\boldsymbol{E}}$ and the uncertain parts $\Delta\boldsymbol{\phi}$, $\Delta\boldsymbol{E}$. The input torque $\boldsymbol{u}$ is divided into the control torque $\boldsymbol{u}_c$ and the unknown disturbance torque $\Delta\boldsymbol{u}$. $\boldsymbol{\phi}$, $\boldsymbol{E}$ and $\boldsymbol{u}$ are rewritten in the following format.

$$\begin{cases} \boldsymbol{\phi} = \overline{\boldsymbol{\phi}} + \Delta\boldsymbol{\phi} \\ \boldsymbol{E} = \overline{\boldsymbol{E}} + \Delta\boldsymbol{E} \\ \boldsymbol{u} = \boldsymbol{u}_c + \Delta\boldsymbol{u} \end{cases} \tag{22}$$

Substituting (22) into (21) yields:

$$\begin{cases} \dot{\boldsymbol{\xi}}_1 = \boldsymbol{\xi}_2 \\ \dot{\boldsymbol{\xi}}_2 = (\overline{\boldsymbol{\phi}} + \overline{\boldsymbol{E}}\boldsymbol{u}_c) + (\Delta\boldsymbol{\phi} + \Delta\boldsymbol{E}\boldsymbol{u}_c + \boldsymbol{E}\Delta\boldsymbol{u}) \end{cases} \tag{23}$$

Define an auxiliary input vector $\boldsymbol{\mu}$ and a lumped disturbance vector $\Delta\boldsymbol{\mu}$ as follows:

$$\begin{cases} \boldsymbol{\mu} = \overline{\boldsymbol{\phi}} + \overline{\boldsymbol{E}}\boldsymbol{u}_c \\ \Delta\boldsymbol{\mu} = \Delta\boldsymbol{\phi} + \Delta\boldsymbol{E}\boldsymbol{u}_c + \boldsymbol{E}\Delta\boldsymbol{u} \end{cases} \tag{24}$$

Substituting (24) into (23) yields:

$$\begin{cases} \dot{\xi}_1 = \xi_2 \\ \dot{\xi}_2 = \mu + \Delta\mu \end{cases} \tag{25}$$

4.2. Fixed-Time Third-Order Sliding Mode Observer-Based Extended State Observer (ESO)

According to (24), the equivalent disturbance torque $\Delta\tau$ is related to $\Delta\mu$ in the format:

$$\Delta\tau = \overline{E}^{-1}\Delta\mu \tag{26}$$

Since $\overline{E}$ is a non-singular matrix, the estimation of $\Delta\tau$ is equivalent to the estimation of $\Delta\mu$. Augment $\Delta\mu$ as an extended state $\xi_3 = \Delta\mu$, the reconstructed linear system is rewritten as:

$$\begin{cases} \dot{\xi}_1 = \xi_2 \\ \dot{\xi}_2 = \xi_3 + \mu \\ \dot{\xi}_3 = \zeta \end{cases} \tag{27}$$

where ζ is the derivative of ξ_3. Assume that each element of ζ is bounded.

$$\zeta_i \in [-L, L], i = 1,2,3 \tag{28}$$

The elements in a 3-dimensional vector ξ_1, ξ_2 and ξ_3 are independent of each other. Motivated by [31], a fixed-time convergent extended state observer (FTESO) for each element i (I = 1,2,3) is designed as:

$$\begin{aligned} \dot{\hat{\xi}}_{1,i} &= -\theta_i\alpha_{1,i}\left\lceil\frac{\hat{\xi}_{1,i}-\xi_{1,i}}{\delta}\right\rfloor^{\frac{2}{3}} - (1-\theta_i)\beta_{1,i}\left\lceil\frac{\hat{\xi}_{1,i}-\xi_{1,i}}{\delta}\right\rfloor^{1+\varepsilon} + \hat{\xi}_{2,i} \\ \dot{\hat{\xi}}_{2,i} &= -\theta_i\frac{\alpha_{2,i}}{\delta}\left\lceil\frac{\hat{\xi}_{1,i}-\xi_{1,i}}{\delta}\right\rfloor^{\frac{1}{3}} - (1-\theta_i)\frac{\beta_{2,i}}{\delta}\left\lceil\frac{\hat{\xi}_{1,i}-\xi_{1,i}}{\delta}\right\rfloor^{1+2\varepsilon} + \hat{\xi}_{3,i} + \mu_i \\ \dot{\hat{\xi}}_{3,i} &= -\theta_i\frac{\alpha_{3,i}}{\delta^2}\left\lceil\frac{\hat{\xi}_{1,i}-\xi_{1,i}}{\delta}\right\rfloor^{0} - (1-\theta_i)\frac{\beta_{3,i}}{\delta^2}\left\lceil\frac{\hat{\xi}_{1,i}-\xi_{1,i}}{\delta}\right\rfloor^{1+3\varepsilon} \\ \theta_i &= \begin{cases} 0, & \hat{\xi}_{1,i}-\xi_{1,i} > \delta \\ 1, & \hat{\xi}_{1,i}-\xi_{1,i} \le \delta \end{cases} \end{aligned} \tag{29}$$

where $\hat{\xi}_{1,i}$, $\hat{\xi}_{2,i}$ and $\hat{\xi}_{3,i}$ are the estimation of $\xi_{1,i}$. $\xi_{2,i}$ and $\xi_{3,i}$ respectively. $\delta > 0$ is used to scale the estimation error. Function $\lceil x\rfloor^k = |x|^k\text{sign}(x)$, $\text{sign}(\cdot)$ represents the signum function. $\varepsilon > 0$ is chosen small enough. θ_i is used to switch between two different exponential functions. $\alpha_{j,i}$ $(j = 1,2,3)$ are selected based on the boundary value L using the formulas for the HOSM differentiator in [28]. $\beta_{j,i}$ $(j = 1,2,3)$ are selected such that the following matrix is Hurwitz as suggested by [31].

$$A = \begin{bmatrix} -\beta_{1,i} & 1 & 0 \\ -\beta_{2,i} & 0 & 1 \\ -\beta_{3,i} & 0 & 0 \end{bmatrix} \tag{30}$$

There are two major improvements between the above observer with the differentiators used in [31,32].

1. The switching is conducted according to the estimation error instead of an arbitrary fixed time.
2. The estimation error $\hat{\xi}_{1,i} - \xi_{1,i}$ is scaled with $1/\delta$.

The switching structure of the observer takes advantage of dealing with different estimation errors with different exponential functions. The function with exponential larger than 1 is more efficient with larger errors, while the function with exponential smaller than 1 is more efficient with smaller errors.

So, we use a switch mechanism controlled by estimation errors to make full use of different exponential functions. The scaling factor is used to adjust flexibly the estimation errors for better performance.

Theorem 1. *Suppose that $\boldsymbol{\xi}_1$ and $\boldsymbol{\mu}$ are available in real time. FTESO (29) converges to the true extended states $\xi_{1,i}$, $\xi_{2,i}$ and $\xi_{3,i}$ in a fixed time with the parameters chosen according to preceding rules.*

Proof of Theorem 1. Define the following notations for observation errors.

$$\begin{cases} \Delta\xi_{1,i} = \hat{\xi}_{1,i} - \xi_{1,i} \\ \Delta\xi_{2,i} = \hat{\xi}_{2,i} - \xi_{2,i} \\ \Delta\xi_{3,i} = \hat{\xi}_{3,i} - \xi_{3,i} \end{cases} \tag{31}$$

Differentiating $\Delta\xi_{1,i}$, $\Delta\xi_{2,i}$ and $\Delta\xi_{3,i}$, one can obtain:

$$\begin{cases} \Delta\dot{\xi}_{1,i} = -\theta_i\alpha_{1,i}\left\lceil\frac{\hat{\xi}_{1,i}-\xi_{1,i}}{\delta}\right\rfloor^{\frac{2}{3}} - (1-\theta_i)\beta_{1,i}\left\lceil\frac{\hat{\xi}_{1,i}-\xi_{1,i}}{\delta}\right\rfloor^{1+\varepsilon} + \Delta\hat{\xi}_{2,i} \\ \Delta\dot{\xi}_{2,i} = -\theta_i\frac{\alpha_{2,i}}{\delta}\left\lceil\frac{\hat{\xi}_{1,i}-\xi_{1,i}}{\delta}\right\rfloor^{\frac{1}{3}} - (1-\theta_i)\frac{\beta_{2,i}}{\delta}\left\lceil\frac{\hat{\xi}_{1,i}-\xi_{1,i}}{\delta}\right\rfloor^{1+2\varepsilon} + \Delta\hat{\xi}_{3,i} \\ \Delta\dot{\xi}_{3,i} = -\theta_i\frac{\alpha_{3,i}}{\delta^2}\left\lceil\frac{\hat{\xi}_{1,i}-\xi_{1,i}}{\delta}\right\rfloor^{0} - (1-\theta_i)\frac{\beta_{3,i}}{\delta^2}\left\lceil\frac{\hat{\xi}_{1,i}-\xi_{1,i}}{\delta}\right\rfloor^{1+3\varepsilon} - \zeta_i \end{cases} \tag{32}$$

Define:

$$\begin{cases} \eta_{1,i} = \Delta\xi_{1,i}(\delta t)/\delta \\ \eta_{2,i} = \Delta\xi_{2,i}(\delta t) \\ \eta_{3,i} = \delta\Delta\xi_{3,i}(\delta t) \end{cases} \tag{33}$$

Substituting (33) into (32) yields:

$$\begin{cases} \dot{\eta}_{1,i} = -\theta_i\alpha_{1,i}\lceil\eta_{1,i}\rfloor^{\frac{2}{3}} - (1-\theta_i)\beta_{1,i}\lceil\eta_{1,i}\rfloor^{1+\varepsilon} + \Delta\hat{\xi}_{2,i} \\ \dot{\eta}_{2,i} = -\theta_i\alpha_{2,i}\lceil\eta_{1,i}\rfloor^{\frac{1}{3}} - (1-\theta_i)\beta_{2,i}\lceil\xi_{1,i}\rfloor^{1+2\varepsilon} + \Delta\hat{\xi}_{3,i} \\ \dot{\xi}_{3,i} = -\theta_i\alpha_{3,i}\lceil\xi_{1,i}\rfloor^{0} - (1-\theta_i)\beta_{3,i}\lceil\xi_{1,i}\rfloor^{1+3\varepsilon} - \delta^2\zeta_i \end{cases} \tag{34}$$

According to Theorem 1 in [31], (34) converges to a 3-sliding surface $S = \{\eta_{1,i} = 0, \eta_{2,i} = 0, \eta_{3,i} = 0\}$ in a fixed time T. Therefore, (29) converges to the true extended states $\xi_{1,i}$, $\xi_{2,i}$ and $\xi_{3,i}$ in a fixed time T/δ. This complete the proof. □

4.2.1. Tracking Differentiator

The tracking differentiator for the desired attitude $\boldsymbol{\rho}_\mathrm{d} = [\rho_{1,\mathrm{d}}, \rho_{2,\mathrm{d}}, \rho_{3,\mathrm{d}}]^\mathrm{T}$ based on the above feedback linearization model is obtained by directly applying the TD in the classical ADRC structure.

$$\begin{cases} \dot{\boldsymbol{\xi}}_{1,\mathrm{d}} = \boldsymbol{\xi}_{2,\mathrm{d}} \\ \dot{\boldsymbol{\xi}}_{2,\mathrm{d}} = \boldsymbol{fhan}_3(\boldsymbol{\xi}_{1,\mathrm{d}} - \boldsymbol{\rho}_\mathrm{d}, \boldsymbol{\xi}_{2,\mathrm{d}}, r, h_0) \end{cases} \tag{35}$$

where function $\boldsymbol{fhan}_3$ is the vector version of tracking function *fhan* in [21]. It performs function *fhan* for each element of the 3-D vectors. The above tracking differentiator outputs the desired linearized states $\boldsymbol{\xi}_{1,\mathrm{d}}$ and $\boldsymbol{\xi}_{2,\mathrm{d}}$. Considering the multi-solution characteristic of MRP in describing rotation, the shadow tracking differentiator for $\boldsymbol{\rho}_\mathrm{d}^\mathrm{S}$ should also be constructed as follows.

$$\begin{cases} \dot{\boldsymbol{\xi}}^\mathrm{S}_{1,\mathrm{d}} = \boldsymbol{\xi}^\mathrm{S}_{2,\mathrm{d}} \\ \dot{\boldsymbol{\xi}}^\mathrm{S}_{2,\mathrm{d}} = \boldsymbol{fhan}_3(\boldsymbol{\xi}^\mathrm{S}_{1,\mathrm{d}} - \boldsymbol{\rho}^\mathrm{S}_\mathrm{d}, \boldsymbol{\xi}^\mathrm{S}_{2,\mathrm{d}}, r, h_0) \end{cases} \tag{36}$$

4.2.2. Multivariable High-Order Sliding Mode (HOSM)-Based Fixed-Time Non-Linear Feedback Law

Define the sliding surface and shadow sliding surface:

$$\begin{aligned} \sigma &= C(\xi_1 - \xi_{1,\text{d}}) + (\xi_2 - \xi_{2,\text{d}}) \\ \sigma^{\text{S}} &= C(\xi_1 - \xi^{\text{S}}_{1,\text{d}}) + (\xi_2 - \xi^{\text{S}}_{2,\text{d}}) \end{aligned} \tag{37}$$

where C is a 3×3 diagonal matrix with positive elements of main diagonal. For tracking the desired attitude described by the MRP set $\{\rho_{\text{d}}, \rho^{\text{S}}_{\text{d}}\}$, there are two candidate sliding mode surfaces to approach. We force the quadrotor moving towards the "nearby candidate" in $\{\sigma, \sigma^{\text{S}}\}$ with minimal 2-norm. Without loss of generality, the fixed-time feedback law is designed and analyzed for σ in the following.

Differentiating σ and substituting (25) into the expression of $\dot{\sigma}$, one can obtain:

$$\dot{\sigma} = C(\xi_2 - \xi_{2,\text{d}}) + \mu + \Delta\mu \tag{38}$$

Inspired by [39], we use 2-norm of the 3-dimension vector σ to apply the high-order sliding mode algorithm on the above multivariable system. A fixed-time second-order sliding mode control law with disturbance rejection is designed as:

$$\mu = -k_1\sigma\|\sigma\|^{-p_1} - k_2\sigma\|\sigma\|^{p_2} - C(\xi_2 - \xi_{2,\text{d}}) - \hat{\xi}_3 \tag{39}$$

where $k_1 > 0$, $k_2 > 0$, $0 < p_1 < 1$, $0 < p_2 < 1$, and the notation $\|\sigma\|$ represents 2-norm of vector σ. According to Theorem 1, $\hat{\xi}_3$ becomes equal to $\Delta\mu$ in a fixed time. Substituting (39) into (38), the close-loop dynamic equation of σ can be obtained.

$$\dot{\sigma} = -k_1\sigma\|\sigma\|^{-p_1} - k_2\sigma\|\sigma\|^{p_2} \tag{40}$$

Theorem 2. *Consider the closed-loop system (40), the sliding variable σ and its derivative $\dot{\sigma}$ reach the origin in fixed time.*

Proof of Theorem 2. Given (40), the 3-dimension vectors $\dot{\sigma}$ and σ are parallel in opposite directions. That means the direction of σ does not change with time, and the following equation holds.

$$\|\dot{\sigma}\| = -\frac{d\|\sigma\|}{dt} \tag{41}$$

Substituting (40) into (41) yields:

$$\frac{d\|\sigma\|}{dt} = -(k_1\|\sigma\|^{-p_1} + k_2\|\sigma\|^{p_2})\|\sigma\| = -k_1\|\sigma\|^{1-p_1} - k_2\|\sigma\|^{1+p_2} \tag{42}$$

When the initial value of the system satisfies the inequation $\|\sigma(0)\| > 1$, according to the above equation, one can obtain:

$$\frac{d\|\sigma\|}{dt} < -k_2\|\sigma\|^{1+p_2} \tag{43}$$

Inequation (43) can be rewritten as:

$$\frac{d\|\sigma\|}{\|\sigma\|^{1+p_2}} < -k_2 dt \tag{44}$$

Denote the time consumed for $\|\sigma(t)\|$ reducing from $\|\sigma(0)\|$ to 1 to T_1, i.e., $\|\sigma(T_1)\| = 1$. Integrate Equation (44), and it can be found that T_1 is bounded by a fixed value that is independent of initial value $\sigma(0)$.

$$T_1 < \frac{1}{k_2 p_2} \tag{45}$$

After the system state reaching $\|\sigma\| = 1$ in time T_1, according to (42), one can obtain:

$$\frac{d\|\sigma\|}{dt} < -k_1 \|\sigma\|^{1-p_1} \tag{46}$$

Denote the time consumed for $\|\sigma(t)\|$ reducing from 1 to 0 to T_2, i.e., $\|\sigma(T_1 + T_2)\| = 0$. Integrate Equation (46), and it can be concluded that if $\|\sigma(T_1)\|$ is in range (0,1], T_2 is bounded by a fixed value.

$$T_2 < \frac{1}{k_1 p_1} \tag{47}$$

For a 3-dimensional vector σ, $\|\sigma\| = 0 \Leftrightarrow \sigma = 0$. According to Equation (40), $\dot{\sigma} = 0$ when $\sigma = 0$. This completes the proof. □

The control torque is calculated according to the first equation in (24).

$$u_c = \overline{\boldsymbol{E}}^{-1} \mu - \overline{\boldsymbol{\Phi}} \tag{48}$$

The feedback law (39) performs exponential functions on the 2-norm of σ instead of three Euler angles used in [15]. In this way, the exponential functions-related computation is reduced. This characteristic is helpful when using a low-cost advanced RISC (reduced instruction set computing) machine (ARM) microprocessor that is ineffective to nonlinear functions such as exponential, trigonometric and inverse trigonometric functions.

By means of input–output feedback linearization, the non-linear attitude system is transformed into a linear system and the parameters in the observer and controller are adjusted in a standard way [28]. More importantly, the matrix B for tuning the disturbance estimation and compensation in traditional ADRC [21] is a model-related parameter even though the whole method is independent of the type of system function. This means the parameter is hard to tune in attitude control of quadrotor that the control torque is not generated directly by the controller output. The HOSM-based FTESO and non-linear feedback law are constructed based on the equivalent linear model, such that the observer and the controller are completely independent without common parameters and can be adjusted individually. In conclusion, the proposed FTADRC method is tuned more easily than the traditional ADRC method.

4.2.3. Non-Linear Control Allocation

The control torque is allocated to four actuators according to the actuator models. After the motors reach the steady state, the PWM signals outputted by the controller are proportional to the rotor speeds. Given the quadratic dependence of thrust and torque upon rotor speed, a simple nonlinear control allocation method is used in this paper.

$$\begin{cases} r_1 = \frac{1}{2}\sqrt{-\frac{2\tau_x}{\sqrt{2}l\lambda_T} + \frac{2\tau_y}{\sqrt{2}l\lambda_T} + \frac{\tau_z}{\lambda_Q} + \frac{F}{\lambda_T}} \\ r_2 = \frac{1}{2}\sqrt{\frac{2\tau_x}{\sqrt{2}l\lambda_T} + \frac{2\tau_y}{\sqrt{2}l\lambda_T} - \frac{\tau_z}{\lambda_Q} + \frac{F}{\lambda_T}} \\ r_3 = \frac{1}{2}\sqrt{\frac{2\tau_x}{\sqrt{2}l\lambda_T} - \frac{2\tau_y}{\sqrt{2}l\lambda_T} + \frac{\tau_z}{\lambda_Q} + \frac{F}{\lambda_T}} \\ r_4 = \frac{1}{2}\sqrt{-\frac{2\tau_x}{\sqrt{2}l\lambda_T} - \frac{2\tau_y}{\sqrt{2}l\lambda_T} - \frac{\tau_z}{\lambda_Q} + \frac{F}{\lambda_T}} \end{cases} \tag{49}$$

where r_i, $(i = 1, 2, 3, 4)$ represents the duty circle of PWM signal for each actuator, l is the distance between the rotor and the center of mass of the quadcopter, F is the whole thrust provided by remote controller, λ_T and λ_Q are the coefficients related to actuator models which can be approximately calculated with maximum thrust T_{max}, maximum torque Q_{max} and maximum duty circle r_{max}.

$$\lambda_T = \frac{T_{max}}{{r_{max}}^2} \quad \lambda_Q = \frac{Q_{max}}{{r_{max}}^2} \tag{50}$$

5. Simulation and Experimental Results

5.1. Simulation Results

A simulation model for quadrotor attitude control is built according to the mathematical models and data acquired in Section 2. The structure of the simulation model is shown in Figure 8. The controller module includes the control method to be verified. The output signals are passed into a zero-order holder before the mathematical models in order to simulate the discrete calculation on the digital processor. The model part includes the motor model, propeller aerodynamic model, and rigid body dynamic model.

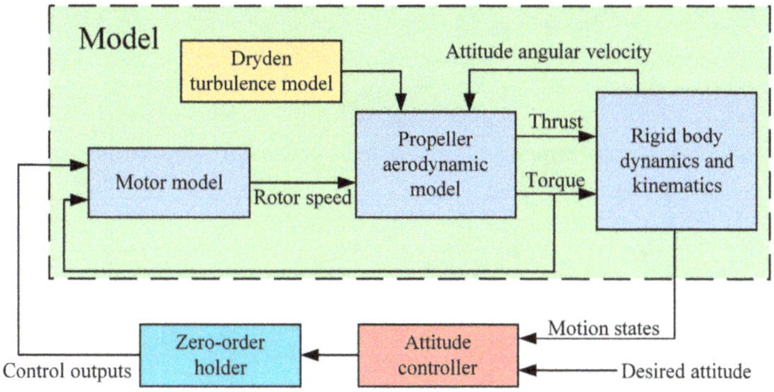

Figure 8. Structure of simulation model.

The unknown wind is added into the model as external disturbance. The rotor thrust and torque are influenced by the wind speed near the propeller. As the quadrotor body is acting as an obstruction to the wind, the wind attenuates while it passes through the body. As a result, the wind speed at the upwind side is larger than the wind speed at the downwind side. The wind attenuation is related to the interaction between the quadrotor body and the wind. In this paper, the wind speed is resolved into a component $\boldsymbol{w}$ parallel to the rotor disk and a component $\boldsymbol{u}$ perpendicular to the rotor disk. The attenuation of $\boldsymbol{u}$ is neglected so that $\boldsymbol{u}$ is constant near the quadrotor body. $\boldsymbol{w}$ decreases proportionally with the distance traveled along the quadrotor body. Figure 9 is a graphical representation of $\boldsymbol{w}$ decreasing with its travel along the quadrotor body. The green dashed line is the component $\boldsymbol{w}$ passing the body centroid. $\boldsymbol{p}_i (i = 1, 2, 3, 4)$ represents four rotor centers. After projecting $\boldsymbol{p}_i$ on the gray line, the projection point closest to the wind direction is selected as the windward point $\boldsymbol{p}_w$. The wind speed near the rotor corresponding to $\boldsymbol{p}_w$ is supposed to be $\boldsymbol{w}$. The wind speed near other three rotors is calculated as follows:

$$\boldsymbol{w}_i = \max(1 - \gamma_w d_i, 0)\boldsymbol{w} \tag{51}$$

where γ_w is the attenuation factor, d_i is the distance between rotor i with $\boldsymbol{p}_w$ along vector $\boldsymbol{w}$.

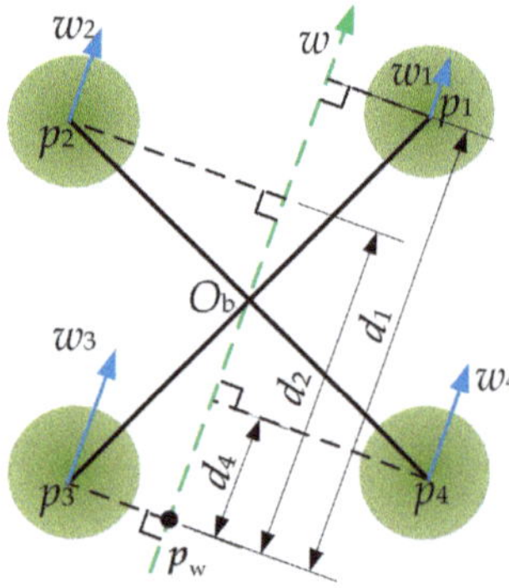

Figure 9. Chart of calculating the wind speed around each propeller.

The above wind attenuation leads to different thrust at the upwind side and downwind side and in turn generates a wind-related overturning torque which acts as the disturbance in attitude control system.

The commonly used Dryden turbulence model is used to generate random wind. The fundamental of the Dryden model is constructing a transfer function based on spectrum functions of atmospheric turbulence velocity and its gradient to convert the white noise signal to a colored noise signal. The Dryden module in aerospace toolbox of Simulink (Version 8.9, MathWorks., Natick, MA, United States, 2017) is used in this paper to generate random wind speed. The low-altitude intensity in the model is set to 10m/s. The other parameters are decided according to MIL-F-8785C standard. The wind speed along three axes of Earth coordinates framed with an altitude of 6m is shown in Figure 10.

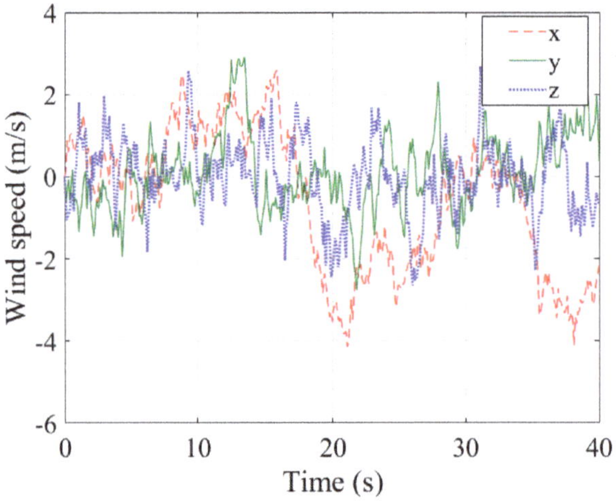

Figure 10. Random wind speed generated by Dryden model.

We present the mathematical simulations of proposed FTADRC attitude control method on a quadrotor model with the parameters in Table 3. The other parameters related to propeller shape and aerodynamics are chosen as Section 2.3. The first four parameters in Table 3 are used in the controller. Practically, l is easy to measure, while the inertias are hard to obtain. So, the nominal parameters are selected as the last column.

In the simulations, the position of the quadrotor is assumed to be free from the forces that include the propeller thrust and gravity. A square wave signal with amplitude of 20°, period of 8s is used as the desired roll angle input to analyze the step response. The random wind generated by the Dryden model is added into the simulation since the 15th second in order to compare the proposed FTADRC with the traditional ADRC in the absence and presence of dynamic wind disturbance. The simulation results of two different attitude control methods are shown in Figures 11–16.

Table 3. Parameters of the quadrotor in simulations.

Parameter	Description	True Value	Nominal Value
I_x	Inertia along x_b-axis	0.1 kgm^2	0.05 kgm^2
I_y	Inertia along y_b-axis	0.1 kgm^2	0.05 kgm^2
I_z	Inertia along z_b-axis	0.22 kgm^2	0.5 kgm^2
l	Distance between rotor and centroid	0.4 m	0.4 m
m	Mass	8 kg	-
K_v	Motor velocity constant	325 rpm/V	-
R_M	Equivalent resistance of motor	0.26 Ω	-

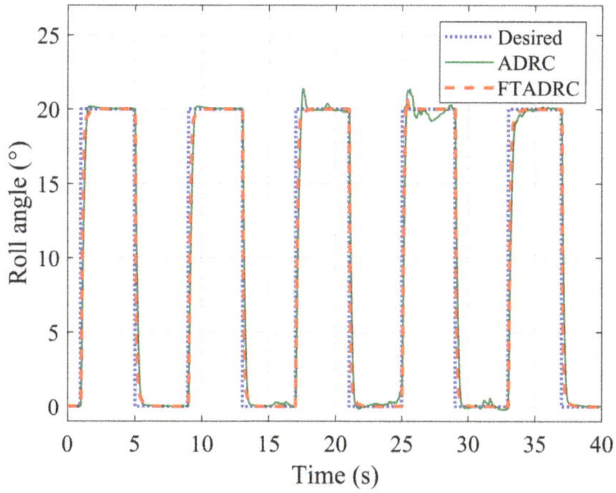

Figure 11. Curves of roll angle.

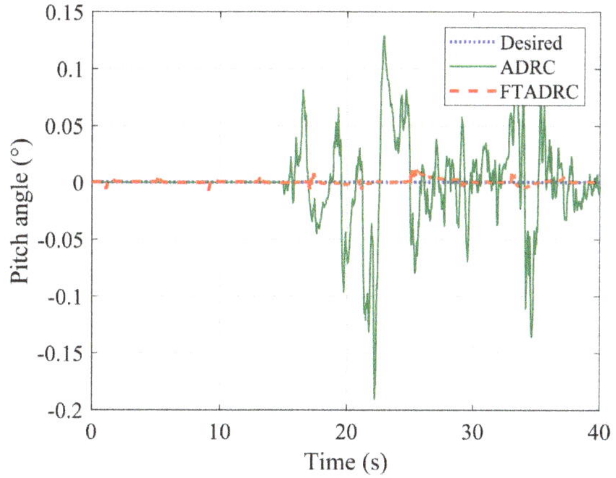

Figure 12. Curves of pitch angle.

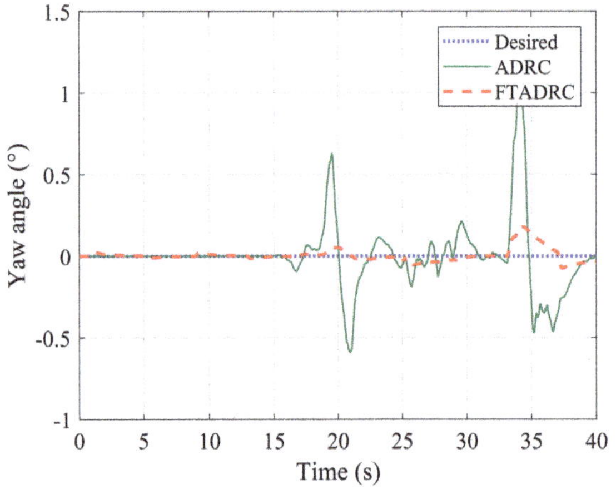

Figure 13. Curves of yaw angle.

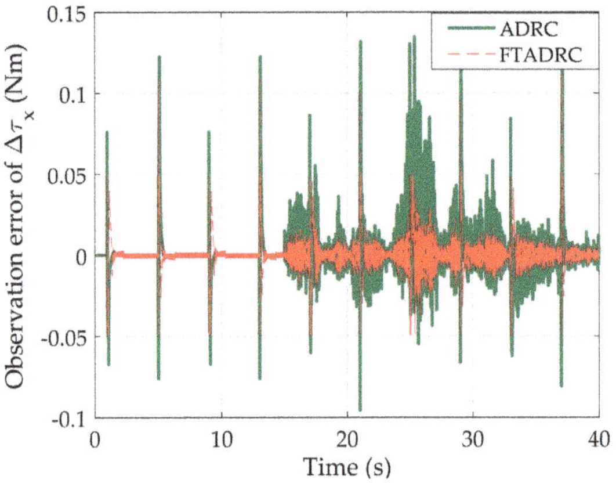

Figure 14. Curves of observation error for $\Delta\tau_x$.

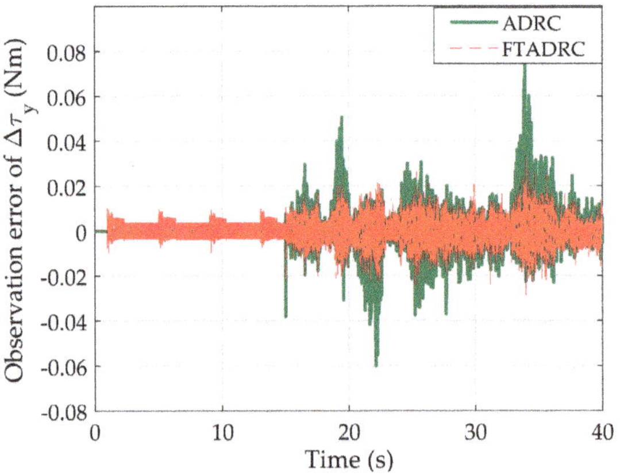

Figure 15. Curves of observation error for $\Delta\tau_y$.

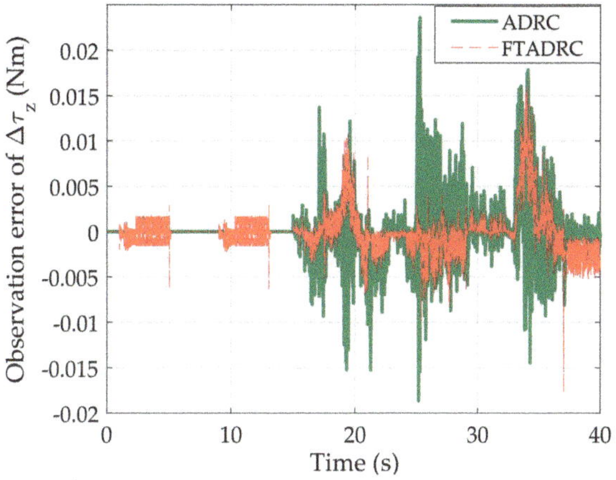

Figure 16. Curves of observation error for $\Delta\tau_z$.

As illustrated in Figure 11, quick convergence of the roll angle is achieved by both methods before 15 s with no disturbances because we tune the parameters of the two different controllers to achieve high-gain feedback control. With the presence of simulated dynamic wind, FTADRC tracks the desired attitude more accurately than traditional ADRC. Figures 14–16 indicate that the disturbance estimation of FTESO is more precise than traditional ESO.

The sharp peaks in Figures 14–16 are caused by motor inertia which is generally neglected in previous researches. The motors are not able to provide the rapidly changing control torque to follow the step input. The deviation of control torque acts as a rapidly changing disturbance which is difficult to track for both FTESO and ESO. However, the sudden disturbance vanishes quickly as the motor tracks the desired speed, and the attitude of the quadrotor is not seriously affected.

5.2. Experimental Results

In order to verify the effectiveness of the proposed FTADRC attitude control method in practical applications, we have developed a flight control unit (FCU) mainly using an STM32F103 micro-processor (STMicroelectronics, Geneva, Switzerland) and two ICM-20689 inertial measurement units (IMUs, InvenSense, San Jose, CA, United States). The angular speed and acceleration information provided by the two IMUs is averaged to generate more accurate measurements. An external HMC5983 magnetometer (Honeywell, Morris Plains, NJ, United States) is used for assisting the attitude determination process. The attitude determination algorithm in [40] is applied to the FCU providing attitude measurements. The core components including micro-processor and sensors cost less than 40 US dollars. In the experiments, the proposed fixed-time disturbance rejection control method consumes less than 1 ms on the 72 MHz ARM processor.

The experiments are conducted with a self-assembled QR450 quadrotor. The experimental setup consists of a remote controller, a ground control station (GCS) and the quadrotor with FCU onboard is shown in Figure 17. A pair of Xbee modules is used for communication between the FCU and GCS.

The quadrotor frame is made by aluminium alloy and carbon fiber to lower costs, and it has been deformed after being used for more than a year in plenty of flight tests, as shown in Figure 18. The deformations result in additional disturbances as the parameters are set ignoring them. The parameters are chosen as Table 4, in which I_x, I_y and I_z are estimated intuitively without precise measurement.

Figure 17. Experimental system.

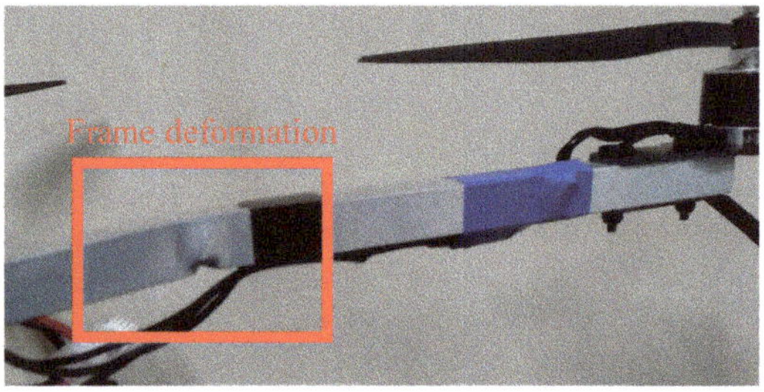

Figure 18. Deformation of quadrotor frame used in the experiments.

Table 4. Control parameters used in the experiments.

Parameters	Nominal Value
I_x	0.01 kgm^2
I_y	0.01 kgm^2
I_z	0.02 kgm^2
$\boldsymbol{C}$	*diag*(5,5,30)
k_1	70
k_2	10
p_1	0.5
p_2	0.5

The experiments are conducted indoors without Global Positioning System (GPS) signals. To assure flight security, the quadrotor is controlled remotely. The external unknown disturbance is simulated in three different ways. The performance of rejecting disturbance is verified respectively in all cases.

Case 1. *Eccentric mass*

In this case, an iron mass block of 194 g acting as the eccentric mass is hung from the right-hand front corner of the quadrotor frame, as shown in Figure 19. The original quadrotor without the eccentric mass is 1195 g. The eccentric mass is equivalent to 16% of the original quadrotor and 64% of individual motor thrust during hovering in weight. In addition, the block waves during flight and generates time-varying disturbance. The iron block hung from a corner of the frame mainly acts as an unknown constant torque in the attitude control system. In a simple error-driven control method, the constant disturbance torque will cause an attitude bias.

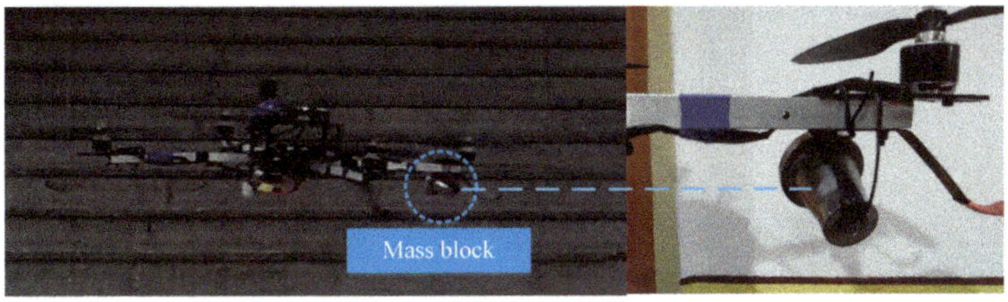

Figure 19. A mass block hung from the right hand front corner is used as the eccentric mass.

The curves of the attitude angles are shown in Figures 20–22, and the curves of the estimated disturbances are shown in Figure 23. The roll and pitch angles are affected by the eccentric mass and the control errors increase to about 2° in 1.5 s after the quadrotor takes off from the ground. The FTESO tracks the unknown disturbances in about 3 s with an initial value of 0. The fast-changing desired roll and pitch angles are produced by remote controller to test the controller's performance. By estimating and attenuating the estimated disturbances, the attitude angles are tracked accurately.

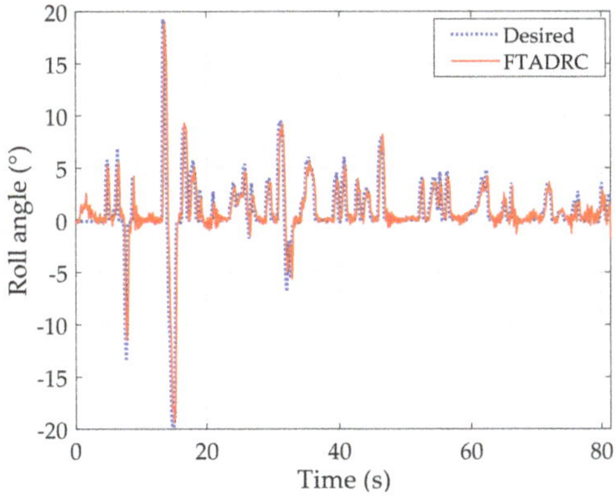

Figure 20. Curves of roll angle.

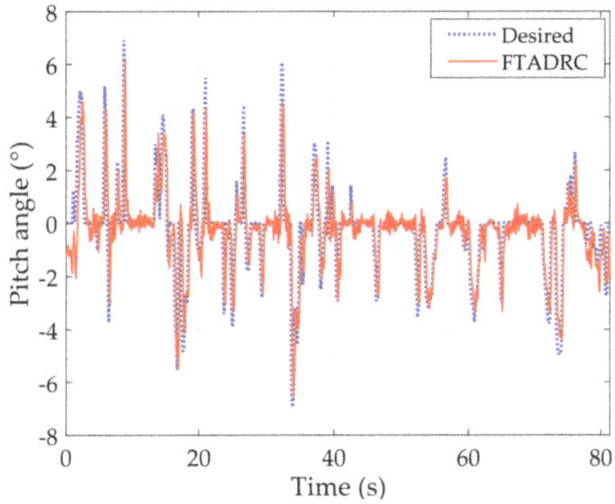

Figure 21. Curves of pitch angle.

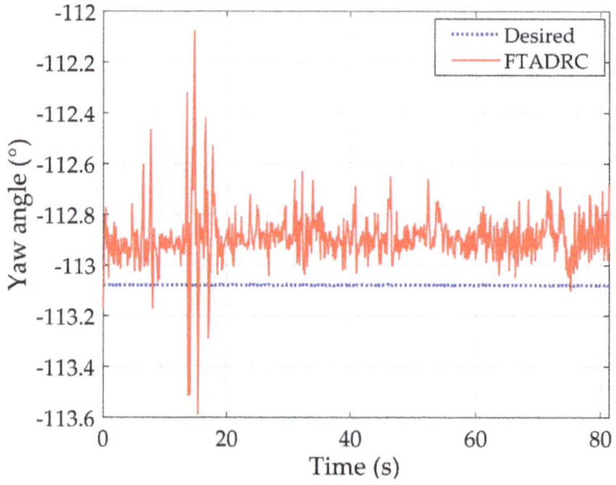

Figure 22. Curves of yaw angle.

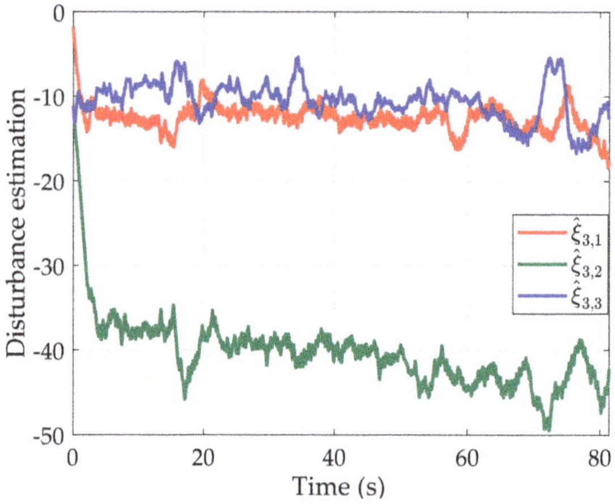

Figure 23. Curves of disturbance estimation.

Case 2. *Sudden fault of a single motor*

In this case, we simulate the actuator fault of motor 1 (right hand front motor) with a programed effectiveness loss of 30%, which means the output duty circle of motor 1 is multiplied by 0.7. The time of fault and recovery is controlled by a digital switch on the remote controller. As a non-redundant system, the quadrotor is sensitive to motor faults. The attitude model is uncontrollable if any of the four motors is completely disabled. So, we employ a programed motor effectiveness loss instead of a complete motor failure to model an in-flight actuator fault.

In the experiment, the fault occurs at 6.4 s and the motor recovers at 32.4 s. The curves of attitude angles are shown in Figures 24–26, and the curves of estimated disturbances are shown in Figure 27. The roll angle quickly changes to 19.8°, and the pitch angle quickly changes to −14.8° as a result of the motor fault. FTESO converges to the disturbances caused by the fault and the attitude angles track the desired values after 16 s.

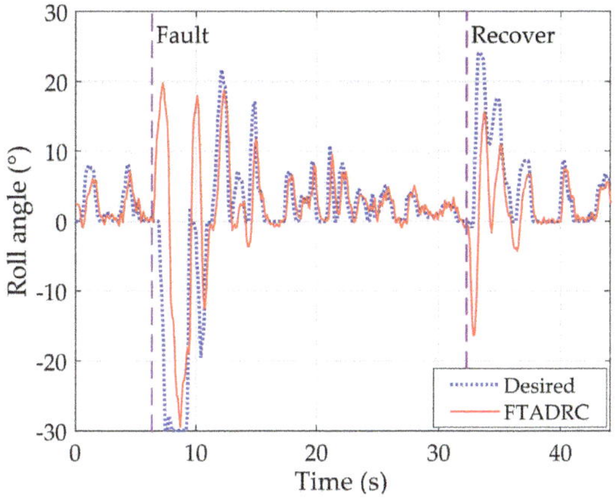

Figure 24. Curves of roll angle.

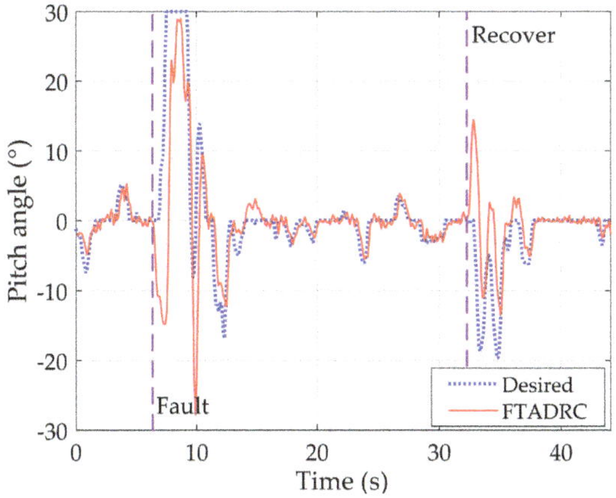

Figure 25. Curves of pitch angle.

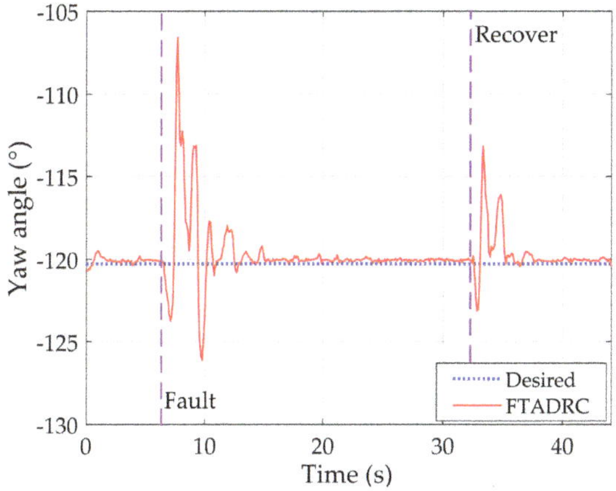

Figure 26. Curves of yaw angle.

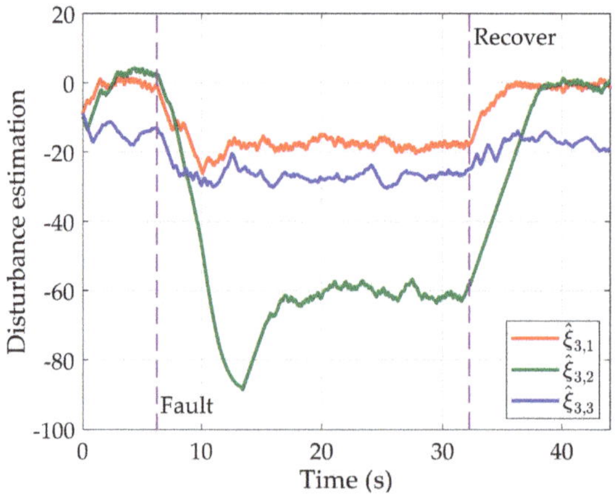

Figure 27. Curves of disturbance estimation.

The recovery of the motor is also treated as an external disturbance. It can be easily known from the motor model that the acceleration of the motor is slower than the deceleration. Thus, the recovery process has less impact on the quadrotor with a bounded disturbance differential-based FTESO. The roll angle changes to $-16.4°$, and the pitch angle changes to $14.4°$ 0.4 s after the motor recovers from the fault. The disturbance estimations of FTESO approximately turn back to the values before the motor fault 5.6 s after the recovery.

Case 3. *Wind and damaged propeller*

In this case, we install a damaged propeller on motor 2 (left hand front motor) and use two electrical fans to generate the external wind as shown in Figure 28. We make the quadrotor hover in the wind by remote control. The average wind speed generated by the two electrical fans is around 4.5 m/s. The main purpose of this case is to show the performance of FTADRC method with common lumped disturbances in quadrotor attitude control such as wind disturbance and propeller damage. Figures 29–32 present the experimental results of the FTADRC and conventional ADRC schemes. In each figure, panel (a) indicates the result of FTADRC, and panel (b) indicates the result of conventional ADRC.

Figures 29–31 present the tracking errors of attitude angles of FTADRC and conventional ADRC control schemes. The tracking errors of roll and pitch angles using FTADRC control scheme keep below $2°$ with average values around $0°$, while the tracking errors of roll and pitch angles using the ADRC control scheme have peak-to-peak values larger than $15°$ and obviously non-zero average values. Figure 32 shows disturbance estimation with FTADRC and traditional ADRC schemes. The disturbance introduced by the wind is hard to evaluate accurately because of the complicated wind dynamics related to the electrical fans. However, it still could be seen from Figure 32 that the disturbance estimation using FTADRC changes faster than that using conventional ADRC. Along with the tracking errors of attitude angles, it can be demonstrated that FTADRC responds more quickly to the external disturbances and tracks the input attitude more accurately than conventional ADRC.

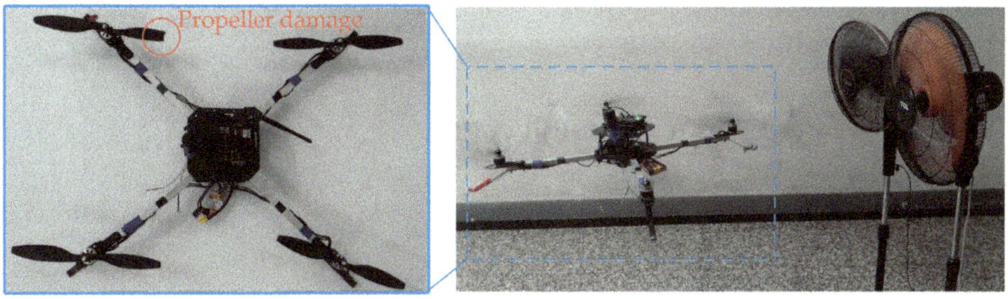

Figure 28. Experimental environment with wind and damaged propeller.

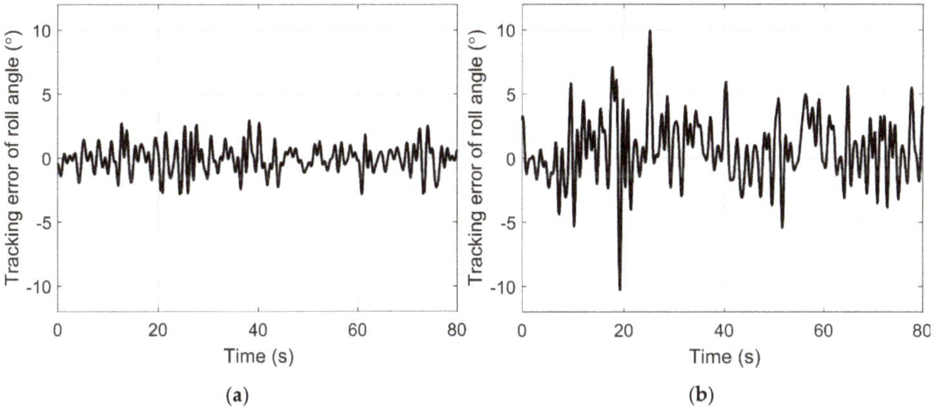

Figure 29. Tracking errors of roll angle using different control schemes: (**a**) FTADRC; (**b**) conventional ADRC.

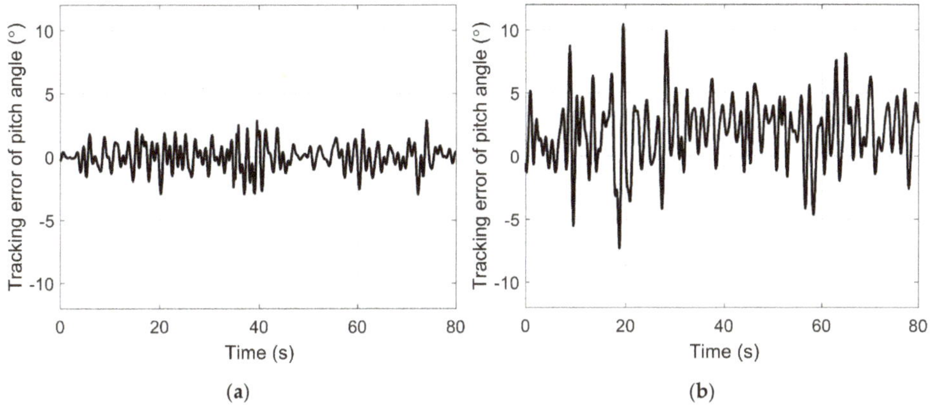

Figure 30. Tracking errors of pitch angle using different control schemes: (**a**) FTADRC; (**b**) conventional ADRC.

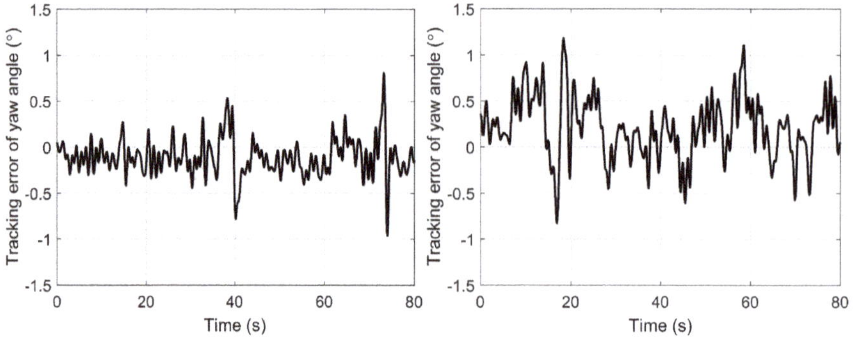

(a) (b)

Figure 31. Tracking errors of yaw angle using different control schemes: (**a**) FTADRC; (**b**) conventional ADRC.

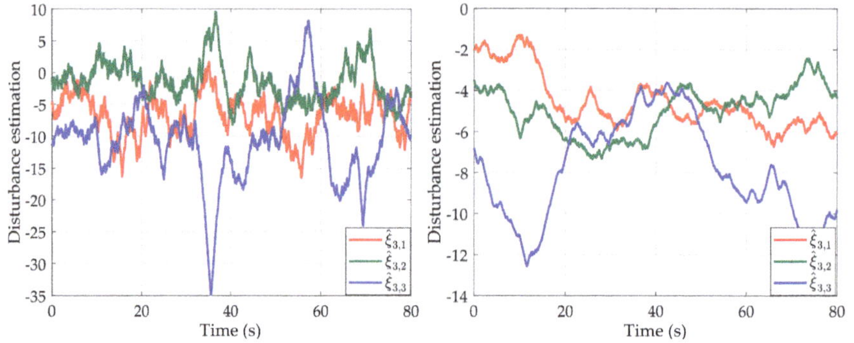

(a) (b)

Figure 32. Curves of disturbance estimation using different control schemes: (**a**) FTADRC; (**b**) conventional ADRC.

6. Discussion

In the experiment with eccentric mass, the quadrotor takes off from the ground slowly so that the eccentric mass could be regard as a disturbance accumulating during take-off. Thus, its influence on attitude angles is not obvious and FTESO converges to the lumped disturbances quickly. During the flight, the attitude of the quadrotor accurately tracks the desired attitude with disturbances rejected by FTESO. This experiment together with the simulations demonstrate the stability and robustness of FTADRC in the presence of slowly varying disturbance such as dynamic wind and eccentric mass.

In the experiment with a motor fault, the disturbance caused by the fault takes effect rapidly. Strictly speaking, the differential of disturbance exceeds the bound set in FTESO, and the proposed FTADRC is not suitable for such a case theoretically. However, the results show that the attitude system of quadrotor is stabled by FTADRC except that FTESO consumes more time for convergence. Therefore, the proposed robust control scheme could be applied to the practical situations beyond the theoretical constraints.

Since the motor merely fails in flight and the centroid of mass is generally adjusted before flight, the conditions in the previous two cases are rigorous, and not normal in the practical operation of quadrotors. The last case with damaged propeller and unknown wind represents a more common operation situation of a low-cost quadrotor. In this experiment, the proposed FTADRC and conventional ADRC attitude control schemes are compared. The performance of FTADRC clearly exceeds the conventional ADRC according to the results.

7. Conclusions

In this paper, the FTADRC scheme is proposed based on a HOSM method for the attitude control of a quadrotor with unknown disturbance. The control scheme employs a fixed-time ESO for observing the lumped disturbance, and uses a multivariable fixed-time second-order sliding mode method to fast approach the sliding surface for a feedback-linearized system. Detailed mathematical models are built for simulation analysis. According to the comparative simulation and experiment results, the attitude control accuracy is improved compared with the traditional ADRC because FTESO achieves higher precision than the traditional ESO, and the control scheme is effective on a low-cost flight control unit. It can be concluded that the proposed FTADRC is robust toward dynamic wind, mass eccentricity, and motor fault. Additionally, FTADRC is practical for a low-cost quadrotor because of its high time-efficiency.

Supplementary Materials: The video of the experiment with eccentric mass is available online at https://youtu.be/hf8jx9WN6Ec. The video of the experiment with a motor fault is available online at https://youtu.be/HxRsjkjuvY4. The code on STM32F103 used in the experiments is available online at https://1drv.ms/f/s!ApzqIuEnpaPHhSaF6lRLL5F_qh0Y.

Author Contributions: Funding acquisition, N.C.; Investigation, F.Y.; Methodology, C.S. and F.Y.; Project administration, C.W.; Resources, N.C.; Software, C.S.; Supervision, C.W.; Writing – original draft, C.S.

Funding: This work is supported by the National Nature Science Fund of China (Grant No. 61403100), the Fundamental Research Funds for the Central Universities (Grant No. HIT.NSRIF.2015037) and the open National Defense Key Disciplines Laboratory of Exploration of Deep Space Landing and Return Control Technology, Harbin Institute of Technology (Grant No. HIT.KLOF.2013.079).

Acknowledgments: We gratefully acknowledge the assistance of Mo He in assembling the quadrotor used in our experiments. We also thank Jinbo Wang for processing videos and images.

Conflicts of Interest: The authors declare no conflict of interest.

Appendix A

Define:

$$a = \frac{\rho_1^2+\rho_2^2+\rho_3^2-1}{2} = -\frac{q_0}{1+q_0} \tag{A1}$$

The three elements of $\boldsymbol{\phi}(\boldsymbol{x})$ are:

$$\begin{aligned}
L_f^2 h_1(\boldsymbol{x}) &= \tfrac{1}{4}(\omega_z+\rho_1\omega_y-\rho_2\omega_x)(\rho_3\omega_x-\rho_1\omega_z+\rho_2^2\omega_y-a\omega_y+\rho_1\rho_2\omega_x+\rho_2\rho_3\omega_z)\\
&\quad -\tfrac{1}{4}(\omega_y-\rho_1\omega_z+\rho_3\omega_x)(\rho_1\omega_y-\rho_2\omega_x+\rho_3^2\omega_z-a\omega_z+\rho_1\rho_3\omega_x+\rho_2\rho_3\omega_y)\\
&\quad +\tfrac{1}{4}(\rho_1\omega_x+\rho_2\omega_y+\rho_3\omega_z)(\rho_2\omega_z-\rho_3\omega_y+\rho_1^2\omega_x-a\omega_x+\rho_1\rho_2\omega_y+\rho_1\rho_3\omega_z)\\
&\quad +\tfrac{1}{4I_x}(I_y-I_z)\omega_y\omega_z(\rho_1^2-\rho_2^2-\rho_3^2+1)\\
&\quad +\tfrac{1}{2I_z}(I_x-I_y)\omega_x\omega_y(\rho_1\rho_3+\rho_2)\\
&\quad +\tfrac{1}{2I_y}(I_z-I_x)\omega_x\omega_z(\rho_1\rho_2-\rho_3)\\
L_f^2 h_2(\boldsymbol{x}) &= \tfrac{1}{4}(\omega_x+\rho_2\omega_z-\rho_3\omega_y)(\rho_1\omega_y-\rho_2\omega_x+\rho_3^2\omega_z-a\omega_z+\rho_1\rho_3\omega_x+\rho_2\rho_3\omega_y)\\
&\quad -\tfrac{1}{4}(\omega_z+\rho_1\omega_y-\rho_2\omega_x)(\rho_2\omega_z-\rho_3\omega_y+\rho_1^2\omega_x-a\omega_x+\rho_1\rho_2\omega_y+\rho_1\rho_3\omega_z)\\
&\quad +\tfrac{1}{4}(\rho_1\omega_x+\rho_2\omega_y+\rho_3\omega_z)(\rho_3\omega_x-\rho_1\omega_z+\rho_2^2\omega_y-a\omega_y+\rho_1\rho_2\omega_x+\rho_2\rho_3\omega_z)\\
&\quad +\tfrac{1}{4I_y}(I_z-I_x)\omega_x\omega_z(\rho_2^2-\rho_1^2-\rho_3^2+1)\\
&\quad +\tfrac{1}{2I_z}(I_x-I_y)\omega_x\omega_y(\rho_2\rho_3-\rho_1)\\
&\quad +\tfrac{1}{2I_x}(I_y-I_z)\omega_y\omega_z(\rho_1\rho_2+\rho_3)\\
L_f^2 h_3(\boldsymbol{x}) &= \tfrac{1}{4}(\omega_y-\rho_1\omega_z+\rho_3\omega_x)(\rho_2\omega_z-\rho_3\omega_y+\rho_1^2\omega_x-a\omega_x+\rho_1\rho_2\omega_y+\rho_1\rho_3\omega_z)\\
&\quad -\tfrac{1}{4}(\omega_x+\rho_2\omega_z-\rho_3\omega_y)(\rho_3\omega_x-\rho_1\omega_z+\rho_2^2\omega_y-a\omega_y+\rho_1\rho_2\omega_x+\rho_2\rho_3\omega_z)\\
&\quad \tfrac{1}{4}(\rho_1\omega_x+\rho_2\omega_y+\rho_3\omega_z)(\rho_1\omega_y-\rho_2\omega_x+\rho_3^2\omega_z-a\omega_z+\rho_1\rho_3\omega_x+\rho_2\rho_3\omega_y)\\
&\quad +\tfrac{1}{4I_z}(I_x-I_y)\omega_x\omega_y(\rho_3^2-\rho_1^2-\rho_2^2+1)\\
&\quad +\tfrac{1}{2I_y}(I_z-I_x)\omega_x\omega_z(\rho_2\rho_3+\rho_1)\\
&\quad +\tfrac{1}{2I_x}(I_y-I_z)\omega_y\omega_z(\rho_1\rho_3-\rho_2)
\end{aligned} \tag{A2}$$

References

1. Bouabdallah, S.; Murrieri, P.; Siegwart, R. Design and control of an indoor micro quadrotor. In Proceedings of the IEEE International Conference on Robotics and Automation, New Orleans, LA, USA, 26 April–1 May 2004; pp. 4393–4398.
2. Özbek, N.S.; Önkol, M.; Efe, M.Ö. Feedback control strategies for quadrotor-type aerial robots: A survey. *Trans. Inst. Meas. Control* **2016**, *38*, 529–554. [CrossRef]
3. Lee, H.; Kim, H.J. Trajectory tracking control of multirotors from modelling to experiments: A survey. *Int. J. Control Autom. Syst.* **2017**, *15*, 281–292. [CrossRef]
4. Tayebi, A.; McGilvray, S. Attitude stabilization of a vtol quadrotor aircraft. *IEEE Trans. Control Syst. Technol.* **2006**, *14*, 562–571. [CrossRef]
5. Mahony, R.; Kumar, V.; Corke, P. Multirotor aerial vehicles modeling, estimation, and control of quadrotor. *IEEE Robot. Autom. Mag.* **2012**, *19*, 20–32. [CrossRef]
6. Pounds, P.; Mahony, R.; Corke, P. Modelling and control of a large quadrotor robot. *Control Eng. Practice* **2010**, *18*, 691–699. [CrossRef]
7. Kendoul, F. Survey of advances in guidance, navigation, and control of unmanned rotorcraft systems. *J. Field Robot.* **2012**, *29*, 315–378. [CrossRef]
8. Lee, D.; Kim, H.J.; Sastry, S. Feedback linearization vs. Adaptive sliding mode control for a quadrotor helicopter. *Int. J. Control Autom. Syst.* **2009**, *7*, 419–428. [CrossRef]
9. Kim, J.; Kang, M.S.; Park, S. Accurate modeling and robust hovering control for a quad-rotor vtol aircraft. In Proceedings of the 2nd International Symposium on UAVs, Reno, NV, USA, June 2009; pp. 9–26.
10. Lyu, X.M.; Zhou, J.N.; Gu, H.W.; Li, Z.X.; Shen, S.J.; Zhang, F. Disturbance observer based hovering control of quadrotor tail-sitter vtol uavs using h-infinity synthesis. *IEEE Robot. Autom. Lett.* **2018**, *3*, 2910–2917. [CrossRef]
11. Dierks, T.; Jagannathan, S. Output feedback control of a quadrotor uav using neural networks. *IEEE Trans. Neural Netw.* **2010**, *21*, 50–66. [CrossRef] [PubMed]
12. Aboudonia, A.; El-Badawy, A.; Rashad, R. Disturbance observer-based feedback linearization control of an unmanned quadrotor helicopter. *Proc. Inst. Mech. Eng. Part I: J. Syst. Control Eng.* **2016**, *230*, 877–891. [CrossRef]
13. Aboudonia, A.; Rashad, R.; El-Badawy, A. Composite hierarchical anti-disturbance control of a quadrotor uav in the presence of matched and mismatched disturbances. *J. Intell. Robot. Syst.* **2018**, *90*, 201–216. [CrossRef]
14. Xiao, B.; Yin, S. A new disturbance attenuation control scheme for quadrotor unmanned aerial vehicles. *IEEE Trans. Ind. Inform.* **2017**, *13*, 2922–2932. [CrossRef]
15. Ahmed, N.; Chen, M. Sliding mode control for quadrotor with disturbance observer. *Adv. Mech. Eng.* **2018**, *10*, 16. [CrossRef]
16. Benallegue, A.; Mokhtari, A.; Fridman, L. High-order sliding-mode observer for a quadrotor uav. *Int. J. Robust Nonlinear Control* **2008**, *18*, 427–440. [CrossRef]
17. Besnard, L.; Shtessel, Y.B.; Landrum, B. Quadrotor vehicle control via sliding mode controller driven by sliding mode disturbance observer. *J. Frankl. Inst.* **2012**, *349*, 658–684. [CrossRef]
18. Fethalla, N.; Saad, M.; Michalska, H.; Ghommam, J. Robust observer-based dynamic sliding mode conroller for a quadrotor uav. *IEEE Access* **2018**, *6*, 45846–45859. [CrossRef]
19. Shi, D.; Wu, Z.; Chou, W. Super-twisting extended state observer and sliding mode controller for quadrotor uav attitude system in presence of wind gust and actuator faults. *Electronics* **2018**, *7*, 128. [CrossRef]
20. Shi, D.; Wu, Z.; Chou, W. Generalized extended state observer based high precision attitude control of quadrotor vehicles subject to wind disturbance. *IEEE Access* **2018**, *6*. [CrossRef]
21. Han, J. From pid to active disturbance rejection control. *IEEE Trans. Ind. Electron.* **2009**, *56*, 900–906. [CrossRef]
22. Wu, Z.-H.; Guo, B.-Z. On convergence of active disturbance rejection control for a class of uncertain stochastic nonlinear systems. *Int. J. Control* **2017**, 1–14. [CrossRef]
23. Guo, B.Z.; Zhao, Z.L. *Active Disturbance Rejection Control for Nonlinear Systems: An Introduction*, 1st ed.; John Wiley & Sons: Hoboken, NJ, USA, 2017; pp. 6–9.

24. Yang, H.J.; Cheng, L.; Xia, Y.Q.; Yuan, Y. Active disturbance rejection attitude control for a dual closed-loop quadrotor under gust wind. *IEEE Trans. Control Syst. Technol.* **2018**, *26*, 1400–1405. [CrossRef]
25. Ma, D.L.; Xia, Y.Q.; Li, T.Y.; Chang, K. Active disturbance rejection and predictive control strategy for a quadrotor helicopter. *IET Contr. Theory Appl.* **2016**, *10*, 2213–2222. [CrossRef]
26. Guo, Y.; Jiang, B.; Zhang, Y. A novel robust attitude control for quadrotor aircraft subject to actuator faults and wind gusts. *IEEE/CAA J. Autom. Sinica* **2018**, *5*, 292–300. [CrossRef]
27. Levant, A. Robust exact differentiation via sliding mode technique. *Automatica* **1998**, *34*, 379–384. [CrossRef]
28. Levant, A. Higher-order sliding modes, differentiation and output-feedback control. *Int. J. Control* **2003**, *76*, 924–941. [CrossRef]
29. Basin, M.; Yu, P.; Shtessel, Y. Finite-and fixed-time differentiators utilising hosm techniques. *IET Control Theory Appl.* **2016**, *11*, 1144–1152. [CrossRef]
30. Menard, T.; Moulay, E.; Perruquetti, W. Fixed-time observer with simple gains for uncertain systems. *Automatica* **2017**, *81*, 438–446. [CrossRef]
31. Angulo, M.T.; Moreno, J.A.; Fridman, L. Robust exact uniformly convergent arbitrary order differentiator. *Automatica* **2013**, *49*, 2489–2495. [CrossRef]
32. Li, P.; Ma, J.; Zheng, Z. Disturbance-observer-based fixed-time second-order sliding mode control of an air-breathing hypersonic vehicle with actuator faults. *Proc. Inst. Mech. Eng. Part G: J. Aerosp. Eng.* **2018**, *232*, 344–361. [CrossRef]
33. Ni, J.; Liu, L.; Chen, M.; Liu, C. Fixed-time disturbance observer design for brunovsky system. *IEEE Trans. Circuit Syst. Part 2: Express Birefs* **2018**, *65*, 341–345. [CrossRef]
34. Yu, P.; Shtessel, Y.; Edwards, C. Continuous higher order sliding mode control with adaptation of air breathing hypersonic missile. *Int. J. Adapt. Control Signal Process.* **2016**, *30*, 1099–1117. [CrossRef]
35. Basin, M.V.; Yu, P.; Shtessel, Y.B. Hypersonic missile adaptive sliding mode control using finite-and fixed-time observers. *IEEE Trans. Ind. Electron.* **2018**, *65*, 930–941. [CrossRef]
36. Yu, X.; Li, P.; Zhang, Y. The design of fixed-time observer and finite-time fault-tolerant control for hypersonic gliding vehicles. *IEEE Trans. Ind. Electron.* **2018**, *65*, 4135–4144. [CrossRef]
37. Schaub, H.; Junkins, J.L. Stereographic orientation parameters for attitude dynamics: A generalization of the rodrigues parameters. *J. Astron. Sci.* **1996**, *44*, 1–19.
38. Leishman, G.J. *Principles of Helicopter Aerodynamics*, 2nd ed.; Cambridge university press: Cambridge, UK, 2006; pp. 55–167.
39. Basin, M.; Panathula, C.B.; Shtessel, Y. Multivariable continuous fixed-time second-order sliding mode control: Design and convergence time estimation. *IET Control Theory Appl.* **2016**, *11*, 1104–1111. [CrossRef]
40. Madgwick, S. An efficient orientation filter for inertial and inertial/magnetic sensor arrays. *Report x-io Univ. Bristol* **2010**, *25*, 113–118.

electronics

MDPI

Article

Super-Twisting Extended State Observer and Sliding Mode Controller for Quadrotor UAV Attitude System in Presence of Wind Gust and Actuator Faults

Di Shi [1], Zhong Wu [1,*] and Wusheng Chou [2,3]

[1] School of Instrumentation Science and Optoelectronics Engineering, Beihang University, Beijing 100191, China; shidi666@buaa.edu.cn

[2] School of Mechanical Engineering and Automation, Beihang University, Beijing 100191, China; wschou@buaa.edu.cn

[3] State Key Laboratory of Virtual Reality Technology and System, Beijing 100191, China

* Correspondence: wuzhong@buaa.edu.cn; Tel.: +86-10-8233-9703

Received: 19 June 2018; Accepted: 24 July 2018; Published: 26 July 2018

Abstract: This article addresses the problem of high precision attitude control for quadrotor unmanned aerial vehicle in presence of wind gust and actuator faults. We consider the effect of those factors as lumped disturbances, and in order to realize the quickly and accurately estimation of the disturbances, we propose a control strategy based on the online disturbance uncertainty estimation and attenuation method. Firstly, an enhanced extended state observer (ESO) is constructed based on the super-twisting (ST) algorithm to estimate and attenuate the impact of wind gust and actuator faults in finite time. And the convergence analysis and parameter selection rule of STESO are given following. Secondly, in order to guarantee the asymptotic convergence of desired attitude timely, a sliding mode control law is derived based on the super-twisting algorithm. And a comprehensive stability analysis for the entire system is presented based on the Lyapunov stability theory. Finally, to demonstrate the efficiency of the proposed solution, numerical simulations and real time experiments are carried out in presences of wind disturbance and actuator faults.

Keywords: quadrotor; super twisting extended state observer (STESO); super twisting sliding mode controller (STSMC); wind disturbance; actuator faults

---

1. Introduction

Quadrotor unmanned aerial vehicles (UAVs) are rapidly growing in popularity due to their wide range of civil and military applications such as surveillance, inspection, search and rescue, and disaster response. As a new kind of UAV, quadrotor is a small rotorcraft with four propellers driven by four direct current (DC) motors respectively [1]. Compared with traditional helicopters, the structure of quadrotor is simpler and more efficient, and has unique features in precise hovering, aggressive maneuver, vertical take-off and landing (VTOL) [2,3], etc. Therefore, the researches on quadrotor unmanned aerial vehicle become more and more popular in recent years, and a lot of achievements have been made [4–6].

The quadrotor is an underactuated and nonlinear coupled system [7]. Traditionally, the integral-type control methods are commonly used in the controller design of quadrotor UAVs and have shown to be effective in the attitude and position stabilization control of them. In [8,9], a proportional integral derivative (PID) control method was developed to obtain the stability of quadrotor. In [10], nonlinear PI/PID controller was designed to regulate the posture of quadrotor and showed robustness to aircraft systems effects. In [11], a motion controller of quadrotor has been derived by using time scale separation ideas, and numerical simulations confirmed that the motion control objective is satisfied with the proposed scheme in presence of the forces saturation of propellers, sensor noise and perturbing forces caused by wind. And in [12], the linear quadratic regulator (LQR) controller was applied to deal with the nominal system of quadrotor obtained by feedback linearization method. In order to meet some mission requirements, high precision attitude control is essential in those applications. However, when operating in outdoor environments, quadrotors would be easily affected by wind gust during the course of flight [13]. In addition to wind gust, the actuating motor-propeller system is prone to faults due to component degradation or damage to the motors or propellers, which may lead to significant performance degradation or even instability of the close-loop system [14]. Therefore, it is difficult for the traditional linear controllers to achieve the high precision control requirement under the influence of these factors. To solve this problem, many approaches have been proposed in literatures. Robust adaptive controller was introduced to eliminate the influence of wind gust in [15]. In [16], robust optimal backstepping control (ROBC) is designed to address the stabilization and trajectory tracking problem of quadrotor in the existence of wind gust. And in [14], a nonlinear robust adaptive fault-tolerant altitude and attitude tracking method was implemented to accommodating actuator faults in quadrotor without the need of a faults diagnosis mechanism. In summery, these methods were designed to improve the robustness of the system.

Alternatively, both wind gust and actuator faults can be considered as lumped disturbances and the online disturbance and uncertainty estimation and attenuation (DUEA) method would be a potential solution to this problem. In recent years, the DUEA methods have been widely studied for some types of real time systems in order to cancel the influence of lumped disturbances at the controller stage [17,18], and the framework of which can be divided into two parts, namely, a disturbance and uncertainty estimator (DUE) and a feedback controller (FC).

In the first part, DUE is designed to estimate the disturbances so that they could be compensated in the feedforward loop. To achieve this aim, a series of observers have been proposed as the DUE so far, such as disturbance observer (DO) [19,20], extended state observer (ESO) [21–24], proportional integral observer (PIO) [25,26] and acceleration based disturbance observer (ABDOB) [27], etc. By the appropriate use of the observer, disturbance rejection performance and robustness of the existing control system could be significantly improved. The extended state observer (ESO), known as the key module of active disturbance rejection control, can estimate both the states of system and the total disturbances with less dependence on model information [28,29]. This method was first proposed by Han in 1990s and the basic idea behind ESO is to view disturbance as an extended state and utilize observer to estimate it [28]. As for the ESO based control structure, the performance of closed-loop system is largely determined by the estimation accuracy of observer. The traditional ESO approaches focus primarily on dealing with slowly changing disturbances. However, it's obvious that the disturbance torque caused by the wind gust and actuator faults happens suddenly, which can not be estimated by traditional ESO thoroughly [30]. Therefore, an enhanced ESO that can quickly estimate the disturbance is necessary in this field. In [31], a higher-order ESO is investigated and the estimation accuracy was improved, however, higher level of the observer order will lead to a higher observer gain which will in return excite the sensor noise and introduce them into the control loop. In [32], a sliding model method was used in disturbance observer to estimate the quadrotor velocities, the external disturbances such as wind and parameter uncertainties, and achieves good results, except for serious chattering. To reduce this problem, super-twisting algorithm have been adopted in design of the observer. In [33], the super-twisting observer (STO) is constructed to reject aperiodic disturbances and input unmatched periodic disturbances with reduced chattering.

In the second part, the FC is designed to guarantee fast convergence of the closed-loop system. Sliding mode control (SMC) has been known as one of the most efficiency controller in fast convergence [34]. In [35], a fixed-time second-order sliding mode control law is designed to guarantee the reaching time, independent of initial conditions. However, the robustness of the SMC is achieved at the cost of a high frequency switching of the control signal, which has a negative effect in the actuator.

To reduce the chattering, a family of continuous sliding mode controllers based on super-twisting algorithm have been developed [36].

Motivated by the above observations and inspired from Ref. [33,36], a high precision attitude control law is developed for quadrotor unmanned aerial vehicle in presence of wind gust and actuator faults. The main contributions of this paper are summarized as follows:

- Propose a STESO to accurately estimate the disturbance torque caused by wind gust and actuator faults in finite time, and give the parameter selection rule of the observer;
- Design a fast convergence attitude control law based on STSMC, and give a comprehensive stability analysis on the entire system.

The remainder of the paper is organized as follows. The mathematical model and control problem is formulated in Section 2. A STESO is designed in Section 3, as well as parament selection rule of the observer is also given in this section. In Section 4, a fast convergence attitude controller is designed based on ST algorithm. Numerical simulation and real time experimental results are presented in Section 5. Finally, we conclude the paper in Section 6.

2. Mathematical Model and Problem Formulation

2.1. Notation

$||\bullet||$ denotes the 2-norm of a vector or a matrix. For a given vector $\boldsymbol{v} = [v_1, ..., v_n]^T \in \mathbb{R}^n$, $||\boldsymbol{v}|| = \sqrt{\boldsymbol{v}^T\boldsymbol{v}}$. For a given matrix $\boldsymbol{A} \in \mathbb{R}^{n\times n}$, $\lambda_{\max}(\boldsymbol{A})$ and $\lambda_{\min}(\boldsymbol{A})$ denote the maximal and minimum eigenvalue of the matrix respectively. In addition, the operator $S(\bullet)$ maps a vector $\boldsymbol{x} = \begin{bmatrix} x_1 & x_2 & x_3 \end{bmatrix}^T$ to a skew symmetric matrix as:

$$S(\boldsymbol{x}) = \begin{bmatrix} 0 & -x_3 & x_2 \\ x_3 & 0 & -x_1 \\ -x_2 & x_1 & 0 \end{bmatrix}$$

sgn($\bullet$) is the sign function, and for a scalar x:

$$\text{sgn} = \begin{cases} \dfrac{x}{|x|}, & |x| \neq 0 \\ 0, & |x| = 0 \end{cases}$$

For a vector $\boldsymbol{x} = \begin{bmatrix} x_1 & x_2 & x_3 \end{bmatrix}^T$ and $r \geq 0$, define:

$$\text{sig}(\boldsymbol{x})^r = \begin{bmatrix} |x_1|^r \,\text{sgn}(x_1) \\ |x_2|^r \,\text{sgn}(x_2) \\ |x_3|^r \,\text{sgn}(x_3) \end{bmatrix}$$

2.2. Quaternion Operations

In order to avoid the singularity problem of trigonometric functions, unit quaternion $\boldsymbol{q} = \begin{bmatrix} q_0 & \boldsymbol{q}_v^T \end{bmatrix}^T \in \mathbb{R}^4$, $||\boldsymbol{q}|| = 1$ is used to represent rotation [37] of the quadrotor. Following are the operations we used.

The quaternion multiplication is:

$$\boldsymbol{q}_1 \otimes \boldsymbol{q}_2 = \begin{bmatrix} q_{01}q_{02} - \boldsymbol{q}_{v1}^T\boldsymbol{q}_{v2} \\ q_{01}\boldsymbol{q}_{v2} + q_{02}\boldsymbol{q}_{v1} - S(\boldsymbol{q}_{v2})\boldsymbol{q}_{v1} \end{bmatrix} \tag{1}$$

The relationship between rotation matrix $\boldsymbol{C}_A^B$ and $\boldsymbol{q}$ is calculated as:

$$C_A^B = (q_0^2 - q_v^T q_v)I_3 + 2q_v q_v^T + 2q_0 S(q_v) \quad (2)$$

$$\dot{C}_A^B = -S(\omega)C_A^B \quad (3)$$

The derivative of a quaternion is given by the quaternion multiplication of the quaternion q and the angular velocity of the system ω :

$$\dot{q} = \begin{bmatrix} \dot{q}_0 \\ \dot{q}_v \end{bmatrix} = \frac{1}{2} q \otimes \begin{bmatrix} 0 \\ \omega \end{bmatrix} = \frac{1}{2} \begin{bmatrix} -q_v^T \\ S(q_v) + q_0 I_3 \end{bmatrix} \omega \quad (4)$$

The quaternion error q_e is given as the quaternion multiplication of the conjugate of the actual quaternion q and the desired quaternion q_d :

$$q_e = q_d^* \otimes q = \begin{bmatrix} q_0 q_{0d} + q_v{}^T q_{vd} \\ q_{0d} q_v - q_0 q_{vd} + S(q_v) q_{vd} \end{bmatrix} \quad (5)$$

2.3. Kinematics and Dynamics of Quadrotor

In this section, the kinematic and dynamic differential equations of the quadrotor are established. The quadrotor can be considered as a rigid cross frame attached with four rotors, and the center of gravity coincides with the body-fixed frame origin.

The simplified model of the quadrotor is presented in Figure 1, rotors $R1$ and $R3$ rotate counterclockwise and rotors $R2$ and $R4$ rotate clockwise. Each propeller rotates at the angular speed $\Omega_i \in [\Omega_{i,min}, \Omega_{i,max}]$ and produces a force F_i (i = 1, 2, 3, 4) along the negative z-direction relative to the body frame [37,38]:

$$F_i = \begin{bmatrix} 0 \\ 0 \\ -k_T \Omega_i^2 \end{bmatrix} \quad (6)$$

where $k_T > 0$ denotes the aerodynamic coefficient which consists formed of the atmospheric density ρ, the radius of the propeller r, and the thrust coefficient c_T. In addition, due to the spinning of the rotors, a reaction torque M_i (i = 1, 2, 3, 4) is generated on the quadrotor body by each rotor:

$$M_i = \begin{bmatrix} 0 \\ 0 \\ (-1)^{i+1} k_D \Omega_i^2 \end{bmatrix} \quad (7)$$

where $k_D > 0$ denotes the drag coefficient of the rotor, which depends on the same factors as $k_T > 0$.

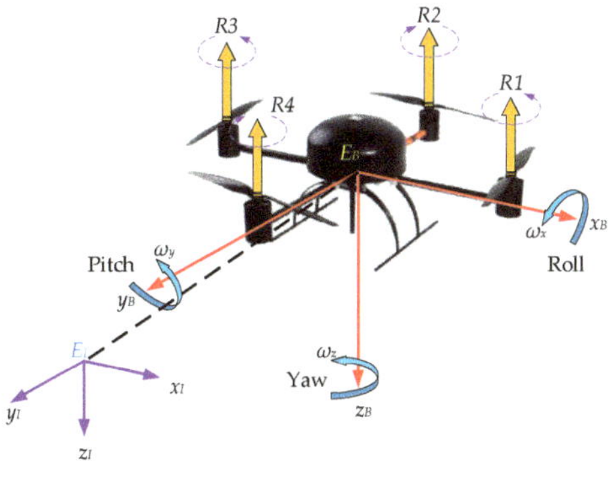

Figure 1. Coordinate systems of the quadrotor.

In the mathematical model of quadrotor, three coordinate frames are considered: the non-moving inertial frame $E_I : \{o_I, x_I, y_I, z_I\}$, the body-fixed frame $E_B : \{o_B, x_B, y_B, z_B\}$ and the desired frame $E_D : \{o_D, x_D, y_D, z_D\}$ to represent the actual attitude and desired attitude of quadrotor respectively. Note that North-East-Down (NED) coordinates are used to define all frames. Attitude angle and angular velocities of the body-fixed frame E_B with respect to the inertial frame E_I are written as $\Theta = \begin{bmatrix} \phi & \theta & \psi \end{bmatrix}^T$ and $\boldsymbol{\omega} = [\ \omega_x \quad \omega_y \quad \omega_z\ ]^T$ respectively, and the quaternion expression of the attitude is $\boldsymbol{q} = [\ q_0 \quad \boldsymbol{q}_v\ ]^T$.

The variation of the orientation is achieved by varying the angular speed of a specific rotor. The torque created around a particular axis with respect to the body-fixed frame is defined as follows:

$$\boldsymbol{u} = \begin{bmatrix} \tau_\phi \\ \tau_\theta \\ \tau_\psi \end{bmatrix} = \overbrace{\begin{bmatrix} 0 & -lk_T & 0 & lk_T \\ lk_T & 0 & -lk_T & 0 \\ -k_D & k_D & -k_D & k_D \end{bmatrix}}^{\boldsymbol{M}} \begin{bmatrix} \Omega_1^2 \\ \Omega_2^2 \\ \Omega_3^2 \\ \Omega_4^2 \end{bmatrix} \tag{8}$$

where l denotes the distance from the rotors to the center of mass and $\boldsymbol{u}$ represents the control signal to be designed.

Assuming a symmetric mass distribution of the quadrotor, the nominal inertia matrix $\boldsymbol{J} = \mathrm{diag}(J_x, J_y, J_z)$ is diagonal. With the disturbances $\boldsymbol{d} = [\ d_x \quad d_y \quad d_z\ ]^T$ caused by wind gust $\boldsymbol{d}_w$ and actuator faults $\boldsymbol{d}_u$ into consideration, the attitude dynamic model of the quadrotor can be obtained as the following differential equations:

$$\begin{bmatrix} J_x\dot{\omega}_x \\ J_y\dot{\omega}_y \\ J_z\dot{\omega}_z \end{bmatrix} + \begin{bmatrix} (J_z - J_y)\omega_y\omega_z \\ (J_x - J_z)\omega_x\omega_z \\ (J_y - J_z)\omega_x\omega_y \end{bmatrix} = \begin{bmatrix} \tau_\phi \\ \tau_\theta \\ \tau_\psi \end{bmatrix} + \begin{bmatrix} d_x \\ d_y \\ d_z \end{bmatrix} \tag{9}$$

According to Equation (4), we summarized the mathematical model of the quadrotor as:

$$\begin{cases} \dot{q} = \frac{1}{2} q \otimes \begin{bmatrix} 0 & \omega \end{bmatrix}^T \\ \dot{\omega} = -\boldsymbol{J}^{-1} S(\boldsymbol{\omega}) \boldsymbol{J}\boldsymbol{\omega} + \boldsymbol{J}^{-1}\boldsymbol{u} + \boldsymbol{J}^{-1}\boldsymbol{d} \end{cases} \tag{10}$$

In practice, we can use micro electro mechanical system (MEMS) inertial measurement unit (IMU) to measure the attitude information $\boldsymbol{\omega}$ and $\boldsymbol{q}$.

2.4. Lumped Disturbaces

2.4.1. Wind Gust

Wing gust produces a strong disturbance torque on the quadrotor. In this article, a Dryden wind gust model is introduced [39]. We assume that the disturbance caused by wind gust $\boldsymbol{d}_w$ is proportional to the speed of wind gust, therefore, $\boldsymbol{d}_w$ can be described based on the random theory [40] and defined as a summation of sinusoidal excitations:

$$d_{w,k}(t) = d_{w,k}^0 + \sum_{i=1}^{n_k} a_{i,k} \sin(\varpi_{i,k} t + \varphi_{i,k}) \tag{11}$$

where $d_{w,k}(t)$ is a time-dependent description of the wind disturbance in $k = x, y, z$ channel in a given time t. $\varpi_{i,k}$ and $\varphi_{i,k}$ are randomly selected frequencies and phase shifts, $n_{i,k}$ is the number of sinusoids, $a_{i,k}$ is the amplitude of the sinusoid, and $d_{w,k}^0$ is the static wind disturbance.

2.4.2. Actuator Faults

In this article, we consider actuator faults represented by partial loss of effectiveness in the rotors. For instance, caused by structural damage to a propeller [14], battery power loss [41], etc. Thus, the actuator faults in this article are modeled as follows, for $i = 1, ..., 4$:

$$\Omega_i^* = \alpha_i \Omega_i \tag{12}$$

where Ω_i represents the commanded rotor angular velocity, Ω_i^* is the loss of angular velocity, and $\alpha_i \in [0, \bar{\alpha})$ is an unknown ratio characterizing the occurrence of a partial loss of effectiveness fault in rotor i with $\bar{\alpha}$ being a known upper bound needed to maintain the controllability of the quadrotor. For instance, in the extreme case of complete failure $b_i = \bar{\alpha}$, the quadrotor becomes uncontrollable. The case of $b_i = 1$ represents a healthy rotor, and $0 \leq \alpha_i < \bar{\alpha} < 1$ represents a faulty rotor with partial loss of effectiveness.

$$\boldsymbol{d}_u = \boldsymbol{M} \begin{bmatrix} \alpha_1^2 \Omega_1^2 \beta_1 (t - T_1) \\ \alpha_2^2 \Omega_2^2 \beta_2 (t - T_2) \\ \alpha_3^2 \Omega_3^2 \beta_3 (t - T_3) \\ \alpha_4^2 \Omega_4^2 \beta_4 (t - T_4) \end{bmatrix} \tag{13}$$

where the fault time profile function $\beta_i(t - T_i)$ is assumed to be a step function with unknown fault occurrence time T_i for $i = 1, ..., 4$, that is:

$$\beta_i (t - T_i) = \begin{cases} 0, & if \quad t < T_i \\ 1, & if \quad t \geq T_i \end{cases} \tag{14}$$

In summary, we can see that the disturbances $\boldsymbol{d} = \boldsymbol{d}_w + \boldsymbol{d}_u$ acting on quadrotor are high-order, non-Gaussian and happen suddenly, furthermore, their randomness and nonlinearity are also very strong.

2.5. Problem Formulation

The purpose of this article is to achieve the high precision tracking to the desired attitude in presence of wind gust and actuator faults. Therefore, the dynamics of attitude error should be introduced. We use $\boldsymbol{\omega}_d = [\ \omega_{d,x} \quad \omega_{d,y} \quad \omega_{d,z}\ ]^T$ and $\boldsymbol{q}_d = [\ q_{0d} \quad \boldsymbol{q}_{vd}\ ]^T$ to denote the desired angular velocities and attitude respectively, thus the tracking error vector of the angular velocities $\boldsymbol{\omega}_e = [\ \omega_{e,x} \quad \omega_{e,y} \quad \omega_{e,z}\ ]^T$ can be expressed as:

$$\boldsymbol{\omega}_e = \boldsymbol{\omega} - \boldsymbol{C}_d^b \boldsymbol{\omega}_d \tag{15}$$

Then, we can obtain the dynamics of $\boldsymbol{\omega}_e$ according to Equations (3), (10) and (15):

$$\dot{\boldsymbol{\omega}}_e = S(\boldsymbol{\omega}_e)\boldsymbol{C}_d^b \boldsymbol{\omega}_d - \boldsymbol{C}_d^b \dot{\boldsymbol{\omega}}_d - \boldsymbol{J}^{-1} S(\boldsymbol{\omega}) \boldsymbol{J} \boldsymbol{\omega} + \boldsymbol{J}^{-1} \boldsymbol{u} + \boldsymbol{J}^{-1} \boldsymbol{d} \tag{16}$$

where $\boldsymbol{C}_d^b$ can be calculated according to Equations (2) and (5). And according to Equations (4), (5), and (15), we can obtain the kinematics of attitude tracking error:

$$\dot{\boldsymbol{q}}_e = \frac{1}{2} \boldsymbol{q}_e \otimes \begin{bmatrix} 0 & \boldsymbol{\omega}_e \end{bmatrix}^T = \frac{1}{2} \begin{bmatrix} -\boldsymbol{q}_{ve}^T \\ S(\boldsymbol{q}_{ve}) + q_{0e} \boldsymbol{I}_3 \end{bmatrix} \boldsymbol{\omega}_e \tag{17}$$

Therefore, the problem we try to tackle in this work is to design a continuous control law $\boldsymbol{u}$, which guarantees errors of attitude angles $\boldsymbol{q}_e$ and angular velocities $\boldsymbol{\omega}_e$ asymptotic converge to zero in the presence of the lumped disturbances $\boldsymbol{d}$.

Figure 2 illustrates the control structure that we designed. Based on the DUEA control methodology, the attitude tracking problem for quadrotor can be divided into two components:

- Design the feedforward loop so that the lumped disturbances are estimated by STESO and compensated this way.
- Design the feedback loop that regulates the orientation of quadrotor to track the desired attitude produced by the commander timely.

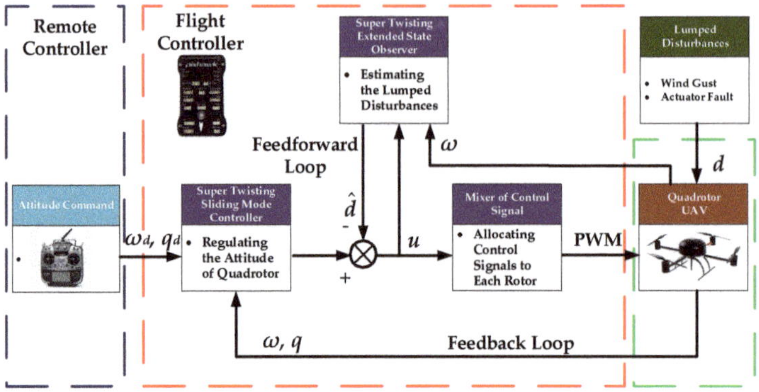

Figure 2. Block diagram of the proposed control scheme.

3. Design and Analysis of Super Twisting Extended State Observer

Through the analysis in previous section, we can see that the lumped disturbances acting on quadrotor are high-order and rapidly changing, therefore, it is difficult for traditional ESO to estimate them thoroughly [30]. Higher-order ESO have been applied in some articles. However, higher observer orders will lead to higher control gains with fixed bandwidth. Which will in return excite the sensor noise and introduce them into the control loop. By introducing super twisting algorithm, excessively high observer gain can be avoided. In this section, the STESO is proposed and the convergence analysis and parameter selection rule of STESO are given.

3.1. Design of STESO

Consider the dynamics Equation (10) of quadrotor, since the angular velocities $\boldsymbol{\omega}$ can be measured by the MEMS gyroscope, the original control input can be reformulated by employing the feedback linearization technique as:

$$\boldsymbol{u} = \boldsymbol{u}^* + S(\boldsymbol{\omega})\boldsymbol{J}\boldsymbol{\omega} \tag{18}$$

Therefore, we can obtain the linearized model of the quadrotor as $\boldsymbol{J}\dot{\boldsymbol{\omega}} = \boldsymbol{u}^* + \boldsymbol{d}$.

It is supposed that each component of the linearized model is independent from each other. Hence, the controller policy developed from one channel can be directly applied to the other two and the description of only the i-th channel is sufficient $(i = x, y, z)$. In this way, the one-dimensional dynamic of the quadrotor is obtained as:

$$J_i\dot{\omega}_i = \tau_i^* + d_i \tag{19}$$

Introducing a new state vector $\boldsymbol{\xi}_i = \begin{bmatrix} \xi_{1,i} & \xi_{2,i} \end{bmatrix}^T$ whose components are defined as $\xi_{1,i} = J_i\omega_i$ and augmenting lumped disturbances d_i as an extended state $\xi_{2,i} = d_i$, the reconstructed system is rewritten as:

$$\begin{cases} \dot{\xi}_{1,i} = \xi_{2,i} + \tau_i^* \\ \dot{\xi}_{2,i} = \delta_i \end{cases} \tag{20}$$

where δ_i is the derivative of d_i. Assume that the system states are bounded, then the existence of a constant f_i^+ is ensured such that the inequality $|\delta_i| < f_i^+$ holds for any time.

It can be verify that the pair $(\boldsymbol{A}, \boldsymbol{C})$ is observable. Then, consider $\hat{\boldsymbol{\xi}}_i$ as the estimation of $\boldsymbol{\xi}_i$, STESO can be designed as follows:

$$\begin{cases} \dot{\hat{\xi}}_{1,i} = \hat{\xi}_{2,i} + \tau_i^* + a_i|\xi_{1,i} - \hat{\xi}_{1,i}|^{\frac{1}{2}} \operatorname{sgn}(\xi_{1,i} - \hat{\xi}_{1,i}) \\ \dot{\hat{\xi}}_{2,i} = b_i \operatorname{sgn}(\xi_{1,i} - \hat{\xi}_{1,i}) \end{cases} \tag{21}$$

where a_i and b_i are the observer gains to be designed.

Define the estimation error variables $\tilde{\boldsymbol{\xi}}_i = \begin{bmatrix} \tilde{\xi}_{1,i} & \tilde{\xi}_{2,i} \end{bmatrix}^T$ as $\tilde{\boldsymbol{\xi}}_i = \boldsymbol{\xi}_i - \hat{\boldsymbol{\xi}}_i$, and the dynamics of the $\tilde{\boldsymbol{\xi}}_i$ can be obtained as follows:

$$\begin{cases} \dot{\tilde{\xi}}_{1,i} = \tilde{\xi}_{2,i} - a_i|\xi_{1,i} - \hat{\xi}_{1,i}|^{\frac{1}{2}} \operatorname{sgn}(\xi_{1,i} - \hat{\xi}_{1,i}) \\ \dot{\tilde{\xi}}_{2,i} = \delta_i - b_i \operatorname{sgn}(\xi_{1,i} - \hat{\xi}_{1,i}) \end{cases} \tag{22}$$

The dynamics of the estimation error, presented in Equation (22), have the form of a non-recursive exact robust differentiator. Therefore, the errors $\tilde{\xi}_{1,i}$ and $\tilde{\xi}_{2,i}$ will converge to zero in a finite time if the gains a_i and b_i are chosen appropriately. The convergence analysis and parameter selection rule will be demonstrated following.

3.2. Convergence Analysis and Parameter Selection Rule

Firstly, introduce a new state vector $\boldsymbol{\eta}_i = \begin{bmatrix} \eta_{1,i} & \eta_{2,i} \end{bmatrix}^T$ as $\eta_{1,i} = |\tilde{\xi}_{1,i}|^{\frac{1}{2}} \operatorname{sgn}(\tilde{\xi}_{1,i})$, $\eta_{2,i} = \tilde{\xi}_{2,i}$ and take the time derivative of $\boldsymbol{\eta}_i$, we have:

$$\begin{cases} \dot{\eta}_{1,i} = \frac{1}{2}|\tilde{\xi}_{1,i}|^{-\frac{1}{2}} \left(-a_i|\tilde{\xi}_{1,i}|^{\frac{1}{2}} \operatorname{sgn}(\tilde{\xi}_{1,i}) + \tilde{\xi}_{2,i} \right) \\ \dot{\eta}_{2,i} = -b_i \operatorname{sgn}(\tilde{\xi}_{1,i}) + \delta_i \end{cases} \tag{23}$$

which can be rewritten as:

$$\dot{\boldsymbol{\eta}}_i = |\tilde{\xi}_{1,i}|^{-\frac{1}{2}} \overbrace{\begin{bmatrix} \frac{-a_i}{2} & \frac{1}{2} \\ -b_i & 0 \end{bmatrix}}^{\boldsymbol{A}} \boldsymbol{\eta}_i + \overbrace{\begin{bmatrix} 0 \\ 1 \end{bmatrix}}^{\boldsymbol{B}} \delta_i \tag{24}$$

Then, introduce a positive definite matrix $\boldsymbol{P} = \frac{1}{2}\begin{bmatrix} 4b_i + a_i^2 & -a_i \\ -a_i & 2 \end{bmatrix}$ and consider the following Lyapunov function:

$$V_i = \boldsymbol{\eta}_i{}^T \boldsymbol{P} \boldsymbol{\eta}_i \tag{25}$$

Notice that in Equation (25), V_i is continuous but is not differentiable at $\tilde{\xi}_{1,i} = 0$, and it is positive definite and radially unbounded if $b_i > 0$, thus we have:

$$\lambda_{\min}(\boldsymbol{P}) \|\boldsymbol{\eta}_i\|^2 \le V_i \le \lambda_{\max}(\boldsymbol{P}) \|\boldsymbol{\eta}_i\|^2 \tag{26}$$

Take the time derivative of V_i and define $\boldsymbol{Q} = \boldsymbol{A}^T\boldsymbol{P} + \boldsymbol{P}\boldsymbol{A}$, we have:

$$\begin{aligned} \dot{V}_i &= |\tilde{\xi}_{1,i}|^{-\frac{1}{2}} \boldsymbol{\eta}_i{}^T \left(\boldsymbol{A}^T\boldsymbol{P} + \boldsymbol{P}\boldsymbol{A} \right) \boldsymbol{\eta}_i + \delta_i \left(\boldsymbol{B}^T\boldsymbol{P}\boldsymbol{\eta}_i + \boldsymbol{\eta}_i{}^T\boldsymbol{P}\boldsymbol{B} \right) \\ &\le -|\tilde{\xi}_{1,i}|^{-\frac{1}{2}} \boldsymbol{\eta}_i{}^T \boldsymbol{Q} \boldsymbol{\eta}_i + 2f_i^+ \boldsymbol{B}^T\boldsymbol{P}\boldsymbol{\eta}_i \end{aligned} \tag{27}$$

where $|\delta_i| < f_i^+$ and $\boldsymbol{Q} = \frac{a_i}{2}\begin{bmatrix} 2b_i + a_i^2 & -a_i \\ -a_i & 1 \end{bmatrix}$ is positive defined. Notice that:

$$
\begin{aligned}
2f_i^{+}\boldsymbol{B}^T\boldsymbol{P}\boldsymbol{\eta_i} &= |\tilde{\xi}_{1,i}|^{-\frac{1}{2}} f_i^{+}\left(|\tilde{\xi}_{1,i}|^{\frac{1}{2}}(-a_i\eta_{1,i}+2\eta_{2,i})\right) \\
&= |\tilde{\xi}_{1,i}|^{-\frac{1}{2}} f_i^{+}\left(|\eta_{1,i}|(-a_i\eta_{1,i}+2\eta_{2,i})\right) \\
&\le |\tilde{\xi}_{1,i}|^{-\frac{1}{2}} f_i^{+}\left(a_i|\eta_{1,i}|^2+2|\eta_{1,i}||\eta_{2,i}|\right) \\
&\le |\tilde{\xi}_{1,i}|^{-\frac{1}{2}} f_i^{+}\left(a_i|\eta_{1,i}|^2+|\eta_{1,i}|^2+|\eta_{2,i}|^2\right) \\
&= |\tilde{\xi}_{1,i}|^{-\frac{1}{2}} \boldsymbol{\eta_i}^T \begin{bmatrix} f_i^{+}(a_i+1) & 0 \\ 0 & f_i^{+} \end{bmatrix} \boldsymbol{\eta_i} \\
&= -|\tilde{\xi}_{1,i}|^{-\frac{1}{2}} \boldsymbol{\eta_i}^T \Delta\boldsymbol{Q}\boldsymbol{\eta_i}
\end{aligned} \tag{28}
$$

Thus according to Equations (27) and (28), we have:

$$
\dot{V}_i \le -|\tilde{\xi}_{1,i}|^{-\frac{1}{2}} \boldsymbol{\eta_i}^T (\boldsymbol{Q}+\Delta\boldsymbol{Q})\boldsymbol{\eta_i} \tag{29}
$$

where $\Delta\boldsymbol{Q} = -\begin{bmatrix} f_i^{+}(a_i+1) & 0 \\ 0 & f_i^{+} \end{bmatrix}$ and $\boldsymbol{Q}+\Delta\boldsymbol{Q} = \frac{a_i}{2}\begin{bmatrix} 2b_i + a_i^2 - 2f_i^{+}\frac{a_i+1}{a_i} & -a_i \\ -a_i & 1-\frac{2f_i^{+}}{a_i} \end{bmatrix}$.

From Equation (29), we can find that $\dot{V}_i$ is negative definite on condition that $\boldsymbol{Q}+\Delta\boldsymbol{Q}$ is positive definite, what is exactly the case if:

$$
\begin{cases} a_i > 2f_i^{+} \\ b_i > f_i^{+}\dfrac{a_i^2}{a_i - 2f_i^{+}} + f_i^{+}\dfrac{a_i+1}{a_i} \end{cases} \tag{30}
$$

Then, analyze the finite time convergence of $\boldsymbol{\eta_i}$, according to Equation (26), we have:

$$
|\tilde{\xi}_{1,i}|^{\frac{1}{2}} < \|\boldsymbol{\eta_i}\| \le \frac{V_i^{\frac{1}{2}}}{\lambda_{\min}^{\frac{1}{2}}(\boldsymbol{P})} \tag{31}
$$

And according to Equations (29) and (31), we can conclude that:

$$
\begin{aligned}
\dot{V}_i &\le -\lambda_{\min}^{\frac{1}{2}}(\boldsymbol{P})\,\lambda_{\min}(\boldsymbol{Q}+\Delta\boldsymbol{Q})\,\|\boldsymbol{\eta_i}\|^2 V_i^{-\frac{1}{2}} \\
&\le -\frac{\lambda_{\min}^{\frac{1}{2}}(\boldsymbol{P})\lambda_{\min}(\boldsymbol{Q}+\Delta\boldsymbol{Q})}{\lambda_{\max}(\boldsymbol{P})} V_i^{\frac{1}{2}} \\
&= -\gamma V_i^{\frac{1}{2}}
\end{aligned} \tag{32}
$$

where $\gamma = \frac{\lambda_{\min}^{\frac{1}{2}}(\boldsymbol{P})\lambda_{\min}(\boldsymbol{Q}+\Delta\boldsymbol{Q})}{\lambda_{\max}(\boldsymbol{P})} > 0$.

Indeed, separating variables and integrating inequality Equation (32) over the time interval $0 < \tau < t < 0$, we obtain:

$$
V_i^{\frac{1}{2}}(t) \le -\frac{1}{2}\gamma t + V_{i,0}^{\frac{1}{2}} \tag{33}
$$

where $V_{i,0}$ is the initial value of $V_i(t)$. Consequently, $V_i(t)$ reaches zero in a finite time T_r that is bounded by:

$$
T_r = 2V_{i,0}^{\frac{1}{2}}\gamma^{-1} \tag{34}
$$

Therefore, accoding to [42], a STESO which is designed to satisfy Equations (21) and (30) will drive the uniformed vector of errors $\boldsymbol{\eta_i}$ and then $\tilde{\boldsymbol{\xi}}_i$ to zero in finite time T_r and will keep it at zero thereafter.

4. Design of Super Twisting Sliding Mode Controller

Sliding mode control (SMC) has been known as one of the most important tools for those systems subjected to disturbances and uncertainties, while chattering is inevitable in those methods. In order to reduce the chattering, supper twisting SMC is introduced in this section. The main objective of the FC is to guarantee that the state of attitude q and ω converge to the reference values q_d and ω_d timely. Thus, the sliding mode manifold in this article is chosen as follows:

$$\boldsymbol{s} = \boldsymbol{k}_1 \boldsymbol{q}_e + \boldsymbol{J}\boldsymbol{\omega}_e \tag{35}$$

where $\boldsymbol{k}_1 = \mathrm{diag}(k_{1,x}, k_{1,y}, k_{1,z})$ is a positive defined three dimensional coefficient matrix to be designed. Take time derivative of $\boldsymbol{s} = \begin{bmatrix} s_x & s_y & s_z \end{bmatrix}^T$, we have:

$$\dot{\boldsymbol{s}} = \boldsymbol{k}_1 \dot{\boldsymbol{q}}_e + \boldsymbol{J}\dot{\boldsymbol{\omega}}_e \tag{36}$$

Then, submitting Equation (16) into (36):

$$\dot{\boldsymbol{s}} = \boldsymbol{k}_1 \dot{\boldsymbol{q}}_e + \boldsymbol{u} + \boldsymbol{d} + \boldsymbol{J}(S(\boldsymbol{\omega}_e)\boldsymbol{C}_d^b \boldsymbol{\omega}_d - \boldsymbol{C}_d^b \dot{\boldsymbol{\omega}}_d) - S(\boldsymbol{\omega})\boldsymbol{J}\boldsymbol{\omega} \tag{37}$$

Define the control signal $\boldsymbol{u}$ as:

$$\boldsymbol{u} = -\left(\boldsymbol{J}(S(\boldsymbol{\omega}_e)\boldsymbol{C}_d^b \boldsymbol{\omega}_d - \boldsymbol{C}_d^b \dot{\boldsymbol{\omega}}_d) - S(\boldsymbol{\omega})\boldsymbol{J}\boldsymbol{\omega}\right) - \boldsymbol{k}_1 \dot{\boldsymbol{q}}_e - \boldsymbol{k}_2 \,\mathrm{sig}\,(\boldsymbol{s})^{\frac{1}{2}} - \int_0^t \boldsymbol{k}_3 \,\mathrm{sig}\,(\boldsymbol{s})^0 d\tau - \hat{\boldsymbol{d}} \tag{38}$$

and plug Equation (38) into (37), we have:

$$\dot{\boldsymbol{s}} = -\boldsymbol{k}_2 \,\mathrm{sig}\,(\boldsymbol{s})^{\frac{1}{2}} - \int_0^t \boldsymbol{k}_3 \,\mathrm{sig}\,(\boldsymbol{s})^0 d\tau + \tilde{\boldsymbol{d}} \tag{39}$$

where $\tilde{\boldsymbol{d}} = \boldsymbol{d} - \hat{\boldsymbol{d}}$ is the estimation error of multiple disturbances, and according to the analysis of the previous section, $\tilde{\boldsymbol{d}}$ is bounded and converges to zero in finite time. Define $\boldsymbol{\sigma} = \begin{bmatrix} \sigma_x & \sigma_y & \sigma_z \end{bmatrix}^T$ as $\boldsymbol{\sigma} = -\int_0^t \boldsymbol{k}_3 \,\mathrm{sig}\,(\boldsymbol{s})^0 d\tau + \tilde{\boldsymbol{d}}$, then Equation (39) can be rewritten as:

$$\begin{cases} \dot{s}_i = \sigma_i - k_{2,i}\|s_i\|^{\frac{1}{2}} \,\mathrm{sgn}\,(s_i) \\ \dot{\sigma}_i = \dot{\tilde{d}}_i - k_{3,i} \,\mathrm{sgn}\,(s_i)\, d\tau \end{cases}, \quad i = x, y, z \tag{40}$$

where the positive defined matrix $\boldsymbol{k}_2 = \mathrm{diag}(k_{2,x}, k_{2,y}, k_{2,z})$ and $\boldsymbol{k}_3 = \mathrm{diag}(k_{3,x}, k_{3,y}, k_{3,z})$ are the controller parameters to be determined.

Subject to the restriction of article length, the convergence analysis and parameter selection rule of the STSMC will not be introduced in detail. Since Equation (40) has the same form as Equation (22), the detailed convergence analysis can refer to the contents of the previous section. Meanwhile according to Equation (30), the controller parameters can be chosen as:

$$\begin{cases} k_{2,i} > 2\left\|\dot{\tilde{d}}_i\right\| \\ k_{3,i} > \left\|\dot{\tilde{d}}_i\right\| \frac{k_{2,i}^2}{k_{2,i} - 2\left\|\dot{\tilde{d}}_i\right\|} + \left\|\dot{\tilde{d}}_i\right\| \frac{k_{2,i}+1}{k_{2,i}} \end{cases} \tag{41}$$

5. Simulation and Experimental Results

In order to evaluate the performance of the proposed control method, numerical simulation and real world experimental results are carried out in this section.

5.1. Simulation Results

We present the numerical simulations of the proposed STESO based DUEA control strategy on a model generated by the online toolbox of Quan and Dai [43], and the values of the nominal model parameters are list in Table 1.

Table 1. Quadrotor parameters used in simulation.

Parameter	Description	Value
m	Mass	1.79 kg
J_x	Roll inertia	1.335×10^{-2} kg · m^2
J_y	Pitch inertia	1.335×10^{-2} kg · m^2
J_z	Yaw inertia	2.465×10^{-2} kg · m^2
l	Motor moment arm	0.18 m
g	Gravity acceleration	9.81 m · s^{-2}
k_T	Aerodynamic coefficient	8.82×10^{-6} N/(rad/s)^2
k_D	Drag coefficient	1.09×10^{-7} N · m/(rad/s)^2
$\Omega_{i,max}$	Maximum rotational speed	8214 r/min
$\Omega_{i,min}$	Minimum rotational speed	100 r/min

In numerical simulations, the position of quadrotor is free and only the attitude of it is controlled. We assume that the rotor $R1$ fails in the 6th second, and loses 20% of effectiveness, which means $T_1 = 6$ s and $\alpha_1 = 0.2$ in Equation (13). Then, according to [15], the disturbance torque caused by the wind field is proportional to the wind speed, and we assume that the three-axis components of $\boldsymbol{d_w} = [\ d_{w,x}\quad d_{w,y}\quad d_{w,z}\ ]^T$ are equal $d_x = d_y = d_z = d_w$ without loss of generality. The values of $\varpi_{k,i}$ are taken between 0.01π rad/s and 2.5π rad/s. The disturbance torque of wind gust used in numerical simulation is Equation (42).

$$\begin{aligned} d_w = \ & 0.01\sin(2.5\pi t - 3) + 0.02\sin(2\pi t + 7) + 0.06\sin(\pi t + 0.6) \\ & + 0.03\sin(0.5\pi t - 9.5) + 0.02\sin(0.3\pi t) + 0.12\sin(0.1\pi t + 4.5) \\ & + 0.01\sin(0.05\pi t + 2) + 0.003\sin(0.01\pi t + 3) + 0.05 \end{aligned} \quad (42)$$

The numerical simulation is carried out in MatLab/Simulink with a fixed-sampling time of 1 ms. And to validate the performance of the proposed control strategy, two simulation cases are presented in this part. The initial conditions of the attitude angles and angular velocities are set to zero, and the desired reference commands are selected as:

$$\Theta_{ref} = \begin{bmatrix} 15\sin(0.4\pi t) & 15\cos(0.4\pi t) & 0 \end{bmatrix}^T \text{deg} \quad (43)$$

5.1.1. Case A: STESO vs. 2nd-Order ESO

In order to verify the enhancement of STESO relative to traditional Higher-order ESO, three comparative simulations are conducted on condition that use nonlinear PD controller as the FC. The controller gains are chosen as $\boldsymbol{K_1} = \boldsymbol{I_3}$ and $\boldsymbol{K_2} = 5\boldsymbol{I_3}$ [44], the observer gains of STESO are chosen as $a_i = 24$ and $b_i = 50$, and the bandwidth of 2nd-order ESO is chosen as 10 rad/s, i.e., the observer gains are $\boldsymbol{L} = [\ 30\quad 300\quad 1000\ ]^T$ [45].

Figures 3–5 show the comparison in attitude tracking results of nonlinear PD controller with STESO, 2nd-order ESO and without DUE. From those figures, we can see that the desired attitude commands can be tracked effectively by the controller with DUE. Moreover, the tracking errors are further reduced by introducing STESO as the DUE instead of 2nd-order ESO. Figure 6 shows the comparison in lumped disturbances estimation results of STESO and 2nd-order ESO. It is obvious that compared with STESO, some phase delay exist in the estimation results of the 2nd-order ESO, which leads to its estimation error convergences into a bounded area. Meanwhile, the estimation errors of STESO almost asymptotically convergence to zero. Especially when the disturbance torque suddenly changes, STESO has more advantages. In general, from the numerical simulation results, we can be conclude that STESO has a higher disturbance estimation accuracy, which in turn improves attitude control accuracy.

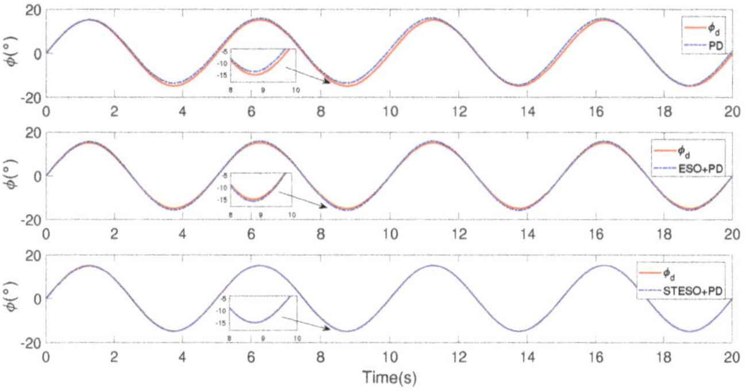

Figure 3. Simulation curves of ϕ in Case A.

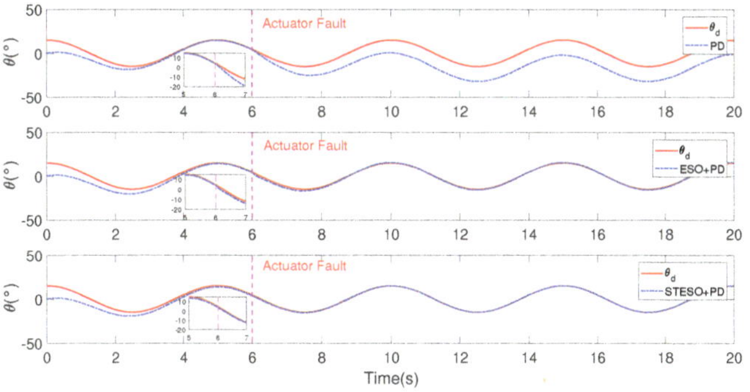

Figure 4. Simulation curves of θ in Case A.

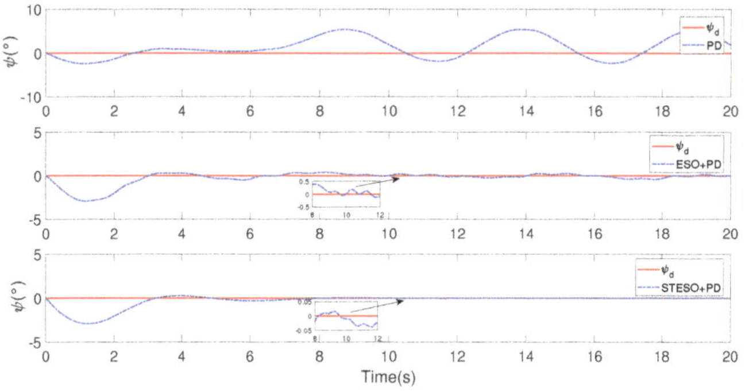

Figure 5. Simulation curves of ψ in Case A.

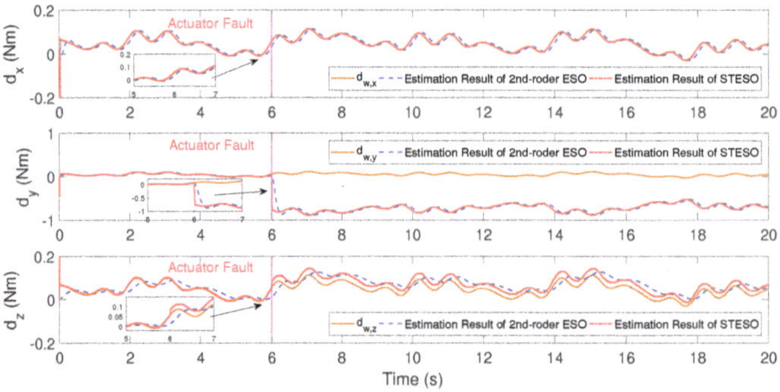

Figure 6. Disturbance estimation results in Case A.

5.1.2. Case B. STSMC vs. Nonlinear PD

In order to verify the fast convergence of STSMC, two comparative simulations are performed with the same STESO of different controllers. The controller gains of STSMC are chosen as $\boldsymbol{k}_1 = \text{diag}(1,1,1)$, $\boldsymbol{k}_2 = \text{diag}(5,5,5)$, $\boldsymbol{k}_3 = \text{diag}(20,20,20)$, the STESO and nonlinear PD parameters are the same with those provided in previous.

The comparison in attitude tracking results between STSMC and Nonlinear PD are illustrated in Figure 7. From this figure, we can see that quickly convergence of the attitude of quadrotor can be achieved by using STSMC as the FC. And by introducing ST algorithm into SMC, the chattering is reduced.

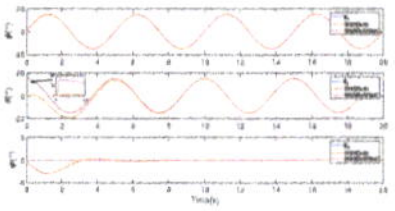

Figure 7. Simulation curves of attitude angle: STSMC vs. nonlinear PD.

5.2. Experimental Results

In order to evaluate the effectiveness of the developed algorithm in practical applications, we have also tested the proposed control scheme on a selfassembled GF360 quadrotor, where an open-source flight controller PIXHAWK [46,47] was used as the autopilot of the quadrotor.

5.2.1. Case A: STESO vs. 2nd-Order ESO

In this case, quadrotor is freely flying and we mainly aim to achieve the fast stabilization of quadrotor attitude on condition that actuator faults occur. According to the simulation results, we can conclude that STESO algorithm has advantage in quick response to the lumped disturbances. And in order to verify its effectiveness in actual flight, three comparative real time experiments are conducted to handle the sudden lose of rotor effectiveness. The experimental setup is shown in Figure 8, and freely flying is performed. As it is dangerous to damage the propeller during flight, we use software to set up a sudden lose of rotor effectiveness in pitch channel at the 10th second. We choose the traditional nonlinear PD controller as the FC in these experiments, where the performances of the PD controller with STESO, 2nd-order ESO or without DUE are compared. The gains of the PD controller are chosen as $\boldsymbol{k}_1 = \text{diag}(7,7,2.8)$ and $\boldsymbol{k}_2 = \text{diag}(0.15,0.15,2)$, the gains of the STESO are $a_i = 1$ and $b_i = 0.24$, the bandwidth of the 2nd-order ESO is 4 rad/s, i.e., the observer gains are $\boldsymbol{L} = [\ 12 \quad 48 \quad 64\ ]^T$.

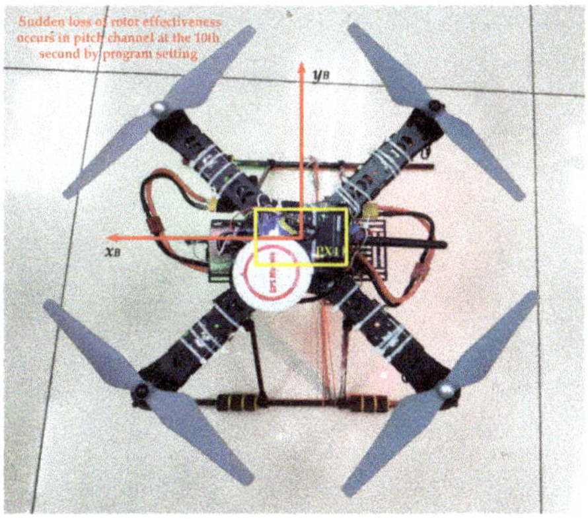

Figure 8. Experimental setup of the quadrotor hovering with sudden lose of rotor effectiveness in Case A.

The experiment curves of the attitude are plotted in Figures 9 and 10. From Figure 9, we can see that when the same loss of rotor effectiveness occurs, the deflections of pitch angle and angular rate in PD controller with STESO method are the smallest and the recovery times are the shortest. Figure 11 show the disturbances estimate curves of proposed STESO and 2nd-order ESO respectively. It can be seen that the convergence time of STESO is shorter that 2nd-order ESO. In general, the comparison of the experiment results are list detailly in Table 2. Furthermore, the corresponding control torques are shown in Figure 12.

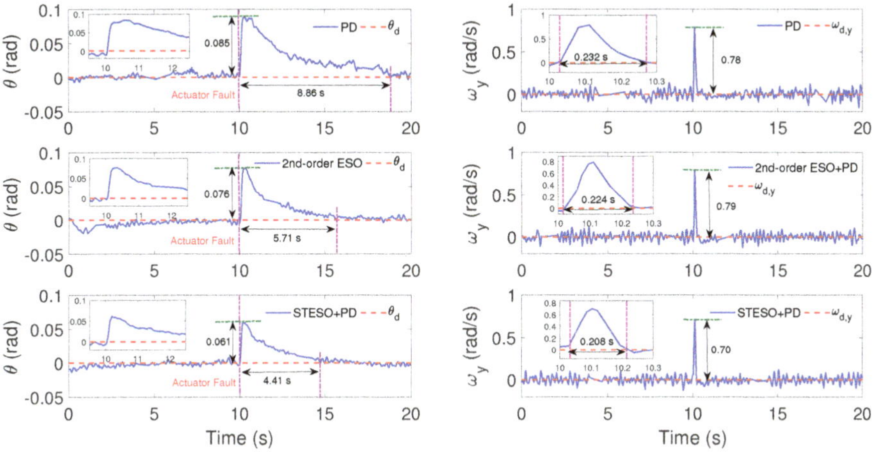

Figure 9. Experimental curves of θ and ω_y with different DUE in Case A.

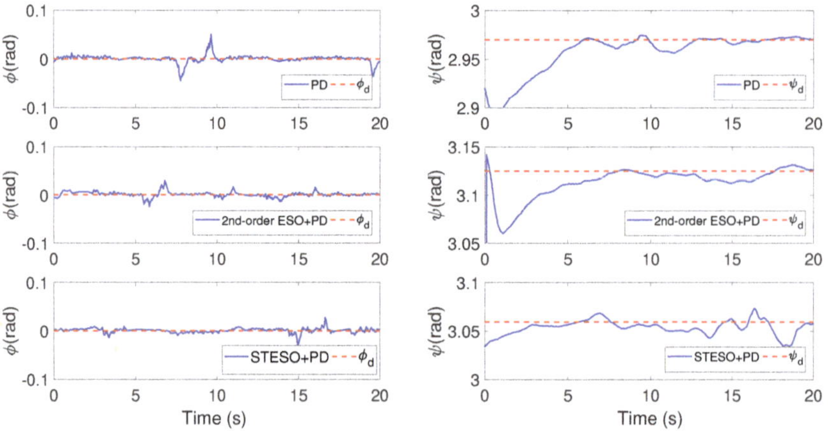

Figure 10. Experimental curves of ϕ and ψ with different DUE in Case A.

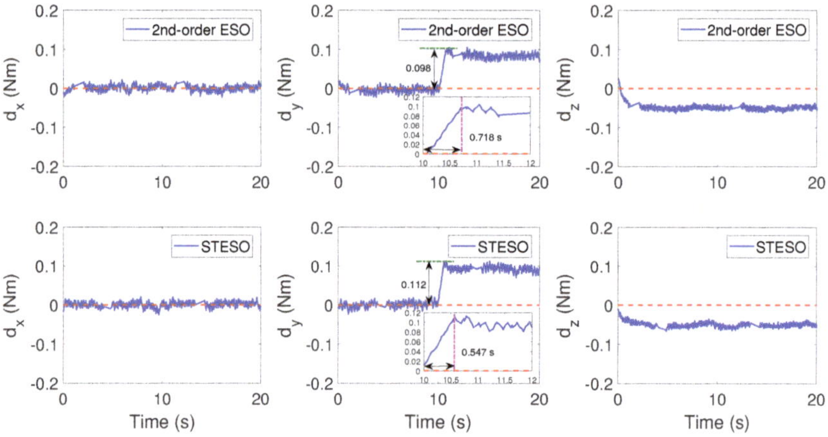

Figure 11. Disturbance estimation results of different DUE with actuator faults in Case A.

Table 2. Comparison of control performances with different observers.

	Deflection		**Recovery Time**		**Convergence Time**
	θ	ω_y	θ	ω_y	
Without Estimator	0.085 rad	0.78 rad/s	8.86 s	0.232 s	/
2nd-order ESO + PD	0.076 rad	0.79 rad/s	5.71 s	0.224 s	0.718 s
STESO + PD	0.061 rad	0.70 rad/s	4.41 s	0.208 s	0.547 s

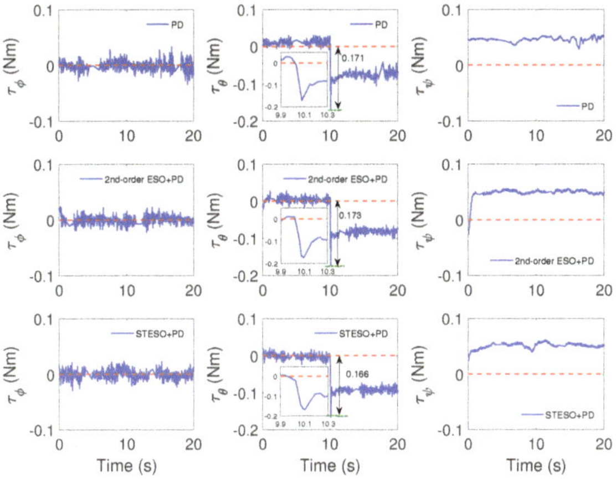

Figure 12. Control input during flight experiments in Case A.

5.2.2. Case B: Proposed Method vs. Nonlinear PD

The main propose in this case is to show the performance of the developed method for quadrotor subject to lumped disturbances such as wind disturbance and actuator fault. As shown in Figure 13, We install a damaged propeller on the $R1$ to perform the fault of the actuator, and then keep the quadrotor hovering in wind gust by remote control. In order to ensure the same experimental conditions, our experiments is run in a controlled indoor environment. We use an electrical fan with adjustable wind speed to generate the disurbance torque acting on the pitch channel of quadrotor. The average wind speed is around 4.5 m/s and turn on the electric fan at the 30 s. The experiments are carried out in our lab without GPS signals.

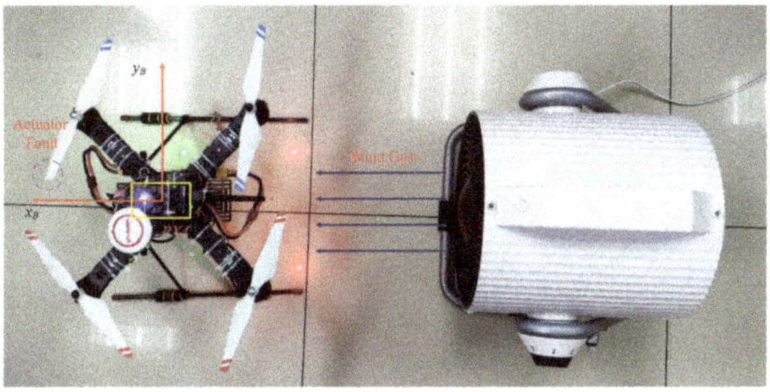

Figure 13. Experimental setup of the quadrotor hovering in the wind field in Case B.

Figure 14 shows the estimation results of the STESO. From this figure, we can see that the actuator fault in $R1$ leads to a steady disturbance torque acting on the quadrotor in hovering flight, i.e., $d_{u,x} \approx 0.04$ Nm, $d_{u,y} \approx -0.03$ Nm and $d_{u,z} \approx -0.07$ Nm. The wind gust mainly leads to the stochastic disturbance torque in each channel, plus a steady torque about -0.01 Nm in pitch channel. The attitude control results in this case are shown in Figures 15 and 16. It can be observed that the control performance is improved by introducing the proposed method compared with the nonlinear PD controller. In addition, the root mean square (RMS) errors of the attitude angles obtained by the proposed controller and PD controller are list in Table 3. And Figure 17 illustrates the control inputs in each channel.

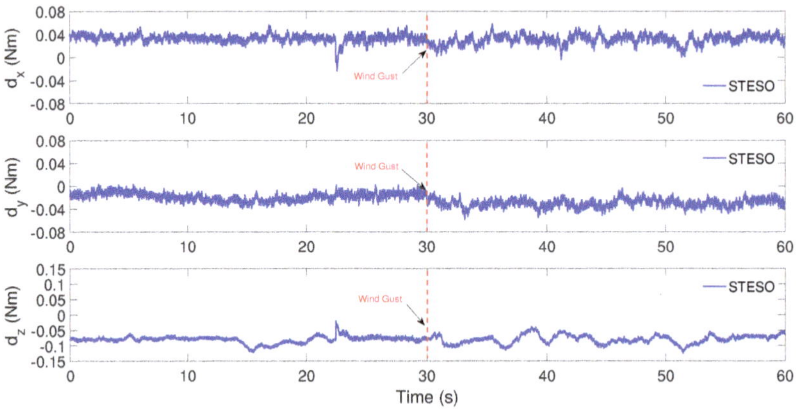

Figure 14. Disturbance estimation results of STESO in Case B.

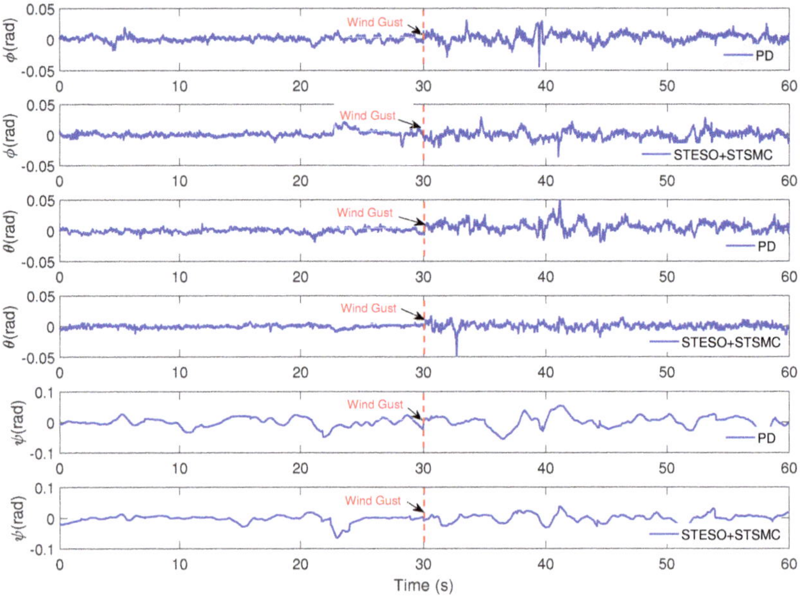

Figure 15. Experimental curves of attitude angle ϕ, θ, ψ in Case B.

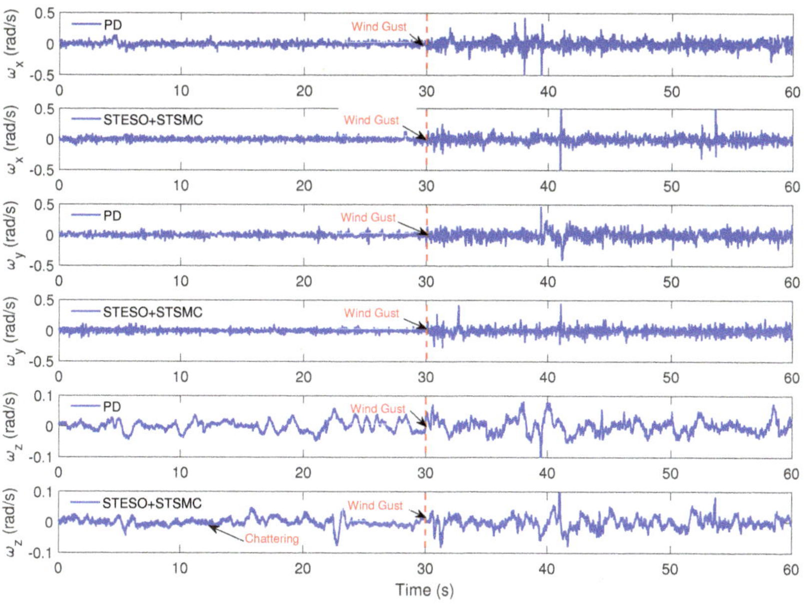

Figure 16. Experimental curves of angular rates ω_x, ω_y, ω_z in Case B.

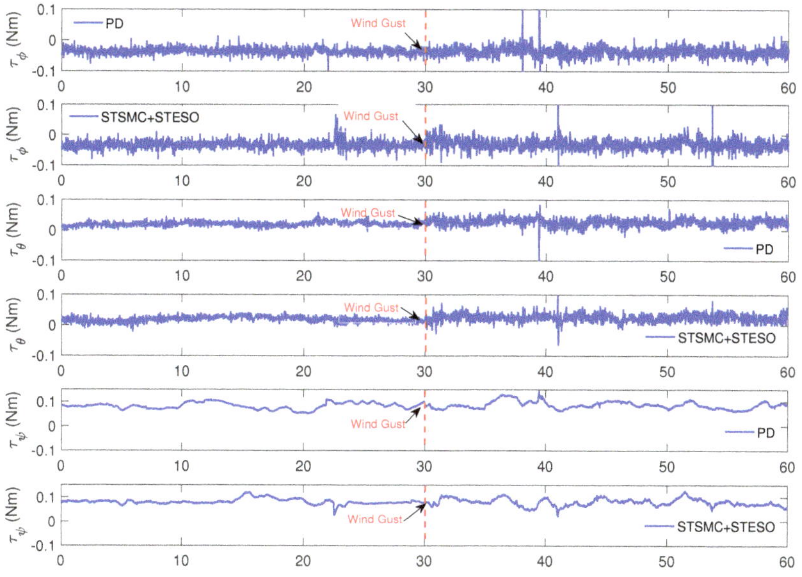

Figure 17. Control input during flight experiments in Case B.

Table 3. Comparison of attitude control performance: RMS error (rad).

	Actuator Fault Only			Actuator Fault+Wind Gust		
	ϕ	θ	ψ	ϕ	θ	ψ
nonlinear PD	0.0082	0.0072	0.0187	0.0160	0.0202	0.0204
STESO + STSMC	0.0094	0.0052	0.0135	0.0136	0.0116	0.0132

6. Conclusions

In this paper, the problem of high precision attitude tracking for quadrotor in the presence of wind gust and actuator fault is investigated. In order to estimate and attenuate the disturbances timely and accurately, a STESO is proposed and successfully implemented as the DUE in experiments. Also, a STSMC is designed as the FC to drive the attitude angle and angular velocity to their desired value in finite time. From the comparative simulation and experiment results, we can conclude that when the parameter selection rule given in this article of is satisfied, the proposed super-twisting algorithm based controller can relize the fast converge to the desired attitude precisely with less chattering. And compared with the traditional Higher-order ESO, STESO has a higher disturbance estimation accuracy, which in turn improves attitude control accuracy.

Supplementary Materials: The Simulinlk simulation algorithm is available online at https://pan.baidu.com/s/1I-OVIDoOKMJhaNP-KbFjsA, password: 113i.

Author Contributions: Conceive and design the algorithm, Z.W. and D.S.; Perform the experiments, D.S.; Software, D.S.; Project administration, Z.W. and W.C.; Supervision, Z.W.

Funding: This work was funded by the State Key Development Program for Basic Research of China (Grant No. 2013CB035503).

Conflicts of Interest: The authors declare no conflicts of interest.

References

1. Stepaniak, M.J.; van Graas, F.; de Haag, M.U. Design of an Electric Propulsion System for a Quadrotor Unmanned Aerial Vehicle. *J. Aircr.* **2009**, *46*, 1050–1058. [CrossRef]
2. Gupte, S.; Mohandas, P.I.T.; Conrad, J.M. A Survey of Quadrotor Unmanned Aerial Vehicles. In Proceedings of the IEEE Southeastcon, Orlando, FL, USA, 15–18 March 2012 .
3. Ozbek, N.S.; Onkol, M.; Efe, M.O. Feedback control strategies for quadrotor-type aerial robots: A survey. *Trans. Inst. Meas. Control* **2016**, *38*, 529–554. [CrossRef]
4. Bouabdallah, S.; Becker, M.; Siegwart, R. Autonomous miniature flying robots: Coming soon! *IEEE Robot Autom Mag.* **2007**, *14*, 88–98. [CrossRef]
5. Han, W.X.; Wang, Z.H.; Shen, Y. Fault estimation for a quadrotor unmanned aerial vehicle by integrating the parity space approach with recursive least squares. *Proc. Inst. Mech. Eng. Part G J. Aerosp. Eng.* **2018**, *232*, 783–796. [CrossRef]
6. Tian, B.L.; Ma, Y.X.; Zong, Q. A Continuous Finite-Time Output Feedback Control Scheme and Its Application in Quadrotor UAVs. *IEEE Access* **2018**, *6*, 19807–19813. [CrossRef]
7. Huang, M.; Xian, B.; Diao, C.; Yang, K.Y.; Feng, Y. Adaptive Tracking Control of Underactuated Quadrotor Unmanned Aerial Vehicles via Backstepping. In Proceedings of the 2010 American Control Conference, Baltimore, MD, USA, 30 June–2 July 2010; pp. 2076–2081.
8. Salih, A.L.; Moghavvemi, M.; Mohamed, H.A.F.; Gaeid, K.S. Flight PID controller design for a UAV quadrotor. *Sci. Res. Essays* **2010**, *5*, 3660–3667.
9. Lee, K.U.; Kim, H.S.; Park, J.B.; Choi, Y.H. Hovering Control of a Quadrotor. In Proceedings of the 2012 12th International Conference on Control, Automation and Systems (Iccas), JeJu Island, Korea, 17–21 October 2012; pp. 162–167.
10. González-Vázquez, S.; Moreno-Valenzuela, J. A New Nonlinear PI/PID Controller for Quadrotor Posture Regulation. In Proceedings of the 2010 IEEE Electronics, Robotics and Automotive Mechanics Conference, Morelos, Mexico, 28 September–1 October 2010; pp. 642–647.

11. González-Vázquez, S.; Moreno-Valenzuela, J. Motion Control of a Quadrotor Aircraft via Singular Perturbations. *Int. J. Adv. Robot. Syst.* **2013**, *10*, 368. [CrossRef]
12. Liu, H.; Lu, G.; Zhong, Y.S. Robust LQR Attitude Control of a 3-DOF Laboratory Helicopter for Aggressive Maneuvers. *IEEE Trans. Ind. Electron.* **2013**, *60*, 4627–4636. [CrossRef]
13. Munoz, F.; Gonzalez-Hernandez, I.; Salazar, S.; Espinoza, E.S.; Lozano, R. Second order sliding mode controllers for altitude control of a quadrotor UAS: Real-time implementation in outdoor environments. *Neurocomputing* **2017**, *233*, 61–71. [CrossRef]
14. Avram, R.C.; Zhang, X.D.; Muse, J. Nonlinear Adaptive Fault-Tolerant Quadrotor Altitude and Attitude Tracking With Multiple Actuator Faults. *IEEE Trans. Control Syst. Technol.* **2018**, *26*, 701–707. [CrossRef]
15. Wang, C.; Song, B.F.; Huang, P.F.; Tang, C.H. Trajectory Tracking Control for Quadrotor Robot Subject to Payload Variation and Wind Gust Disturbance. *J. Intell. Robot. Syst.* **2016**, *83*, 315–333. [CrossRef]
16. Basri, M.A.M.; Husain, A.R.; Danapalasingam, K.A. Stabilization and trajectory tracking control for underactuated quadrotor helicopter subject to wind-gust disturbance. *Sadhana Acad. Proc. Eng. Sci.* **2015**, *40*, 1531–1553. [CrossRef]
17. Chen, W.H.; Ohnishi, K.; Guo, L. Advances in Disturbance/Uncertainty Estimation and Attenuation. *IEEE Trans. Ind. Electron.* **2015**, *62*, 5758–5762. [CrossRef]
18. Yang, J.; Chen, W.H.; Li, S.H.; Guo, L.; Yan, Y.D. Disturbance/Uncertainty Estimation and Attenuation Techniques in PMSM Drives-A Survey. *IEEE Trans. Ind. Electron.* **2017**, *64*, 3273–3285. [CrossRef]
19. Wang, Z.; Wu, Z. Nonlinear attitude control scheme with disturbance observer for flexible spacecrafts. *Nonlinear Dyn.* **2015**, *81*, 257–264. [CrossRef]
20. Wang, Z.; Wu, Z.; Du, Y.J. Adaptive sliding mode backstepping control for entry reusable launch vehicles based on nonlinear disturbance observer. *Proc. Inst. Mech. Eng. Part G J. Aerosp. Eng.* **2016**, *230*, 19–29. [CrossRef]
21. Pu, Z.Q.; Yuan, R.Y.; Yi, J.Q.; Tan, X.M. A Class of Adaptive Extended State Observers for Nonlinear Disturbed Systems. *IEEE Trans. Ind. Electron.* **2015**, *62*, 5858–5869. [CrossRef]
22. Wang, H.D.; Huang, Y.B.; Xu, C. ADRC Methodology for a Quadrotor UAV Transporting Hanged Payload. In Proceedings of the 2016 IEEE International Conference on Information and Automation (ICIA), Ningbo, China, 1–3 August 2016; pp. 1641–1646.
23. Xia, Y.Q.; Pu, F.; Li, S.F.; Gao, Y. Lateral Path Tracking Control of Autonomous Land Vehicle Based on ADRC and Differential Flatness. *IEEE Trans. Ind. Electron.* **2016**, *63*, 3091–3099. [CrossRef]
24. Dou, J.X.; Kong, X.X.; Wen, B.C. Altitude and attitude active disturbance rejection controller design of a quadrotor unmanned aerial vehicle. *Proc. Inst. Mech. Eng. Part G J. Aerosp. Eng.* **2017**, *231*, 1732–1745. [CrossRef]
25. Busawon, K.K.; Kabore, P. Disturbance attenuation using proportional integral observers. *Int. J. Control* **2001**, *74*, 618–627. [CrossRef]
26. Gao, Z.W.; Breikin, T.; Nang, H. Discrete-time proportional and integral observer and observer-based controller for systems with both unknown input and output disturbances. *Opt. Control Appl. Meth.* **2008**, *29*, 171–189. [CrossRef]
27. Alcan, G.; Unel, M. Robust Hovering Control of a Quadrotor Using Acceleration Feedback. In Proceedings of the 2017 International Conference on Unmanned Aircraft Systems (Icuas'17), Miami, FL, USA, 13–16 June 2017; pp. 1455–1462.
28. Han, J.Q. From PID to Active Disturbance Rejection Control. *IEEE Trans. Ind. Electron.* **2009**, *56*, 900–906. [CrossRef]
29. Wang, W.; Gao, Z. A comparison study of advanced state observer design techniques. In Proceedings of the American Control Conference, Denver, CO, USA, 4–6 June 2003; Volume 6, pp. 4754–4759
30. Madonski, R.; Herman, P. Survey on methods of increasing the efficiency of extended state disturbance observers. *ISA Trans.* **2015**, *56*, 18–27. [CrossRef] [PubMed]
31. Godbole, A.A.; Kolhe, J.P.; Talole, S.E. Performance Analysis of Generalized Extended State Observer in Tackling Sinusoidal Disturbances. *IEEE Trans. Control Syst. Technol.* **2013**, *21*, 2212–2223. [CrossRef]
32. Madani, T.; Benallegue, A. Sliding mode observer and backstepping control for a quadrotor unmanned aerial vehicles. In Proceedings of the 2007 American Control Conference, New York, NY, USA, 9–13 July 2007; Volume 1–13, pp. 2462–2467.

33. Chuei, R.; Cao, Z.W.; Man, Z.H. Super Twisting Observer based Repetitive Control for Aperiodic Disturbance Rejection in a Brushless DC Servo Motor. *Int. J. Control Autom. Syst.* **2017**, *15*, 2063–2071. [CrossRef]
34. Utkin, V. On Convergence Time and Disturbance Rejection of Super-Twisting Control. *IEEE Trans. Autom. Control* **2013**, *58*, 2013–2017. [CrossRef]
35. Li, P.; Ma, J.J.; Zheng, Z.Q. Disturbance-observer-based fixed-time second-order sliding mode control of an air-breathing hypersonic vehicle with actuator faults. *Proc. Inst. Mech. Eng. Part G J. Aerosp. Eng.* **2018**, *232*, 344–361. [CrossRef]
36. Tian, B.L.; Liu, L.H.; Lu, H.C.; Zuo, Z.Y.; Zong, Q.; Zhang, Y.P. Multivariable Finite Time Attitude Control for Quadrotor UAV: Theory and Experimentation. *IEEE Trans. Ind. Electron.* **2018**, *65*, 2567–2577. [CrossRef]
37. Chovancova, A.; Fico, T.; Hubinsky, P.; Duchon, F. Comparison of various quaternion-based control methods applied to quadrotor with disturbance observer and position estimator. *Robot. Auton. Syst.* **2016**, *79*, 87–98. [CrossRef]
38. Bouabdallah, S.; Siegwart, R. Full control of a quadrotor. In Proceedings of the 2007 IEEE/RSJ International Conference on Intelligent Robots and Systems, San Diego, CA, USA, 29 October–2 November 2007; Volume 9, pp. 153–158.
39. Waslander, S.; Wang, C. Wind Disturbance Estimation and Rejection for Quadrotor Position Control. In Proceedings of the AIAA Infotech@Aerospace Conference and AIAA Unmanned...Unlimited Conference, Seattle, WA, USA, 6–9 April 2009.
40. Chen, Y.M.; He, Y.L.; Zhou, M.F. Decentralized PID neural network control for a quadrotor helicopter subjected to wind disturbance. *J. Cent. South Univ.* **2015**, *22*, 168–179. [CrossRef]
41. Önder, E.M. Battery power loss compensated fractional order sliding mode control of a quadrotor UAV. *Asian J. Control 14*, 413–425.
42. Shtessel, Y.; Edwards, C.; Fridman, L.; Levant, A. *Sliding Mode Control and Observation*; Springer: New York, NY, USA, 2014.
43. Quan, Q.; Dai, X. Flight Performance Evaluation of UAVs. Available online: http://flyeval.com/ (accessed on 20 April 2018).
44. Shi, D.; Wu, Z.; Chou, W. Harmonic Extended State Observer Based Anti-Swing Attitude Control for Quadrotor with Slung Load. *Electronics* **2018**, *7*, 83. [CrossRef]
45. Shi, D.; Wu, Z.; Chou, W. Generalized Extended State Observer Based High Precision Attitude Control of Quadrotor Vehicles Subject to Wind Disturbance. *IEEE Access* **2018**, *6*, 2169–3536. [CrossRef]
46. Meier, L.; Tanskanen, P.; Heng, L.; Lee, G.H.; Fraundorfer, F.; Pollefeys, M. PIXHAWK: A micro aerial vehicle design for autonomous flight using onboard computer vision. *Auton. Robot.* **2012**, *33*, 21–39. [CrossRef]
47. Meier, L.; Tanskanen, P.; Fraundorfer, F.; Pollefeys, M. The Pixhawk Open-Source Computer Vision Framework for Mavs. In Proceedings of the International Conference on Unmanned Aerial Vehicle in Geomatics (UAV-G), New York, NY, USA, 14–16 September 2011; Volume 38-1, pp. 13–18.

electronics

MDPI

Article

Harmonic Extended State Observer Based Anti-Swing Attitude Control for Quadrotor with Slung Load

Di Shi [1], Zhong Wu [1,*] and Wusheng Chou [2,3]

[1] School of Instrumentation Science and Optoelectronics Engineering, Beihang University, Beijing 100191, China; shidi666@buaa.edu.cn
[2] School of Mechanical Engineering and Automation, Beihang University, Beijing 100191, China; wschou@buaa.edu.cn
[3] State Key Laboratory of Virtual Reality Technology and System, Beijing 100191, China
* Correspondence: wuzhong@buaa.edu.cn; Tel.: +86-10-8233-9703

Received: 8 May 2018; Accepted: 28 May 2018; Published: 29 May 2018

Abstract: During the flight of the quadrotor, the existence of a slung load will exert a swing effect on the system and the motion of which will significantly change the dynamics of the quadrotor. The external torque caused by the slung load can be considered as a kind of disturbance and it is a threat to the attitude control stability of the system. In order to solve this problem, a high precision disturbance compensation method is presented in this paper, based on the harmonic extended state observer (HESO). Firstly, a generic mathematical model for the quadrotor-slung load system is obtained via the Lagrangian mechanics, and according to the analysis of the slung load motion, we obtain the disturbance as a form of periodic equation. Secondly, based on the dynamic model of the disturbance, we propose a HESO to achieve high precision disturbance estimation and its stability is proved by Lyapunov theory. Thirdly, we designed an attitude tracking controller based on backstepping method, and discussed the stability of the entire system. Finally, numerical simulations and real time experiments are carried out to evaluate the performance of the proposed method. Our results show that the robustness of the quadrotor subject to slung load has been improved.

Keywords: quadrotor; slung load; disturbance; harmonic extended state observer

---

1. Introduction

In recent years, the research on quadrotor vehicles has attracted great interest due to the wide range of civil and military applications, and many achievements have been made [1,2]. As a new kind of unmanned aerial vehicle (UAV), the quadrotor is a small rotorcraft with four propellers driven by four direct current (DC) motors respectively [3]. Compared with traditional helicopters, the structure of the quadrotor is simpler and more efficient, and has significant advantages in precise hovering, aggressive maneuver, vertical take-off and landing (VTOL) [4,5], etc.

Like traditional helicopters, quadrotor vehicles have many important applications in carrying slung load, such as deploying supplies in military operations, or delivering first-aid kits for personal assistance to the victims in disasters like floods, earthquakes, fires, industrial accidents, etc. [6,7], and the research work addressing quadrotor vehicles with slung load becomes an attractive topic. On this application, the major difficulty in modeling and control study is the coupled effects between the quadrotor vehicles and the slung load [8]. The external slung load behaves like a pendulum and the motion of which will significantly change the dynamics of the quadrotor. Moreover, the quadrotor vehicle is a typical underactuated, strong coupled, nonlinear system [9], and inherently unstable without close-loop controller [10]. If the pendulous motion of the load exceeds certain limits, the stability of control system will be broken because of the changes in dynamic characteristics of the plant. Therefore, improving the robustness of the controller subject to the oscillation of slung load is very necessary.

In the relevant literatures on the quadrotor, the problem of addressing quadrotor-slung load system has been discussed in some publications. The mathematical model of the entire system is derived by the Newton-Euler formulation in [8,11], however, there is no further analysis on the dynamic characteristics of the slung load motion in detail. In [12], the slung load is modeled as a point mass spherical pendulum, and a related adaptive control method is proposed to handle the additional forces and torques acting on the quadrotor. In [13], a generalized approach is presented using an iterative optimal control algorithm, and a series of complex tasks are thus solved without the need for manual manipulation of the system dynamics, heuristic simplifications, or manual trajectory generation. In [14], a nonlinear dynamic model is presented, and an interconnection and damping assignment-passivity based Control (IDA-PBC) methodology is used for precise payload's positioning with stabilization of the swing angles to the minimum of the desired energy function.

Alternatively, the effect of slung load on the quadrotor can be considered as a kind of disturbance [15]. Therefore, in order to improve system reliability and achieve the requirements of high precision control, the disturbance and uncertainty estimation and attenuation (DUEA) method would be a potential solution. In recent years, this method has been widely used and achieved good results [16,17] The framework of DUEA can be divided into two parts, namely, a disturbance and uncertainty estimator (DUE) and a feedback controller (FC). In the first part, DUE is designed to estimate the disturbances so that they could be compensated in the feedforward loop. Then the FC in the second part is designed to guarantee fast convergence of the closed-loop system. In this framework the DUE plays an important role because the performance of the closed-loop system is largely determined by the estimation accuracy of it. Therefore, a series of observers have been proposed as the DUE so far to improve estimation accuracy under different conditions, such as disturbance observer (DO) [18,19], extended state observer (ESO) [15,20,21] and proportional integral observer (PIO) [22,23] etc. By the appropriate use of the observer, disturbance rejection performance and robustness of the existing control system could be significantly improved.

The extended state observer (ESO), known as the key module of active disturbance rejection control, can estimate both the states of system and the total disturbances with less dependence on model information [24,25]. This method was first proposed by Han in 1990s and the basic idea behind ESO is to view disturbance as an extended state and utilize observer to estimate it [24]. Traditionally, ESO approaches focus primarily on dealing with slowly changing disturbances. However, it's obvious that the disturbance caused by the slung load is periodic, which cannot be estimated by traditional ESO thoroughly [26]. Therefore, an enhancement of ESO that can handle periodic disturbance is necessary in this field. In [27], a higher-order ESO is investigated, from the results, it can be seen that the higher-order ESO can improve the estimation accuracy of sinusoidal external disturbances more or less, while, there still exists a periodic estimation error that will in turn decrease the control accuracy of the closed-loop system. Furthermore, the higher level of the observer order will lead to a higher observer gain, which will in return excite the sensor noise and introduce them into the control loop. The internal model principle is applied for generalized ESO in [28] and harmonic disturbance observer (HDO) in [29] by embedding the disturbance dynamics into the observer, and the disturbance estimation performance is improved.

Motivated by these methods, a harmonic extended state observer based anti-swing attitude control method is proposed for a quadrotor subject to slung load in this article, where the periodic disturbance caused by slung load motion is compensated by the estimated state signals produced of the HESO. The main contributions of this paper are summarized as follows: (1) Build the generic mathematical model for the quadrotor-slung load system by the Lagrangian mechanics, and analyze the characteristics of the slung load motion in detail. (2) Propose a HESO based on the characteristics of the slung load motion.

The outline of this paper is as follows: The mathematical model and the control problems of quadrotor-slung load system are formulated in Section 2. A HESO is designed in Section 3. In Section 4, an attitude tracking controller is designed via backstepping method. Numerical simulation and real time experimental results are presented in Section 5, and the conclusions are summarized in Section 6.

2. Mathematical Model and Problem Formulation

2.1. Preliminaries

2.1.1. Notations and Assumptions

Throughout this paper, the following notations will be used. $\mathbb{R}$ is the set of real numbers. Let $||\cdot||$ denote the 2-norm of a vector or a matrix. For a given vector $\boldsymbol{v} = [v_1, \ldots, v_n]^T \in \mathbb{R}^n$, $\left|\left|\boldsymbol{v}\right|\right| = \sqrt{\boldsymbol{v}^T\boldsymbol{v}}$, and for a given matrix $\boldsymbol{A} \in \mathbb{R}^{n\times n}$, $\left|\left|\boldsymbol{A}\right|\right| = \sqrt{\lambda_{\max}(\boldsymbol{A}^T\boldsymbol{A})}$, where $\lambda_{\max}(\cdot)$ is the maximal eigenvalue of the matrix. In addition, the operator $S(\cdot)$ maps a vector $\boldsymbol{x} = \begin{bmatrix} x_1 & x_2 & x_3 \end{bmatrix}^T$ to a skew symmetric matrix as:

$$S(\boldsymbol{x}) = \begin{bmatrix} 0 & -x_3 & x_2 \\ x_3 & 0 & -x_1 \\ -x_2 & x_1 & 0 \end{bmatrix} \tag{1}$$

In this section, the mathematical model of the quadrotor-slung load system is established. As shown in Figure 1, we consider this system consists of three parts: a quadrotor, a cable and a payload. And before our work, the following assumptions are made:

(1) The slung load is considered as a particle and only swings in a plane. The length of the connecting cable is constant and known.
(2) The inelastic cable is massless and always tight and no consideration of the energy loss caused by the friction force in the swing.
(3) The aerodynamic effects on the load are neglected.

2.1.2. Quaternion Operations

In order to avoid the singularity problem of trigonometric functions, unit quaternion $\boldsymbol{q} = \begin{bmatrix} q_0 & \boldsymbol{q}_v^T \end{bmatrix}^T \in \mathbb{R}^4, ||\boldsymbol{q}|| = 1$ is used to represent rotation [30]. Following are the operations we used.

The quaternion multiplication is:

$$\boldsymbol{q}_1 \otimes \boldsymbol{q}_2 = \begin{bmatrix} q_{01}q_{02} - \boldsymbol{q}_{v1}^T\boldsymbol{q}_{v2} \\ q_{01}\boldsymbol{q}_{v2} + q_{02}\boldsymbol{q}_{v1} - S(\boldsymbol{q}_{v2})\boldsymbol{q}_{v1} \end{bmatrix} \tag{2}$$

The relationship between rotation matrix $\boldsymbol{R}_A^B$ and $\boldsymbol{q}$ is calculated as:

$$\boldsymbol{R}_A^B = (q_0^2 - \boldsymbol{q}_v^T\boldsymbol{q}_v)\boldsymbol{I}_3 + 2\boldsymbol{q}_v\boldsymbol{q}_v^T + 2q_0S(\boldsymbol{q}_v) \tag{3}$$

$$\dot{\boldsymbol{R}}_A^B = -S(\boldsymbol{\omega})\boldsymbol{R}_A^B \tag{4}$$

The derivative of a quaternion is given by the quaternion multiplication of the quaternion $\boldsymbol{q}$ and the angular velocity of the plant $\boldsymbol{\omega}$:

$$\dot{\boldsymbol{q}} = \begin{bmatrix} \dot{q}_0 \\ \dot{\boldsymbol{q}}_v \end{bmatrix} = \frac{1}{2}\boldsymbol{q} \otimes \begin{bmatrix} 0 \\ \boldsymbol{\omega} \end{bmatrix} = \frac{1}{2}\begin{bmatrix} -\boldsymbol{q}_v^T \\ S(\boldsymbol{q}_v) + q_0\boldsymbol{I}_3 \end{bmatrix}\boldsymbol{\omega} \tag{5}$$

The quaternion error $\boldsymbol{q}_e$ is given as the quaternion multiplication of the conjugate of the desired quaternion $\boldsymbol{q}_d$ and the actual quaternion $\boldsymbol{q}$:

$$\boldsymbol{q}_e = \boldsymbol{q}_d^* \otimes \boldsymbol{q} = \begin{bmatrix} q_{0e} \\ \boldsymbol{q}_{ve} \end{bmatrix} = \begin{bmatrix} q_0 q_{0d} + \boldsymbol{q}_v{}^T \boldsymbol{q}_{vd} \\ q_{0d}\boldsymbol{q}_v - q_0 \boldsymbol{q}_{vd} + S(\boldsymbol{q}_v)\boldsymbol{q}_{vd} \end{bmatrix} \tag{6}$$

where $\boldsymbol{q}^* = \begin{bmatrix} q_0 & -\boldsymbol{q}_v \end{bmatrix}^T$ is the conjugate of the quaternion $\boldsymbol{q}$.

2.2. Mathmetical Model of Quadrotor-Slung Load system

Firstly, the quadrotor can be considered as a rigid cross frame attached with four rotors, and the center of gravity coincides with the body-fixed frame origin. The simplified model of the quadrotor is presented in Figure 1, rotors $R1$ and $R3$ rotate counterclockwise, and rotors $R2$ and $R4$ rotate clockwise, each propeller rotates at the angular speed Ω_i and produces a force F_i (i = 1,2,3,4) along the negative z-direction relative to the body frame [30,31]:

$$F_i = -k_T \Omega_i^2 \tag{7}$$

where $k_T > 0$ denotes the aerodynamic coefficient which consists formed of the atmospheric density ρ, the radius of the propeller r, and the thrust coefficient c_T. In addition, due to the spinning of the rotors, a reaction torque M_i (i = 1,2,3,4) is generated on the quadrotor body by each rotor:

$$M_i = (-1)^{i+1} k_D {\Omega_i}^2 \tag{8}$$

where $k_D > 0$ denotes the drag coefficient of the rotor, which depends on the same factors as k_T.

The variation of the orientation is achieved by varying the angular speed of a specific rotor. The total force and torque acting on the quadrotor are defined as follows:

$$\boldsymbol{u} = \begin{bmatrix} \tau_\phi \\ \tau_\theta \\ \tau_\psi \end{bmatrix} = \begin{bmatrix} 0 & -lk_T & 0 & lk_T \\ lk_T & 0 & -lk_T & 0 \\ -k_D & k_D & -k_D & k_D \end{bmatrix} \begin{bmatrix} \Omega_1^2 \\ \Omega_2^2 \\ \Omega_3^2 \\ \Omega_4^2 \end{bmatrix} \tag{9}$$

where $\boldsymbol{u}$ represents the attitude control signal to be designed.

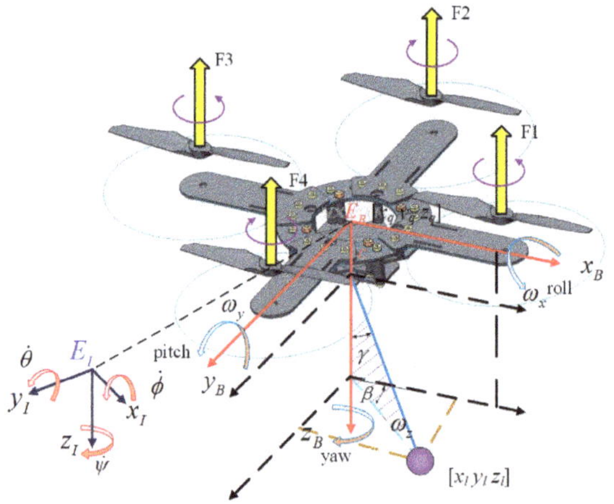

Figure 1. Quadrotor-slung load system geometry and coordinate systems.

In the mathematical model of quadrotor, three coordinate frames are considered: the non-moving inertial coordinate frame $E_I : \{o_I, x_I, y_I, z_I\}$, the body-fixed coordinate frame $E_B : \{o_B, x_B, y_B, z_B\}$

and the desired frame $E_D : \{o_D, x_D, y_D, z_D\}$ to represent the actual attitude and desired attitude of quadrotor respectively. Note that NED coordinates are used to define all frames. In the inertial frame E_I, the position of the quadrotor is $\boldsymbol{X}_q = \begin{bmatrix} x_q & y_q & z_q \end{bmatrix}^T \in \mathbb{R}^3$, the attitude angle is $\Theta = \begin{bmatrix} \phi & \theta & \psi \end{bmatrix}^T \in \mathbb{R}^3$ and the quaternion expression of the attitude is $\boldsymbol{q} = [\begin{matrix} q_0 & \boldsymbol{q}_v \end{matrix}]^T \in \mathbb{R}^4$. In the body-fixed frame E_B, the angular velocity of is $\boldsymbol{\omega} = [\begin{matrix} \omega_x & \omega_y & \omega_z \end{matrix}]^T \in \mathbb{R}^3$. Thus, the rotation matrix from E_B to E_I can be represented as:

$$\boldsymbol{R}_{\boldsymbol{B}}^{\boldsymbol{I}} = \begin{bmatrix} \cos\theta\cos\psi & \sin\phi\sin\theta\cos\psi - \cos\phi\sin\psi & \cos\phi\sin\theta\cos\psi + \sin\phi\sin\psi \\ \cos\theta\sin\psi & \sin\phi\sin\theta\sin\psi + \cos\phi\cos\psi & \cos\phi\sin\theta\sin\psi - \sin\phi\cos\psi \\ -\sin\theta & \sin\phi\cos\theta & \cos\theta\cos\psi \end{bmatrix} \tag{10}$$

and the rotation velocity transfer matrix can be given as:

$$\boldsymbol{P} = \begin{bmatrix} 1 & \sin\phi\tan\theta & \cos\phi\tan\theta \\ 0 & \cos\phi & -\sin\phi \\ 0 & \sin\phi\cos^{-1}\theta & \cos\phi\cos^{-1}\theta \end{bmatrix} \tag{11}$$

Thus, the rotational kinematic equations with respect to the inertial frame E_I can be expressed as:

$$\dot{\Theta} = \boldsymbol{P}\boldsymbol{\omega} \tag{12}$$

Secondly, we study the slung load. The position of the hook point is defined as $\boldsymbol{c} = \begin{bmatrix} 0 & 0 & c \end{bmatrix}^T$ in body frame E_B. According to previous assumptions, the relationship between the position of the slung load $\boldsymbol{X}_l = \begin{bmatrix} x_l & y_l & z_l \end{bmatrix}^T \in \mathbb{R}^3$ and $\boldsymbol{X}_q$ are:

$$\boldsymbol{X}_l = \boldsymbol{X}_q + \boldsymbol{R}_{\boldsymbol{B}}^{\boldsymbol{I}}\boldsymbol{c} + L_c \begin{bmatrix} \sin\gamma\cos\beta \\ \sin\gamma\sin\beta \\ \cos\gamma \end{bmatrix} \tag{13}$$

where $\gamma \in (-90^\circ, 90^\circ)$ is the pendulum angle between the cable and the positive orientation of $o_I z_I$, β is the angle between the pendulum plane and the $x_I o_I z_I$ plane, and L_c is the length of the cable. Moreover, according to the former assumptions, β and L_c are constant.

Finally, we use Lagrangian mechanics to summarize the dynamics of the quadrotor-slung load system. Compared with the Newtonian mechanics, this choice eliminates the need for the constraint forces to enter into the resultant system of equations, so that fewer equations are needed. As shown in Figure 1, the entire system has seven degrees of freedom (DOF): pendulum angle of the slung load γ, position of the quadrotor $\boldsymbol{X}_q = \begin{bmatrix} x_q & y_q & z_q \end{bmatrix}^T$ and attitude of the quadrotor $\Theta = \begin{bmatrix} \phi & \theta & \psi \end{bmatrix}^T$. Therefore, we defined the generalized coordinate of the quadrotor-slung load system as $\boldsymbol{\eta} = \begin{bmatrix} x_q & y_q & z_q & \phi & \theta & \psi & \gamma \end{bmatrix}^T$. The expressions for the kinetic and potential energies will be presented in order to obtain the Lagrangian of the system. The total kinetic energy function of the quadrotor-slung load system, resulting from the translational and rotational motions can be portioned as the sum of the translational kinetic energy:

$$T_t = \frac{1}{2}m_q\dot{\boldsymbol{X}}_q^2 + \frac{1}{2}m_l\dot{\boldsymbol{X}}_l^2 \tag{14}$$

And the rotational kinetic energy of the entire system:

$$T_r = \frac{1}{2}\boldsymbol{\omega}^T\boldsymbol{J}\boldsymbol{\omega} = \frac{1}{2}\dot{\Theta}^T\boldsymbol{P}^{-T}\boldsymbol{J}\boldsymbol{P}^{-1}\dot{\Theta} \tag{15}$$

where the inertia matrix $\boldsymbol{J} = \mathrm{diag}(J_x, J_y, J_z)$ is diagonal.

And the total potential energy function of the system results from the sum of the potential energies of the quadrotor and the slung load:

$$V = -(m_q z_q + m_l z_l)g \tag{16}$$

Thus, the Lagrangian of the generalized system can be defined as:

$$L(\eta, \dot{\eta}) = T_t + T_r - V \tag{17}$$

Notice that gravity $\boldsymbol{G}$ is the conservative force in this system, while total lift $\boldsymbol{F} = \boldsymbol{R}_B^I \begin{bmatrix} 0 & 0 & \sum F_i \end{bmatrix}^T$ and torque $\boldsymbol{u}$ are non-conservative forces, the Euler Lagrangian equations of the second kind can be obtained as:

$$\frac{d}{dt}\left(\frac{\partial L(\eta, \dot{\eta})}{\partial \dot{\eta}}\right) - \frac{\partial L(\eta, \dot{\eta})}{\partial \eta} = \boldsymbol{Q} \tag{18}$$

where $\boldsymbol{Q} = \begin{bmatrix} F_x & F_y & F_z & \tau_\phi & \tau_\theta & \tau_\psi & 0 \end{bmatrix}^T \in \mathbb{R}^7$ is the non-conservative generalized force.

When the quadrotor is carrying a slung load, Large-scale maneuvers should be avoided, therefore, the coupling of angular velocity between different channels can be ignored. Then, we assume $\boldsymbol{P} \approx \boldsymbol{I}_3$ when calculating the dynamic quadrotor-slung load system. According to Equations (14)–(18), the dynamic model can be expressed as:

$$\sum_T : \begin{cases} \ddot{x}_q = \frac{1}{m_q + m_l}\left(F_x - m_l L_c\left(\ddot{\gamma}\cos\gamma - \dot{\gamma}^2 \sin\gamma\right)\cos\beta - m_l c\ddot{\theta}\right) \\ \ddot{y}_q = \frac{1}{m_q + m_l}\left(F_y - m_l L_c\left(\ddot{\gamma}\cos\gamma - \dot{\gamma}^2 \sin\gamma\right)\sin\beta + m_l c\ddot{\phi}\right) \\ \ddot{z}_q = \frac{1}{m_q + m_l}\left(F_z + m_l L_c\left(\ddot{\gamma}\sin\gamma + \dot{\gamma}^2\cos\gamma\right)\right) + g \end{cases} \tag{19}$$

$$\sum_R : \begin{cases} \ddot{\phi} = \frac{1}{J_x + m_l c^2}\left(\tau_\phi + m_l c\left(\ddot{y}_q + L_c\left(\ddot{\gamma}\cos\gamma - \dot{\gamma}^2\sin\gamma\right)\sin\beta\right)\right) \\ \ddot{\theta} = \frac{1}{J_y + m_l c^2}\left(\tau_\theta - m_l c\left(\ddot{x}_q + L_c\left(\ddot{\gamma}\cos\gamma - \dot{\gamma}^2\sin\gamma\right)\cos\beta\right)\right) \\ \ddot{\psi} = \frac{1}{J_z}\tau_\psi \end{cases} \tag{20}$$

$$\ddot{\gamma} = -\frac{1}{L_c}\left(\left(\ddot{x}_q\cos\beta + \ddot{y}_q\sin\beta\right)\cos\gamma + (-\ddot{z}_Q + g)\sin\gamma\right) + \frac{c}{L_c}\Delta(\phi, \theta, \gamma) \tag{21}$$

where $\Delta(\phi, \theta, \gamma)$ represents the residual term of $\ddot{\gamma}$.

According to [30], we can obtain the attitude kinematic of the quadrotor as:

$$\dot{q} = \frac{1}{2} q \otimes \begin{bmatrix} 0 & \omega \end{bmatrix}^T \tag{22}$$

2.3. Analysis of Slung Load Motion

In this article, we consider the effect of slung load on the quadrotor as a kind of disturbance. In order to realize high precision estimation of the disturbance, the characteristics of the disturbance need to be analyzed.

The derived equations of motion of the load and quadrotor are highly nonlinear and strongly coupled, thus, they are difficult to be used for motion analysis. Therefore, we trimmed the mathematical near hovering or uniform linear flight. In this condition, we have $\ddot{\boldsymbol{X}}_q \approx 0$. And in actual flight $c \ll L_c$, then we have $cL_c^{-1} \approx 0$. From Equation (20), we can see that the load movement can hardly affect ψ, therefore, disturbance in pitch and roll channels are considered in this condition. The trimmed results of the rotational dynamic are:

$$\begin{cases} \ddot{\phi} = \frac{1}{J_y + m_l c^2}(\tau_\phi + d_\phi) \\ \ddot{\theta} = \frac{1}{J_y + m_l c^2}(\tau_\theta + d_\theta) \\ \ddot{\gamma} = -\frac{1}{L_c} g \sin\gamma \end{cases} \tag{23}$$

where $d_\phi = m_l c L_c \left(\ddot{\gamma} \cos \gamma - \dot{\gamma}^2 \sin \gamma \right) \sin \beta$ and $d_\theta = -m_l c L_c \left(\ddot{\gamma} \cos \gamma - \dot{\gamma}^2 \sin \gamma \right) \cos \beta$ are the torque disturbance caused by the slung load in roll and pitch channel respectively.

Then, analyze the motion of slung load. When γ is a small angle, we have $\sin \gamma \approx \gamma$ and $\cos \gamma \approx 1$. Thus, we can rewrite Equation (23) as:

$$\ddot{\gamma} = -\frac{1}{L_c} g \gamma \tag{24}$$

According to Equation (24), we can calculate that:

$$\gamma = A_0 \sin\left(\sqrt{g L_c^{-1}} t + \chi_0 \right) \tag{25}$$

where A_0 and χ_0 are the unknown amplitude and phase related to initial conditions.

Substituting Equation (25) into Equation (23), we have:

$$d = \begin{bmatrix} d_\phi \\ d_\theta \\ d_\psi \end{bmatrix} = \begin{bmatrix} A_{1,\phi} \sin\left(\sqrt{g L_c^{-1}} t + \chi_{1,\phi} \right) + A_{2,\phi} \sin\left(3\sqrt{g L_c^{-1}} t + \chi_{2,\phi} \right) + \Delta_\phi \\ A_{1,\theta} \sin\left(\sqrt{g L_c^{-1}} t + \chi_{1,\theta} \right) + A_{2,\theta} \sin\left(3\sqrt{g L_c^{-1}} t + \chi_{2,\theta} \right) + \Delta_\phi \\ 0 \end{bmatrix} \tag{26}$$

where Δ_i $(i = \phi, \theta)$ are the unmodeled residual of the disturbance in each channel.

According to the above analysis, the disturbance torque caused by the slung load is a sum-of-sinusoids function which has two frequencies $\varpi_1 = \sqrt{g L_c^{-1}}$, $\varpi_2 = 3\sqrt{g L_c^{-1}}$, while the amplitude $A_{1,i}$, $A_{2,i}$ and phase $\chi_{1,i}$, $\chi_{2,i}$ $(i = \phi, \theta)$ are unknown.

2.4. Problem Formulation

In order to study the transient and steady-state characteristics of the quadrotor, the dynamics of attitude error are introduced. We use $\boldsymbol{\omega}_d = \begin{bmatrix} \omega_{d,x} & \omega_{d,y} & \omega_{d,z} \end{bmatrix}^T$ and $\boldsymbol{q}_d = \begin{bmatrix} q_{0d} & \boldsymbol{q}_{vd} \end{bmatrix}^T$ to denote the desired angular velocities and attitude respectively, thus:

$$\boldsymbol{\omega}_e = \boldsymbol{\omega} - \boldsymbol{R}_D^B \boldsymbol{\omega}_d \tag{27}$$

where $\boldsymbol{\omega}_e = \begin{bmatrix} \omega_{e,x} & \omega_{e,y} & \omega_{e,z} \end{bmatrix}^T$ is the tracking error vector of the angular velocities. Then, we can obtain the dynamics of $\boldsymbol{\omega}_e$ according to Equations (4), (20) and (27):

$$\dot{\boldsymbol{\omega}}_e = S(\boldsymbol{\omega}_d) \boldsymbol{R}_D^B \boldsymbol{\omega}_d - \boldsymbol{R}_D^B \dot{\boldsymbol{\omega}}_d + \boldsymbol{J}_0^{-1} \boldsymbol{u} + \boldsymbol{J}_0^{-1} \boldsymbol{d} \tag{28}$$

where $\boldsymbol{J}_0 = \mathrm{diag}(J_x + m_l c^2, J_y + m_l c^2, J_z)$, $\boldsymbol{R}_D^B$ can be calculated according to Equations (3) and (6). And according to Equations (5), (6) and (27), we can obtain the kinematics of attitude tracking error

$$\dot{q}_e = \frac{1}{2} q_e \otimes \begin{bmatrix} 0 & \omega_e \end{bmatrix}^T = \frac{1}{2} \begin{bmatrix} -\boldsymbol{q}_{ve}^T \\ S(\boldsymbol{q}_{ve}) + q_{0e} \boldsymbol{I}_3 \end{bmatrix} \omega_e \tag{29}$$

The problem we try to tackle in this work is to design a continuous control law $\boldsymbol{u}$ using only the measurable system output $\boldsymbol{\omega}$ and $\boldsymbol{q}$ such that the error of attitude $\boldsymbol{\omega}_e$ and $\boldsymbol{q}_e$ converge to zero in presence of the slung load. In order to ensure the robustness of the controller, the DUEA strategy is necessary

Figure 2 illustrates the control scheme that we designed. Based on the DUEA control methodology, the attitude control problem for quadrotor can be divided into two components: design the feedforward loop so that the periodic disturbance is estimated by HESO and compensated this way; and design the feedback loop that regulates the orientation to track the desired attitude produced by the commander timely. Therefore, the control signal $\boldsymbol{u}$ contains two parts as:

$$\boldsymbol{u} = \boldsymbol{u}^N + \boldsymbol{u}^E \tag{30}$$

where $\boldsymbol{u}^N$ is the nominal control input vector and $\boldsymbol{u}^E$ is the disturbances attenuation input vector.

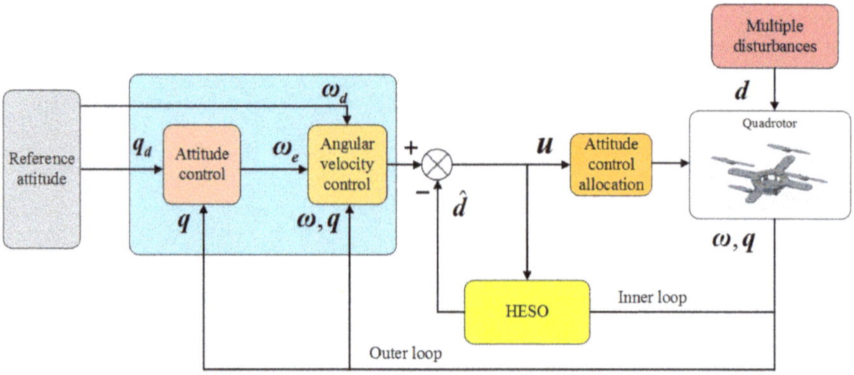

Figure 2. Block diagram of the proposed control scheme.

3. Design and Stability Analysis of HESO

In this section, the design of HESO, which provides the disturbance estimate for the controller, is described in detail. As for the DUEA control methodology, the control performance of closed loop system will be largely determined by the observation performance. However, the disturbances acting on quadrotor are periodic, which cannot be accurately estimated by traditional ESO thoroughly [28]. Therefore, in order to enhance the performance of feedback controller, a HESO is designed to estimate the periodic disturbances, and the stability analysis is carried out afterwards.

3.1. Design of HESO

According to the analysis of the disturbances in previous section, we can conclude that the external disturbances caused by the slung load have the same characteristic in roll and pitch channel. Without loss of generality, the design process in $i(i = \phi, \theta)$ channel is demonstrated in detail. Firstly, we can rewrite the time varying disturbance as:

$$d_i = \overbrace{A_{1,i}\sin(\varpi_1 t + \chi_{1,i})}^{d_{1,i}} + \overbrace{A_{2,i}\sin(\varpi_2 t + \chi_{2,i})}^{d_{2,i}} + \Delta_i \tag{31}$$

where $\dot{\Delta}_i = \delta_i$ is upper bounded $\|\delta_i\| \leq \sigma_i$. In order to establish the model of equivalent disturbance, we write Equation (31) as a form of state equation

$$\begin{cases} \dot{\zeta}_i = A_d\zeta_i + \delta_{d,i} \\ d_i = C_d\zeta_i \end{cases} \tag{32}$$

where

$$A_d = \begin{bmatrix} 0 & 1 & & & & \\ -\varpi_1^2 & 0 & & & & \\ & & 0 & 1 & & \\ & & -\varpi_2^2 & 0 & & \\ & & & & 0 & 1 \\ & & & & 0 & 0 \end{bmatrix}, \delta_{d,\theta} = \begin{bmatrix} 0 \\ 0 \\ 0 \\ 0 \\ 0 \\ \delta_\theta \end{bmatrix}, C_d = \begin{bmatrix} 1 \\ 0 \\ 1 \\ 0 \\ 1 \\ 0 \end{bmatrix}^T$$

$\zeta_i = \begin{bmatrix} d_{1,i} & \dot{d}_{1,i} & d_{2,i} & \dot{d}_{2,i} & d_{3,i} & \dot{d}_{3,i} \end{bmatrix}^T$ is the system state and $\delta_{d,i}$ is the perturbation term of the disturbance.

Then, consider the SISO nonlinear Equation (23), we have

$$\left(J_i + m_l c^2\right)\dot{\omega}_i = \tau_i + d_i \tag{33}$$

Define $x_i = \left[\begin{array}{cc}(J_i + m_l c^2)\omega_i & \zeta_i^T\end{array}\right]^T$ as the new state vector, the reconstructed system can be written as:

$$\begin{cases} \dot{x}_i = \overbrace{\left[\begin{array}{cc} 0 & C_d \\ 0_{6\times 1} & A_d \end{array}\right]}^{A} x_i + \overbrace{\left[\begin{array}{c} 1 \\ 0_{6\times 1} \end{array}\right]}^{B} \tau_i + \overbrace{\left[\begin{array}{c} 0 \\ \delta_{d,i} \end{array}\right]}^{\delta_i} \\ y_i = \underbrace{\left[\begin{array}{cc} 1 & 0_{1\times 6} \end{array}\right]}_{C} x_i \end{cases} \tag{34}$$

Note that in the reconstructed model of quadrotor, it can be verified that the pair (A, C) is observable [28].

From the reconstructed system model in Equation (34), the observer is designed as follows:

$$\begin{cases} \dot{\hat{x}}_i = A\hat{x}_i + B\tau_i + L_{e,i}(y_i - \hat{y}_i) \\ \hat{y}_i = C\hat{x}_i \end{cases} \tag{35}$$

where $\hat{x}_i = \left[\begin{array}{cc}(J_i + m_l c^2)\hat{\omega}_i & \hat{\zeta}_i^T\end{array}\right]^T$ is the stats of the observer, $\hat{y}_i$ is the estimate of the output y_i, and $L_{e,i} = \left[\begin{array}{ccccccc} l_0 & l_1 & l_2 & l_3 & l_4 & l_5 & l_6 \end{array}\right]^T$ is the observer gain to be designed.

By defining the estimation error of the observer as $\tilde{x}_i = x_i - \hat{x}_i$, we have the following equation:

$$\dot{\tilde{x}}_i = \overbrace{(AL_{e,i}C)}^{\overline{A}}\tilde{x}_i + \overbrace{\left[\begin{array}{c} 0 \\ \delta_{d,i} \end{array}\right]}^{\delta_i} \tag{36}$$

From Equation (36), we can find that the estimation accuracy is partly determined by the perturbation term δ_i, thus in order to increase the estimation accuracy of the observer, δ_i should be decreased. Compared with the traditional higher order ESO, the equivalent disturbance model is introduced in HESO which makes the model more precise and the perturbation $\delta_{d,i}$ smaller [28].

3.2. Convergency Analysis of HESO

Frist, define the Lyapunov candidate function $V_{0,i}$ $(i = \phi, \theta)$ as:

$$V_{0,i} = \tilde{x}_i^T P_0 \tilde{x}_i \tag{37}$$

Take the time derivative of $V_{0,i}$, we have:

$$\dot{V}_{0,i} = \tilde{x}_i^T((A - L_{e,i}C)^T P_0 + P_0(A - L_{e,i}C))\tilde{x}_i + 2\tilde{x}_i^T P_0 \delta_i \tag{38}$$

Then, consider Equation (38). Under the condition that $A - L_{e,i}C$ is Hurwitz, for any $k_0 > 0$ there exists a positive definite symmetric matrix P_0 satisfying:

$$(A - L_{e,i}C)^T P_0 + P_0(A - L_{e,i}C) = -k_0 I_3 \tag{39}$$

Substituting Equation (39) into Equation (38), we can get:

$$\dot{V}_{0,i} = -k_0\tilde{x}_i^T\tilde{x}_i + 2\tilde{x}_i^T P_0\delta_i \leq -k_0\|\tilde{x}_i\|^2 + 2\|\tilde{x}_i\|\|P_0\|\|\delta_i\| \leq -\|\tilde{x}_i\|(\|\tilde{x}_i\|k_0 - 2\sigma_i\lambda_{\max}(P_0)) \tag{40}$$

where $\|\delta_{d,i}\|$ is bounded with $\|\delta_{d,i}\| \leq \sigma_i$, and $\lambda_{\max}(P_0)$ is the maximum eigenvalue of P_0. It is obvious that $\dot{V}_{0,i} < 0$ whenever $\|\tilde{x}_{d,i}\| > 2\sigma_i\lambda_{\max}(P_0)k_0^{-1}$. Therefore, the upper bound for estimation error $\|\tilde{x}_i\|$ will be constrained by the bounded ball $B_r = \left\{ r \middle| \|r\| \leq 2\sigma_i\lambda_{\max}(P_0)k_0^{-1} \right\}$. Moreover, the estimation error decreases with the increase of the model accuracy.

4. Design of Attitude Controller

In this section, the main procedures of attitude controller integrated with HESO are presented for effectively handling disturbance caused by slung load. The quadrotor UAV is an underactuated system with six DOF and four control inputs. In order to derive its model, backstepping method is used in the design of attitude controller.

Step 1. Design the control strategy to ensure that $\boldsymbol{q}_e(t)$ converges to zero.

According to the attitude error kinematics subsystem Equation (29), we select the candidate Lyapunov function as:

$$V_1 = \boldsymbol{q}_e{}^T \boldsymbol{q}_e + (1 - q_{0e})^2 > 0 \tag{41}$$

Take the time derivative of V_1, we have:

$$\begin{aligned} \dot{V}_1 &= 2\boldsymbol{q}_{ve}^T \dot{\boldsymbol{q}}_{ve} - 2(1 - q_{0e})\dot{q}_{0e} = (\boldsymbol{q}_{ve}^T S(\boldsymbol{q}_{ve}) + q_{0e}\boldsymbol{q}_{ve}^T \boldsymbol{I}_3 + (1 - q_{0e})\boldsymbol{q}_{ve}^T)\boldsymbol{\omega}_e \\ &= \boldsymbol{q}_{ve}^T \boldsymbol{\omega}_e \end{aligned} \tag{42}$$

Then, we design a virtual control scheme as:

$$\boldsymbol{\omega}_{ed} = -\boldsymbol{K}_1 \boldsymbol{q}_e \tag{43}$$

where $\boldsymbol{K}_1$ is the gain matrix of the controller, which is diagonal positive definite. If the angular velocity tracking error $\boldsymbol{\omega}_e$ is equal to the virtual control input $\boldsymbol{\omega}_{ed}$, $\dot{V}_1$ is negative semidefinite definite:

$$\dot{V}_1 = -\boldsymbol{q}_e^T \boldsymbol{K}_1 \boldsymbol{q}_e \tag{44}$$

According to the Lyapunov stability theorem, we can conclude that $\boldsymbol{q}_e$ converges to zero, under the condition that the virtual control $\boldsymbol{\omega}_e$ converges to $-\boldsymbol{K}_1\boldsymbol{q}_e$.

Step 2. Design the control signal $\boldsymbol{u}$ to ensure that $\boldsymbol{\omega}_e$ track the desired virtual control input $\boldsymbol{\omega}_{ed}$.

We define the error between $\boldsymbol{\omega}_e$ and $\boldsymbol{\omega}_{ed}$ as:

$$\widetilde{\boldsymbol{\omega}}_e = \boldsymbol{\omega}_e + \boldsymbol{K}_1 \boldsymbol{q}_e \tag{45}$$

In order to discuss the stability of the entire system including DUE and FC, we define the following candidate Lyapunov function V_2 as:

$$V_2 = V_1 + \frac{1}{2}\widetilde{\boldsymbol{\omega}}_e{}^T \boldsymbol{J}_0 \widetilde{\boldsymbol{\omega}}_e + V_{0,\phi} + V_{0,\theta} = \left[\boldsymbol{q}_e{}^T \boldsymbol{q}_e + (1 - q_{0e})^2\right] + \frac{1}{2}\widetilde{\boldsymbol{\omega}}_e{}^T \boldsymbol{J}_0 \widetilde{\boldsymbol{\omega}}_e + \widetilde{\boldsymbol{x}}_\phi^T \boldsymbol{P}_0 \widetilde{\boldsymbol{x}}_\phi + \widetilde{\boldsymbol{x}}_\theta^T \boldsymbol{P}_0 \widetilde{\boldsymbol{x}}_\theta \tag{46}$$

Take the time derivative of V_2, and substitute Equations (28), (45), and (47) into $\dot{V}_2$, then we have

$$\begin{aligned} \dot{V}_2 &= \boldsymbol{q}_{ve}^T \widetilde{\boldsymbol{\omega}}_e - \boldsymbol{q}_{ve}^T \boldsymbol{K}_1 \boldsymbol{q}_{ve} + \widetilde{\boldsymbol{\omega}}_e^T (\boldsymbol{J}_0 \dot{\widetilde{\boldsymbol{\omega}}}_e + \boldsymbol{J}_0 \boldsymbol{K}_1 \dot{\boldsymbol{q}}_{ve}) - k_0 \widetilde{\boldsymbol{x}}_\phi^T \widetilde{\boldsymbol{x}}_\phi + 2\widetilde{\boldsymbol{x}}_\phi^T \boldsymbol{P}_0 \boldsymbol{\delta}_\phi - k_0 \widetilde{\boldsymbol{x}}_\theta^T \widetilde{\boldsymbol{x}}_\theta + 2\widetilde{\boldsymbol{x}}_\theta^T \boldsymbol{P}_0 \boldsymbol{\delta}_\theta \\ &= -\boldsymbol{q}_{ve}^T \boldsymbol{K}_1 \boldsymbol{q}_{ve} + \widetilde{\boldsymbol{\omega}}_e^T (\boldsymbol{J}_0 (S(\boldsymbol{\omega}_d) \boldsymbol{R}_D^B \boldsymbol{\omega}_d - \boldsymbol{R}_D^B \dot{\boldsymbol{\omega}}_d) + \boldsymbol{u} + \boldsymbol{d} + \boldsymbol{J}_0 \boldsymbol{K}_1 \dot{\boldsymbol{q}}_{ve} + \boldsymbol{q}_{ve}) - k_0 \left(\widetilde{\boldsymbol{x}}_\phi^T \widetilde{\boldsymbol{x}}_\phi + \widetilde{\boldsymbol{x}}_\theta^T \widetilde{\boldsymbol{x}}_\theta\right) + 2\left(\widetilde{\boldsymbol{x}}_\phi^T \boldsymbol{P}_0 \boldsymbol{\delta}_\phi + \widetilde{\boldsymbol{x}}_\theta^T \boldsymbol{P}_0 \boldsymbol{\delta}_\theta\right) \end{aligned} \tag{47}$$

Define the nominal control input vector as:

$$\boldsymbol{u} = -\boldsymbol{J}_0 \left(S(\boldsymbol{\omega}_d) \boldsymbol{R}_D^B \boldsymbol{\omega}_d - \boldsymbol{R}_D^B \dot{\boldsymbol{\omega}}_d\right) - \boldsymbol{J}_0 \boldsymbol{K}_1 \dot{\boldsymbol{q}}_{ve} - \boldsymbol{q}_{ve} - \boldsymbol{K}_2 \widetilde{\boldsymbol{\omega}}_e - \hat{\boldsymbol{d}} \tag{48}$$

where $\hat{\boldsymbol{d}} = \begin{bmatrix} C_d \hat{\zeta}_\phi & C_d \hat{\zeta}_\theta & 0 \end{bmatrix}^T$ is the estimation results of disturbance. Then, plug Equations (40) and (48) into Equation (47):

$$\begin{aligned} \dot{V}_2 &= -\boldsymbol{q}_{ve}^T \boldsymbol{K}_1 \boldsymbol{q}_{ve} - k_0 \left(\widetilde{\boldsymbol{x}}_\phi^T \widetilde{\boldsymbol{x}}_\phi + \widetilde{\boldsymbol{x}}_\theta^T \widetilde{\boldsymbol{x}}_\theta\right) - \widetilde{\boldsymbol{\omega}}_e^T \boldsymbol{K}_2 \widetilde{\boldsymbol{\omega}}_e + \widetilde{\boldsymbol{\omega}}_e^T \widetilde{\boldsymbol{d}} + 2\left(\widetilde{\boldsymbol{x}}_\phi^T \boldsymbol{P}_0 \boldsymbol{\delta}_\phi + \widetilde{\boldsymbol{x}}_\theta^T \boldsymbol{P}_0 \boldsymbol{\delta}_\theta\right) \\ &\leq -\lambda_{\min}(\boldsymbol{K}_1) \|\boldsymbol{q}_{ve}\|^2 - k_0 \left(\|\widetilde{\boldsymbol{x}}_\phi\|^2 + \|\widetilde{\boldsymbol{x}}_\theta\|^2\right) - \lambda_{\min}(\boldsymbol{K}_2) \|\widetilde{\boldsymbol{\omega}}_e\|^2 + \|\widetilde{\boldsymbol{\omega}}_e\| \|\widetilde{\boldsymbol{d}}\| + 2\sigma \lambda_{\max}(\boldsymbol{P}_0) (\|\widetilde{\boldsymbol{x}}_\phi\| + \|\widetilde{\boldsymbol{x}}_\theta\|) \\ &\leq -\lambda_{\min}(\boldsymbol{K}_1) \|\boldsymbol{q}_{ve}\|^2 - k_0 \left(\|\widetilde{\boldsymbol{x}}_\phi\|^2 + \|\widetilde{\boldsymbol{x}}_\theta\|^2\right) - \lambda_{\min}(\boldsymbol{K}_2) \|\widetilde{\boldsymbol{\omega}}_e\|^2 + \sqrt{3} \|\widetilde{\boldsymbol{\omega}}_e\| (\|\widetilde{\boldsymbol{x}}_\phi\| + \|\widetilde{\boldsymbol{x}}_\theta\|) + 2\sigma \lambda_{\max}(\boldsymbol{P}_0) (\|\widetilde{\boldsymbol{x}}_\phi\| + \|\widetilde{\boldsymbol{x}}_\theta\|) \\ &\leq -\lambda_{\min}(\boldsymbol{K}_1) \|\boldsymbol{q}_{ve}\|^2 - \left(k_0 - \tfrac{\sqrt{3}}{2}\right) \left(\|\widetilde{\boldsymbol{x}}_\phi\|^2 + \|\widetilde{\boldsymbol{x}}_\theta\|^2\right) - \left(\lambda_{\min}(\boldsymbol{K}_2) - \sqrt{3}\right) \|\widetilde{\boldsymbol{\omega}}_e\|^2 + 2\sigma \lambda_{\max}(\boldsymbol{P}_0) (\|\widetilde{\boldsymbol{x}}_\phi\| + \|\widetilde{\boldsymbol{x}}_\theta\|) \end{aligned} \tag{49}$$

where $\|\tilde{d}\| = \sqrt{\|C_d\tilde{\zeta}_\phi\|^2 + \|C_d\tilde{\zeta}_\theta\|^2} \le \|C_d\tilde{\zeta}_\phi\| + \|C_d\tilde{\zeta}_\theta\| \le \sqrt{3}(\|\tilde{x}_\phi\| + \|\tilde{x}_\theta\|)$, $\lambda_{\min}(K_1)$ and $\lambda_{\min}(K_2)$ are the minimal eigenvalue of K_1 and K_2 respectively. Define $z^T = (q_e^T, \tilde{\omega}_e^T, \tilde{x}_\phi^T, \tilde{x}_\theta^T)$ as the uniformed vector of errors, and $\xi = \min\left(\lambda_{\min}(K_1), \lambda_{\min}(K_2) - \sqrt{3}, k_0 - \frac{\sqrt{3}}{2}\right) > 0$, then Equation (49) can be reduced to

$$\dot{V}_3 \le -\xi\left(\|q_{ve}\|^2 + \|\tilde{\omega}_e\|^2 + \|\tilde{x}_\phi\|^2 + \|\tilde{x}_\theta\|^2\right) + 2\sigma\lambda_1(\|\tilde{x}_\phi\| + \|\tilde{x}_\theta\|) \le -\xi\|z\|^2 + 2\sigma\lambda_{\max}(P_0)\|z\| \quad (50)$$

Thus $\dot{V}_3 < 0$ whenever $\|z\| > 2\sigma\lambda_{\max}(P_0)\xi^{-1}$. Notice that $(q_e, \tilde{\omega}_e)$ is a linear diffeomorphism of (q_e, ω_e), hence (q_e, ω_e) can converge into a compact set. We can conclude that, the attitude error (q_e, ω_e), virtual control input error $\tilde{\omega}_e$ and the estimation error $\tilde{x}_\phi$ and $\tilde{x}_\theta$ are uniformly ultimately bounded and exponentially converges to the bounded ball $B_z = \{ z \mid \|z\| \le 2\sigma\lambda_{\max}(P_0)\xi^{-1}\}$.

In general, when we chose a larger ξ, the bounded ball B_z will become smaller and consequently, $\|z\|$ will also become smaller, so a larger ξ is preferred. However, larger ξ will lead to larger control gains which can excite the sensor noise and undesirable high frequency dynamics of the system. Thus, the tuning of controller parameters is a tradeoff between the demand of performance and the real conditions. Moreover, by improving the model accuracy of the disturbance, σ will be reduced, therefore, the accuracy of attitude control can be improved.

5. Simulation and Experimental Results

In order to evaluate the performance of the proposed HESO, simulation and real world experimental results are presented in this section.

5.1. Simulation Results

We present the numerical simulations of the proposed GESO based DUEA control strategy on a model generated by the online toolbox of Quan and Dai [32], and the values of the nominal model parameters are list in Table 1.

Table 1. Quadrotor parameters used in simulation.

Parameter	Description	Value
m_q	Mass of the quadrotor	2 kg
m_l	Mass of the slung load	1 kg
c	Hook position	0.2 m
L_c	Length of the cable	1 m
J_x	Roll inertia	1.335×10^{-2} kg·m^2
J_y	Pitch inertia	1.335×10^{-2} kg·m^2
J_z	Yaw inertia	2.465×10^{-2} kg·m^2
l	Motor moment arm	0.18 m
g	Gravity acceleration	9.81 g·s^2
k_T	Aerodynamic coefficient	8.54×10^{-6} kg·m
k_D	Drag coefficient	1.36×10^{-7} kg·m^2

The values of gain parameters used in our controller are given as $K_1 = 7I_3$ and $K_2 = 2I_3$, and the observer gains of HESO are $L_{e,i} = [\ 6 \quad 27 \quad 27 \quad 8 \quad 8 \quad 2 \quad 2\ ]^T$. Traditional 2nd order ESO is used as the comparison, and the bandwidth is fix at 3 rad/s, the observer gain is $L = \begin{bmatrix} 9 & 27 & 27 \end{bmatrix}^T$.

5.1.1. Comparison in Attitude Stabilization Performance

This part involves attitude stabilization control in the presence of slung load. The initial condition of the slung load is $\gamma_0 = 30°$ and $\beta = 20°$. The initial conditions used of the quadrotor in the simulation are zero, and we select the desired attitude signal as follows $\Omega_d = [\ 0° \quad 0° \quad 0°\ ]^T$.

Three comparative simulations were conducted, and the performances of the proposed controller with different ESOs are compared in pitch and roll channel. Figures 3 and 4 show the curves of the vehicle's attitude response during its flight. We can see that although the proposed controller (without GESO) was able to ensure the stabilization of the attitude angles, the control accuracy was reduced under the influence of the slung load motion. When we introduced Traditional 2nd order ESO and HESO as the DUE, the performances of the proposed controller were improved, and through the comparison of attitude control results, we can conclude that HESO performed better than traditional ESO. Moreover, Figures 5 and 6 show the disturbance estimation result of HESO and Traditional ESO. Obviously, we can see that the estimation error decreases when we adopt HESO.

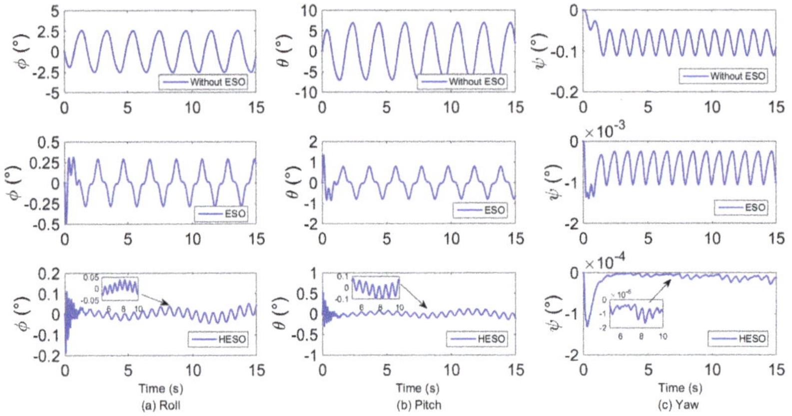

Figure 3. Simulation curves of the attitude angle in attitude stabilization: (**a**) roll channel; (**b**) pitch channel; (**c**) yaw channel.

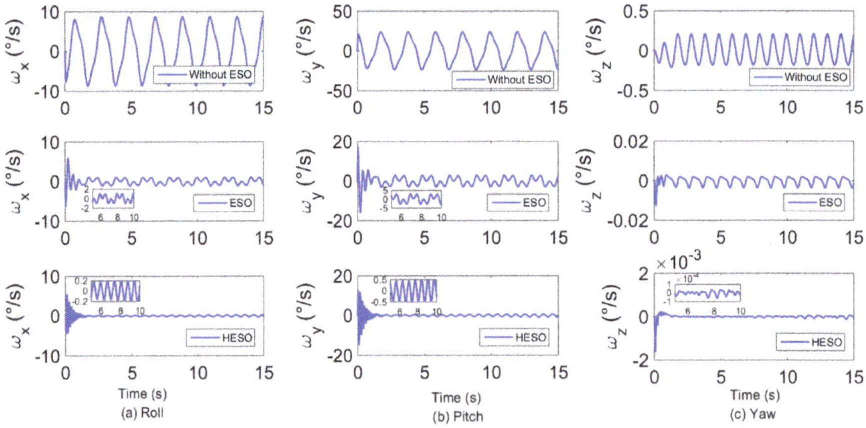

Figure 4. Simulation curves of the attitude rate in attitude stabilization: (**a**) roll channel; (**b**) pitch channel; (**c**) yaw channel.

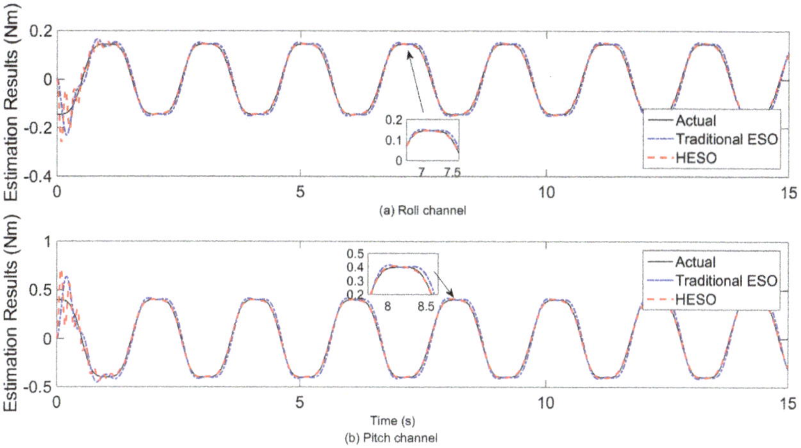

Figure 5. Estimation results comparison of harmonic extended state observer (HESO) and traditional extended state observer (ESO) in attitude stabilization.

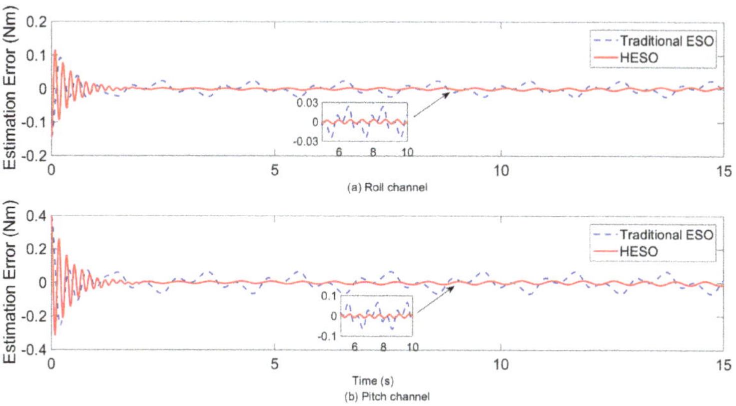

Figure 6. Estimation errors comparison of HESO and traditional ESO in attitude stabilization.

5.1.2. Comparison in Attitude Tracking Performance

In this case, the numerical simulation demonstrates the effectiveness of the proposed control scheme for attitude tracking. The initial condition of the slung load is $\gamma_0 = 0°$. We chose desired attitude signal as $\phi_d = 0°$ and

$$\theta_d = \begin{cases} 10°, & 1 < t < 9 \\ 0°, & 0 < t \leq 1 \cup 9 \leq t \end{cases} \tag{51}$$

The attitude tracking performances of the proposed controller with different ESOs are illustrated in Figures 7 and 8. We can see that the proposed controller alone was hard to ensure the tracking of the desired attitude angles, and the control accuracy was severely affected by the slung load. When we introduced Traditional 2nd ESO and HESO as the DUE, the performances of the proposed controller were improved, and the simulation results show that HESO performed better than traditional ESO. Moreover, as shown in Figure 9, we can see that the estimation error decreases when we adopt HESO.

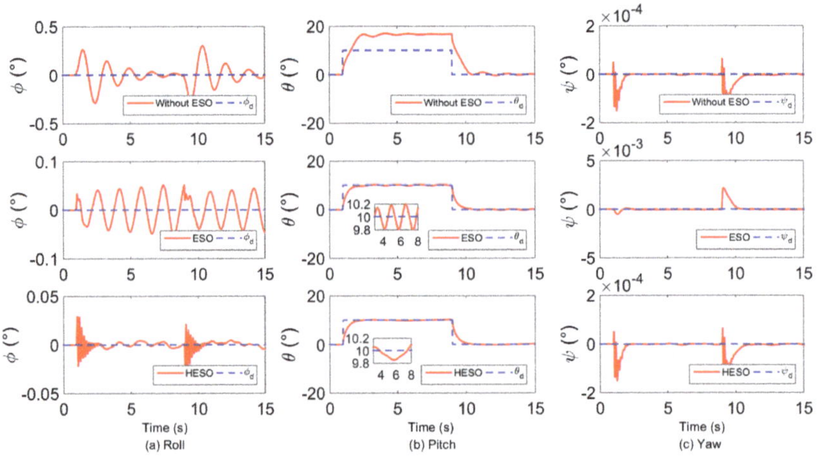

Figure 7. Simulation curves of the attitude angle in attitude tracking: (a) roll channel; (b) pitch channel; (c) yaw channel.

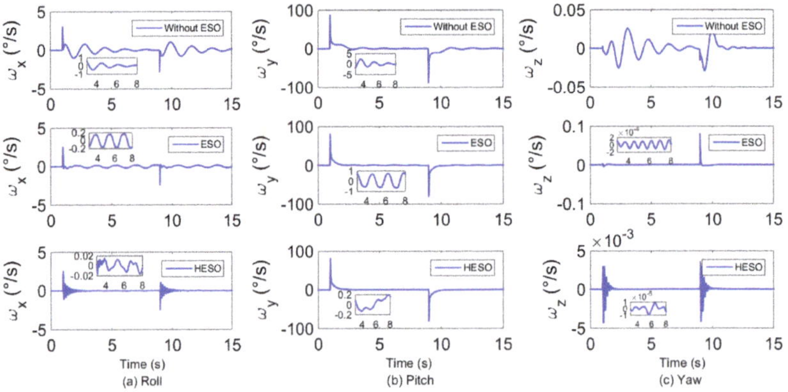

Figure 8. Simulation curves of the attitude rate in attitude tracking: (a) roll channel; (b) pitch channel; (c) yaw channel.

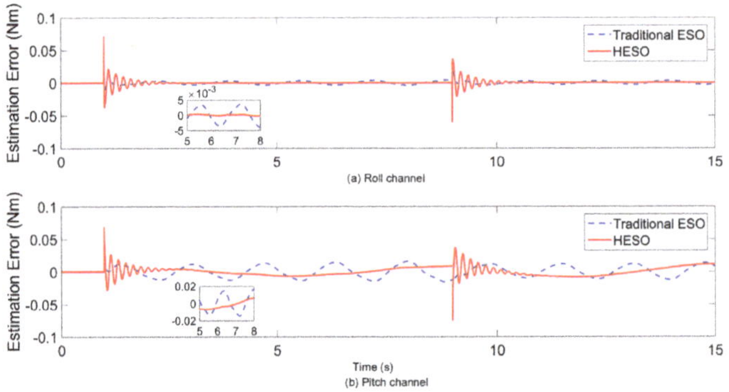

Figure 9. Estimation errors comparison of HESO and traditional ESO in attitude tracking.

5.2. Experimental Results

We have also tested the proposed control scheme on a self-assembled GF360 quadrotor, and PIXHAWK [33,34] was used as the autopilot of the quadrotor. To evaluate the stability and robustness of the proposed control scheme, the experiments were carried out as follows. The weight of the slung load m_l is 0.5 kg and cable length L_c is 1 m, the distance between the hook of the cable and the mass center of quadrotor c is about 20 cm.

Subject to the sampling frequency and noise of MEMS gyro, the observer gain of HESO is chosen as $\boldsymbol{L}_e = \begin{bmatrix} 6 & 12 & 8 & 12 & 8 & 2 & 2 \end{bmatrix}^T$. And subject to the limitations of computing power, HESO is only used in pitch channel, and the slung load only swing in longitudinal plane. The controller gains were chosen as $\boldsymbol{K}_1 = \mathrm{diag}[\ 7 \quad 7 \quad 2.8\ ]$ and $\boldsymbol{K}_2 = \mathrm{diag}[\ 0.15 \quad 0.15 \quad 0.2\ ]$. Three comparative experiments with different ESO are presented.

Figure 10 presents the history of pitch angle tracking errors $\theta_e = \theta - \theta_d$ obtained by the proposed controller with different ESO in the presence of slung load. Figure 11 shows the experiment results of pitch rate ω_y in these three cases. It can be observed that the control performance is improved by HESO compared with the one without ESO and the one with traditional 2nd order ESO. In addition. The root mean square (RMS) errors obtained by proposed controller with different ESO are given in Table 2. It is illustrated that HESO can achieve 50.11% reduction on the RMS error of pitch angle.

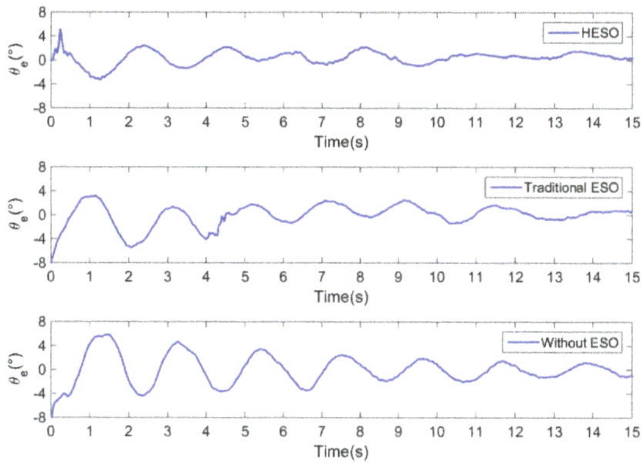

Figure 10. Experimental curves of θ with different ESO.

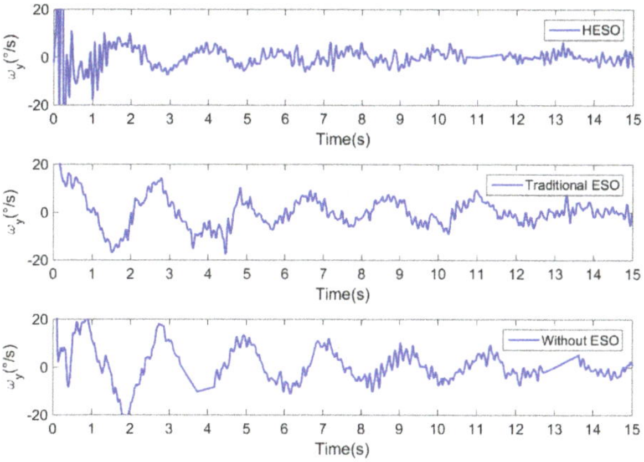

Figure 11. Experimental curves of ω_y with different ESO.

Table 2. Comparison of control performance in the tracking error of pitch angle with different ESO (RMS error).

	Without ESO	Traditional ESO	HESO
θ_e (°)	2.4696	1.8890	1.2320

6. Conclusions

In this paper, the problem of anti-swing attitude control for quadrotor in the presence of slung load is investigated. In the proposed approach, we developed a HESO based DUEA control scheme to accurately attenuate the disturbance torque caused by slung load. Through the analysis of the quadrotor-slung load system, the external disturbance torque can be described as a periodic function with two frequencies $\varpi_1 = \sqrt{gL_c^{-1}}$,$\varpi_2 = 3\sqrt{gL_c^{-1}}$, while the amplitude and phase are unknown. Then the harmonic extended state observer is designed according to the model of the disturbance in propose of enhancing the robustness of feedback nonlinear controller. Subsequently, an attitude tracking controller is designed based on the backstepping method. From the simulation and experimental results, we can conclude that compared to the traditional 2nd order ESO, the HESO can achieve a better performance in estimating the periodic disturbances, so that the robustness of the proposed controller can be further improved.

Author Contributions: Formal analysis, D.S.; Project administration, Z.W. and W.C.; Software, D.S.; Supervision, Z.W.

Acknowledgments: This work was supported by the State Key Development Program for Basic Research of China (Grant No. 2013CB035503).

Conflicts of Interest: The authors declare no conflicts of interest.

References

1. Bouabdallah, S.; Becker, M.; Siegwart, R. Autonomous miniature flying robots: Coming soon! *IEEE Robot. Autom. Mag.* **2007**, *14*, 88–98. [CrossRef]
2. Satici, A.C.; Poonawala, H.; Spong, M.W. Robust optimal control of quadrotor uavs. *IEEE Access* **2013**, *1*, 79–93. [CrossRef]
3. Stepaniak, M.J.; van Graas, F.; de Haag, M.U. Design of an electric propulsion system for a quadrotor unmanned aerial vehicle. *J. Aircr.* **2009**, *46*, 1050–1058. [CrossRef]
4. Gupte, S.; Mohandas, P.I.T.; Conrad, J.M. A survey of quadrotor unmanned aerial vehicles. In Proceedings of the IEEE Southeastcon, Orlando, FL, USA, 15–18 March 2012.
5. Ozbek, N.S.; Onkol, M.; Efe, M.O. Feedback control strategies for quadrotor-type aerial robots: A survey. *Trans. Inst. Meas. Control* **2016**, *38*, 529–554. [CrossRef]
6. Bisgaard, M.; la Cour-Harbo, A.; Bendtsen, J.D. Adaptive control system for autonomous helicopter slung load operations. *Control Eng. Pract.* **2010**, *18*, 800–811. [CrossRef]
7. Omar, H.M. Designing anti-swing fuzzy controller for helicopter slung-load system near hover by particle swarms. *Aerosp. Sci. Technol.* **2013**, *29*, 223–234. [CrossRef]
8. Feng, Y.; Rabbath, C.A.; Rakheja, S.; Su, C.Y. Adaptive controller design for generic quadrotor aircraft platform subject to slung load. In Proceedings of the IEEE 28th Canadian Conference on Electrical and Computer Engineering (CCECE), Halifax, NS, Canada, 3–6 May 2015; pp. 1135–1139.
9. Huang, M.; Xian, B.; Diao, C.; Yang, K.Y.; Feng, Y. Adaptive tracking control of underactuated quadrotor unmanned aerial vehicles via backstepping. In Proceedings of the American Control Conference (ACC), Baltimore, MD, USA, 30 June–2 July 2010; pp. 2076–2081.
10. Alexis, K.; Nikolakopoulos, G.; Tzes, A. Model predictive quadrotor control: Attitude, altitude and position experimental studies. *IET Control Theory Appl.* **2012**, *6*, 1812–1827. [CrossRef]
11. Sadr, S.; Moosavian, S.A.A.; Zarafshan, P. Dynamics modeling and control of a quadrotor with swing load. *J. Robot.* **2014**, *2014*, 265897. [CrossRef]

12. Palunko, I.; Cruz, P.; Fierro, R. Agile load transportation: Safe and efficient load manipulation with aerial robots. *IEEE Robot. Autom. Mag.* **2012**, *19*, 69–79. [CrossRef]
13. De Crousaz, C.; Farshidian, F.; Neunert, M.; Buchli, J. Unified motion control for dynamic quadrotor maneuvers demonstrated on slung load and rotor failure tasks. In Proceedings of the IEEE International Conference on Robotics and Automation (ICRA), Seattle, WA, USA, 26–30 May 2015; pp. 2223–2229.
14. Guerrero-Sanchez, M.E.; Mercado-Ravell, D.A.; Lozano, R.; Garcia-Beltran, C.D. Swing-attenuation for a quadrotor transporting a cable-suspended payload. *ISA Trans.* **2017**, *68*, 433–449. [CrossRef] [PubMed]
15. Wang, H.D.; Huang, Y.B.; Xu, C. Adrc methodology for a quadrotor uav transporting hanged payload. In Proceedings of the IEEE International Conference on Information and Automation (ICIA), Ningbo, China, 1–3 August 2016; pp. 1641–1646.
16. Chen, W.H.; Ohnishi, K.; Guo, L. Advances in disturbance/uncertainty estimation and attenuation. *IEEE Trans. Ind. Electron.* **2015**, *62*, 5758–5762. [CrossRef]
17. Yang, J.; Chen, W.H.; Li, S.H.; Guo, L.; Yan, Y.D. Disturbance/uncertainty estimation and attenuation techniques in pmsm drives-a survey. *IEEE Trans. Ind. Electron.* **2017**, *64*, 3273–3285. [CrossRef]
18. Wang, Z.; Wu, Z. Nonlinear attitude control scheme with disturbance observer for flexible spacecrafts. *Nonlinear Dyn.* **2015**, *81*, 257–264. [CrossRef]
19. Wang, Z.; Wu, Z.; Du, Y.J. Adaptive sliding mode backstepping control for entry reusable launch vehicles based on nonlinear disturbance observer. *Proc. Inst. Mech. Eng. Part G J. Aerosp. Eng.* **2016**, *230*, 19–29. [CrossRef]
20. Pu, Z.Q.; Yuan, R.Y.; Yi, J.Q.; Tan, X.M. A class of adaptive extended state observers for nonlinear disturbed systems. *IEEE Trans. Ind. Electron.* **2015**, *62*, 5858–5869. [CrossRef]
21. Xia, Y.Q.; Pu, F.; Li, S.F.; Gao, Y. Lateral path tracking control of autonomous land vehicle based on adrc and differential flatness. *IEEE Trans. Ind. Electron.* **2016**, *63*, 3091–3099. [CrossRef]
22. Busawon, K.K.; Kabore, P. Disturbance attenuation using proportional integral observers. *Int. J. Control* **2001**, *74*, 618–627. [CrossRef]
23. Gao, Z.W.; Breikin, T.; Nang, H. Discrete-time proportional and integral observer and observer-based controller for systems with both unknown input and output disturbances. *Optim. Control Appl. Methods* **2008**, *29*, 171–189. [CrossRef]
24. Han, J.Q. From pid to active disturbance rejection control. *IEEE Trans. Ind. Electron.* **2009**, *56*, 900–906. [CrossRef]
25. Wang, W.W.; Gao, Z.Q. A comparison study of advanced state observer design techniques. In Proceedings of the 2003 American Control Conference, Denver, CO, USA, 4–6 June 2003; pp. 4754–4759.
26. Madonski, R.; Herman, P. Survey on methods of increasing the efficiency of extended state disturbance observers. *ISA Trans.* **2015**, *56*, 18–27. [CrossRef] [PubMed]
27. Godbole, A.A.; Kolhe, J.P.; Talole, S.E. Performance analysis of generalized extended state observer in tackling sinusoidal disturbances. *IEEE Trans. Control Syst. Technol.* **2013**, *21*, 2212–2223. [CrossRef]
28. Zhang, Y.J.; Zhang, J.; Wang, L.; Su, J.B. Composite disturbance rejection control based on generalized extended state observer. *ISA Trans.* **2016**, *63*, 377–386. [CrossRef] [PubMed]
29. Chen, W.H. Harmonic disturbance observer for nonlinear systems. *J. Dyn. Syst. Meas. Control* **2003**, *125*, 114–117. [CrossRef]
30. Chovancova, A.; Fico, T.; Hubinsky, P.; Duchon, F. Comparison of various quaternion-based control methods applied to quadrotor with disturbance observer and position estimator. *Robot. Auton. Syst.* **2016**, *79*, 87–98. [CrossRef]
31. Bouabdallah, S.; Siegwart, R. Full control of a quadrotor. In Proceedings of the IEEE/RSJ International Conference on Intelligent Robots and Systems, San Diego, CA, USA, 29 October–2 November 2007; pp. 153–158.
32. Quan, Q.; Dai, X. Flight Performance Evaluation of Uavs. Available online: http://flyeval.com/ (accessed on 20 April 2018).

33. Meier, L.; Tanskanen, P.; Heng, L.; Lee, G.H.; Fraundorfer, F.; Pollefeys, M. Pixhawk: A micro aerial vehicle design for autonomous flight using onboard computer vision. *Auton. Robot.* **2012**, *33*, 21–39. [CrossRef]
34. Meier, L.; Tanskanen, P.; Fraundorfer, F.; Pollefeys, M. The pixhawk open-source computer vision framework for mavs. *Int. Arch. Photogramm.* **2011**, *38*, 13–18. [CrossRef]

electronics

MDPI

Article

Recursive Rewarding Modified Adaptive Cell Decomposition (*RR-MACD*): A Dynamic Path Planning Algorithm for UAVs

Franklin Samaniego *, Javier Sanchis, Sergio García-Nieto and Raúl Simarro

Instituto Universitario de Automática e Informática Industrial, Universitat Politècnica de València, 46022 Valencia, Spain; jsanchis@isa.upv.es (J.S.); sgnieto@isa.upv.es (S.G.-N.); rausifer@upvnet.upv.es (R.S.)

* Correspondence: frank7083@gmail.com; Tel.: +34-637-682907

Received: 28 December 2018; Accepted: 5 March 2019; Published: 8 March 2019

Abstract: A relevant task in unmanned aerial vehicles (UAV) flight is path planning in 3*D* environments. This task must be completed using the least possible computing time. The aim of this article is to combine methodologies to optimise the task in time and offer a complete 3*D* trajectory. The flight environment will be considered as a 3*D* adaptive discrete mesh, where grids are created with minimal refinement in the search for collision-free spaces. The proposed path planning algorithm for UAV saves computational time and memory resources compared with classical techniques. With the construction of the discrete meshing, a cost response methodology is applied as a discrete deterministic finite automaton (DDFA). A set of optimal partial responses, calculated recursively, indicates the collision-free spaces in the final path for the UAV flight.

Keywords: UAV; path planning; adaptive discrete mesh; octree

1. Introduction

The world market for unmanned aerial vehicles (UAVs) is expanding rapidly, and there are various forecasts and projections regarding the market for unmanned vehicles. The economic impact of integrating UAVs into the National Airspace System in the United States will grow substantially and reach more than $82.1 billion between 2015 and 2025 [1].

A wide diversity of air missions can be completed by UAVs [2,3] in various scenarios and including outdoor/indoor and water/ground/air/space environments [4]. The types of missions include military (missile launching drones, bomb-dropping drones, flying camouflaged drones) and civilian (video-graph/photography, disaster response, environment and climate) [5–9]. A highly demanded task for UAVs is 3*D* autonomous navigation (either in static or dynamic environments) that optimises the route and minimises the computational cost. Thus, path planning defines the methodology that an autonomous robot must complete to move from an initial location to a final location, deploying its own resources as sensors, actuators, and strategies, while avoiding obstacles during the trip. Several path planning and obstacle avoidance techniques are being used in unmanned ground vehicles (UGVs), autonomous underwater vehicles (AUVs), and unmanned aerial vehicles (UAVs).

From the traditional robotics point of view, numerous works have been developed in which the path planning and obstacle avoidance algorithms perform searches in continuous or discrete Euclidean [10] dimensional movement environments. It is important to mention that LaValle in [11] has done significant work on sampling-based path planning algorithms. However, although his analysis is complete from a two-dimensional perspective, 3*D* planning analysis is not completely addressed. An exhaustive study of the growing work on sampling-based algorithms is presented in [12]. It must be remembered that the 2*D* path planning problem is NP-hard; and so environmental dimensional increases and UAV kinematics affect problem complexity.

An in-depth review of the current literature shows several works focus on two-dimensional ($2D$) scenarios [13] that limit vehicle behaviour to just a flat surface and consider its height as constant by making a dimensional analysis ($2.5D$) [14]. However, in complex unstructured situations (including, for example, forests, urban, or underwater environments) a simple $2D$ algorithm is insufficient and $3D$ path planning is needed.

A diversity of methodological paradigms have been developed to complete the task of $3D$ path planning. These are based on sampling, node/based algorithms, bio-inspired algorithms, and mathematical models, among other techniques. A brief bibliographic review focused on $3D$ trajectory planning is presented below.

Some representative techniques used in path planning methods and based on continuous and discrete environment sampling include: RRT (rapidly-exploring random tree) [15–18]; PRM (probabilistic road maps) [19–23]; Voronoi diagrams [24–26]; and artificial potential [27–30]. Nevertheless, it is important to note that RRT and PRM make random explorations (continuous sampling) of the defined environment. RRT is an expensive algorithm in terms of computational cost when searching for feasible solutions in cluttered environments. It should be emphasised that once the PRM road map is made, a methodological base built on nodes must be invoked to define the lowest cost path. The main disadvantage of the Voronoi diagram is that it is an offline method. Finally, artificial potential algorithms present little computational complexity—although they tend to fall into local minimums.

Node-based algorithms (discrete space) are mathematical structures used to model pairwise relations (in this context, the structures are made with vertices and edges) and the aim is to calculate the cost of exploring nodes to find the optimal path. Various methodologies and subsequent variations, such as Dijkstra's algorithm [31,32], A* [33,34], D* [35,36], and Theta* [37], present these characteristics in their results. In [38] the characteristics and approximations of various methodologies of * (Star) search algorithms are studied.

In recent years, these classical techniques have been improved with new learning machine techniques. ANN (Artificial Neural Networks) [39–41], fuzzy logic [42,43], ACO (Ant Colony Optimisation) [44,45], and PSO (Particle Swarm Optimisation) [46,47], among others [48–50], are examples of these heuristic methodologies. Hence, these biological algorithms attempt to optimise the path by mimicking animal behaviour. The weaknesses and strengths of a set of heuristic techniques are discussed in [51]. The implementation relevance of these methodologies does not present significant experimental results. In addition, the different techniques presented in this section have a particular computational cost and complexity based on the different approaches [52] (see Table 1).

Table 1. Computational cost and complexity in the graph structure, where n is the number of vertex and m is the number of edges.

Method	Time Complexity	Memory	Real Time
Sampling based algorithms	$0(nlogn)$	$0(n^2)$	On-line
Node based algorithms	$0(mlogn)$	$0(n^2)$	On-line
Bioinspired algoritms	$0(nlogn)$	$0(n^2)$	Off-line

A summary of the above mentioned methodologies is shown in Table 2, which details the approximation methodology, authors and reference, type of obstacle avoidance (static or dynamic), type of implementation (simulation or real), and publication year.

The $3D$ path planning problem is still an open issue in this field. The general approach is to combine several of the above mentioned techniques to improve overall performance.

Several planners optimise the path planning distance. However, this paper attempts to include distance as an objective, as well as the geometrical characteristics of the UAV and flight constraints (velocity, turning capacity, battery, flight distance, etc.). All of these constraints are evaluated as potential cost and the path planning result is based on the sum of contributions for each cost (see

Section 4). It is important to highlight that planning results do not attempt to arrive at an optimal path in a shorter distance. Furthermore, unlike other path planning methodologies in which pruning of the results is necessary, this paper attempts to minimise such pruning.

Table 2. 3D path planning methodologies studied list.

Approach	Authors	Static Obstacle	Dynamic Obstacle	Simulation	Real	Year
RRT	Abbadi, A. [15]	x	x	x	o	[2012]
	Aguilar, W. [16]	o	x	x	x	[2016]
	Aguilar, W. [17]	x	o	x	x	[2017]
	Yao, P. [18]	x	o	x	o	[2017]
PRM	Yan, F. [19]	x	o	o	x	[2013]
	Yeh, H. [20]	x	o	x	o	[2012]
	Denny, J. [21]	x	o	x	o	[2013]
	Li, Q. [22]	x	o	x	o	[2014]
	Ortiz-Arroyo, D. [23]	x	o	x	o	[2015]
Voronoi	Thanou, M. [24]	x	o	x	o	[2014]
	Qu, Y. [25]	x	o	x	o	[2014]
	Fang, Z. [26]	x	o	x	x	[2017]
Artificial Potencial	Khuswendi, T. [27]	x	x	x	o	[2011]
	Chen, X. [28]	x	x	x	o	[2013]
	Rivera, D. [29]	x	x	x	o	[2012]
	Liu L. [30]	x	x	x	o	[2016]
ANN	Kroumov, V. [39]	x	o	x	o	[2010]
	Gautam, S. [40]	x	o	x	o	[2014]
	Maturana, D. [41]	x	o	o	x	[2015]
Fuzzy Logic	Iswanto, I. [42]	x	x	x	o	[2016]
	LIU, S. [43]	x	o	x	o	[2012]
ACO	Duan, H. [44]	x	o	x	o	[2010]
	He, Y. [45]	x	o	x	o	[2013]
PSO	Zhang, Y. [46]	x	x	x	o	[2013]
	Goel, U. [47]	x	x	x	o	[2018]
Others	YongBo, C. [48]	x	o	x	o	[2017]
	Wang, G. [49]	x	o	x	o	[2016]
	Aghababa, M. [50]	x	o	x	o	[2012]

An adaptive cell decomposition (ACD) is a strong methodology for solving physical systems led by partial differential equations [53,54]. Such techniques offer a substantial improvement in computational time and discretisation is not governed by a dominant equations system. This methodology is used in accurate complex 3*D* Cartesian geometry reconstructions [55,56]. The approach presented in this work does not seek a refined environment reconstruction, and only tries to determine occupied and free spaces within the 3*D* Cartesian space. The savings in computational and memory effort is significant. The computational structure that constructs the algorithm makes a rapid labelling of the geometric figure of the environment as a 3*D* solid with a rectangular shape.

In this paper, a functional 3*D* UAV path planning algorithm is proposed that is based on an evolution of the (ACD) method. The proposal attempts to achieve a linear speedup, exploring and decomposing the 3*D* environment under a recursive reward cost paradigm, and building an efficient and simple 3*D* path detection. The aim is not to generate a large scale reconstruction of the environment, nor start the procedure with a defined cloud of points [57]. In the presented paper, the UAV just receives the obstacle information from the control station and generates a trajectory. Over time, physical phenomena often generate unknown space distributions between the UAV and obstacles, and so adaptation according these changes and spatial constraints might exist. In the event of obstacle collision with the previously calculated path, a new estimated path is generated.

This paper is organised as follows. Section 2 defines the terrain representation and codification obstacles, and the general problem of path planning under static and dynamic environments is stated. In Section 3, the basis of the adaptive cell decomposition technique is revisited and modified to be ready for our proposal. In Section 4, the new algorithm for planned paths is then explained, and finally, several application examples are shown in Section 5. Conclusions and future works are considered in the final section.

2. Problem Definition

At the moment when obstacles in the 3*D* real world are represented, they do not possess exact geometries, and for simplicity in this paper, obstacles have been modelled as cuboids in direct relation to their dimensional characteristics and location. When cuboids increase or decrease, special care is taken to ensure that they do not collide with each other and there is a free flight route during the environment tests.

In a 3*D* environment where a UAV performs a continuous flight from an *init* point (q_i) to a *goal* point (q_f), a set of various manoeuvres to complete this mission are deployed. The UAV had previously defined the trajectory to follow after taking into account several considerations and constraints.

Let us assume that a complete description of the possible operating environment as an urban space in which buildings of different dimensions are defined. The UAV in flight receives data from its control station about the environmental conditions and it makes the necessary calculations to determine the best trajectory. The relevant data includes the goal point q_f, the current location, and the size of the obstacles (static or dynamic), as well as speed and movement directions. Since q_i is related directly with the current UAV location, the aim is to apply a discrete decomposition (partial and recursive) of the environment to find the set of collision-free spaces for the UAV flight and head towards the middle of those collision-free spaces until it arrives at q_f. Hence, the final trajectory result of this methodology generates a vector (x_i, y_i, z_i) of three-dimensional points (system coordinates) translated as waypoints. Furthermore, it is important to emphasise that the resulting vector indicates spatial positions, and therefore the UAV that performs the trajectory tracking must possess these tracking skills. Thus, a UAV that has a quadropter-type holonomic system (the number of controllable degrees of freedom of the UAV system is equal to the total degrees of freedom) can complete the 3*D* waypoint tracking. Hence, a non-holonomic (the system is described by a set of parameters subject to differential constraints) (fixed wing) UAV does not have to be able to follow these trajectories.

Figure 1 shows an environmental example, where the discrete decomposition is built around the obstacles that interfere with the UAV flight. Hence, the three-dimensional characteristics of the obstacle might be considered to determine an escape trajectory that can surround the obstacle (including its sides and above and below) in continuous flight. Therefore, the 3*D* environment decomposition will take advantage of the 3*D* displacement capabilities of the UAV.

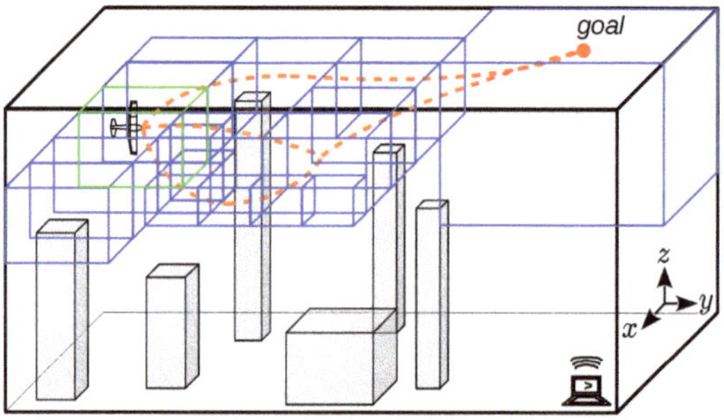

Figure 1. General scenario for 3*D* unmanned aerial vehicles (UAV) path planning. The grey cubes depict the environmental obstacles. Blue cubes are generated by the environmental discretisation and represent collision-free spaces. Orange lines are possible paths.

3. Modified Adaptive Cell Decomposition (*MACD*)

A standard 3*D* ACD algorithm attempts to make a discrete approximation of the environment, typically in a tree data structure known as octree (Octree is a tree data structure in which each internal node has exactly eight children) [58]. This process requires considerable computational resources and time. The modified adaptive cell decomposition *(MACD)* [59] does not make an exhaustive routing for each little space in the environment. If an obstacle exists, the routing in the three dimensions is in direct relation to the obstacle characteristics. The parameterised decomposition level is fixed in direct relation to the UAV manoeuvrability. This means that whenever the UAV needs a minimal space to complete a movement, this value will be defined in the level of decomposition.

Let us say $Y = (x, y, z)$ denotes a discrete 3*D* environment, where the set of collision-free voxels are defined as S_{free}, the set of occupied voxels are defined as S_{occup}, and the optimal path (metric term of distance) between q_i and q_f as ρ. The aim is to find the optimal path ρ compound with a set of the nearest voxels in the S_{free} space that enclose the obstacles.

The procedure is summarised in the flowchart shown in Figure 2, beginning with a specification of the number of decomposition levels. For the first level ($i = 0$), search limits are set to the environmental dimensions Y. Partition of the octree divides the environment in 8 equal parts in relation to the previous boundaries. In every single octree, an obstacle search is performed that determines the possibility of a later decomposition. Once this level decomposition is finished, the complete information of S_{occup} and S_{free} space is completed. Finally, a planner determines the best trajectory in distance terms.

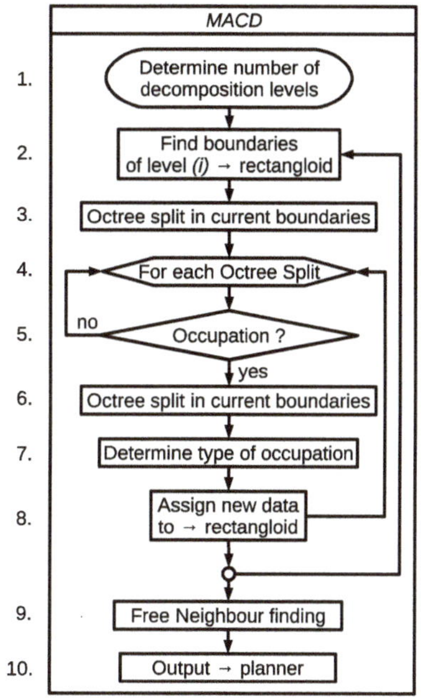

Figure 2. Flowchart of modified adaptive cell decomposition (*MACD*).

A simple example in a 2*D* environment is shown in Figure 3, using *quadtree* decomposition. The decomposition level n is predefined, n being the total levels which define the tree growth as a hierarchical computational structure. In a first segmentation, with $n = 1$, the environment is partitioned in $(2^n)^2$ underlying discrete spaces (blue lines). A second division of the environment ($n = 2$) will generate 16 spaces (green lines).

Figure 3 shows an example of a *quadtree* decomposition, denoted by q_k with current level $n = 0$, with reference to its own neighbourhood. On its left side, there are neighbours with smaller dimensional characteristics than the present q_k. There is a total of 8 neighbours, from $(2^n)^2$ with $n = 3 \rightarrow (2^3)^2 = 64$, of which only 8 are related directly with the q_k neighbourhood. The lower and upper face of q_k has the same level of decomposition, being $n = 0$ and producing a single q_{k+1} neighbour. On its right side, as the dimensions of the neighbour q_{k+1} are larger than q_k, the number of neighbours is equal to 1, counting a total of 11 neighbours and possible movements in this case.

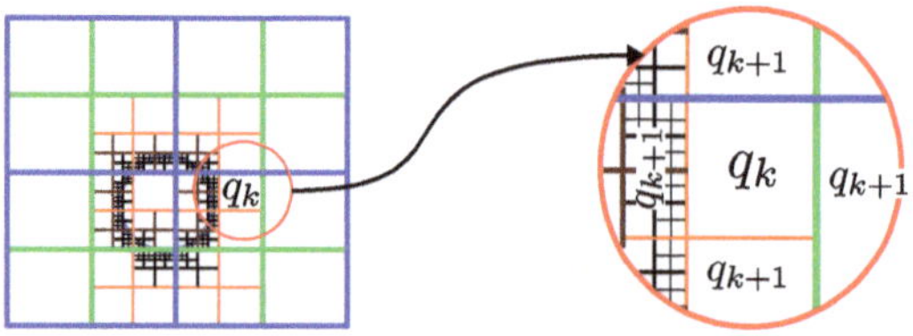

Figure 3. Neighbourhood structure. Quadtree decomposition neighbourhood.

The $3D$ decomposition methodology has two relevant variations. Firstly, the definition and location of each voxel boundary, and secondly, the definition of the number of neighbours per each voxel belonging to S_{free}. The number of neighbours q_k is bounded by at least $q_{k+1} = 3$ neighbours, and the maximum number of each q_k voxel face is a multiple of $(2^{0,1,2,\ldots,n})^2$ with n decomposition levels (shown in Figure 4).

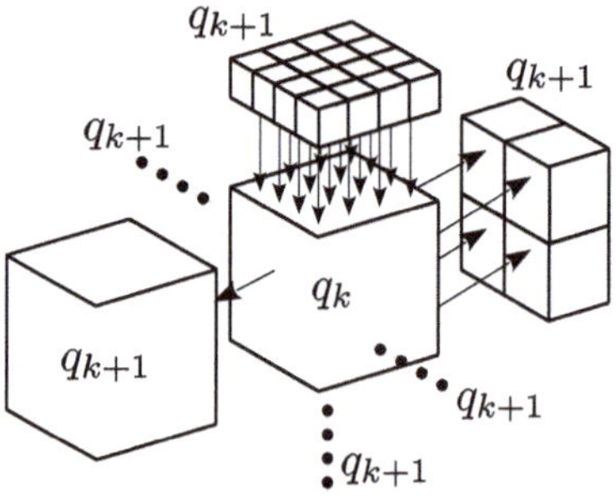

Figure 4. Neighbourhood structure. Octree decomposition neighbourhood.

At this point, the procedure is partially complete, since the decomposition just finds the set S_{free} and additional calculations to determine if ρ are needed. Hence, *MACD* uses *Dijkstra's* [31] algorithm to calculate the optimal path in distance terms.

Algorithm 1 shows the pseudo-code to generate a structure called a *rectangloid* which contains all the information compiled throughout the cell decomposition process. The algorithm performs a recursive searching of the free S_{free} space and the occupied S_{occup} space in the discrete $3D$ environment to determine each voxel property (including its set of neighbours), and it simultaneously builds the computational structure that joins the voxels. For a better understanding, a brief description of several algorithm steps is offered here.

Line 9: Boundaries of the current voxel are calculated.
Line 12: Boundaries of the sub-voxel are assigned to *rectangloid*.
Line 13: This step does a total routing by searching the environment for obstacle collisions.
Line 24: The *vertex* variable collects each (S_{free}) of *rectangloid* structure.
Line 25: The *edges* variable determines the structure that joins every *vertex*.
Line 26: *Dijkstra's* algorithm is used to determine ρ.

Algorithm 1 Modified Adaptive Cell Decomposition (*MACD*)

1: $n \rightarrow decomposition\ level, nextRow = 0$
2: $[boundary] = environment.boundaries$
3: $rectangloid.add = boundary$
4: **for** $i = 0 : n$ **do**

5:     $rows = rectangloid.size;$
6:     $nextRow = nextRow + 1;$
7:     **for** $j = nextRow : rows$ **do**

8:         **if** $rectangloid(j).ocup == S_{occup}$ **then**

9:             $[boundary] = boundaryOctree(rectangloid(j).boundaries);$
10:             **for** $k = 1 : boundary.size$ **do**

11:                 $newRow = newRow + 1;$
12:                 $rectangloid(newRow).add = boundary(k,:);$
13:                 $obj = obstacle.find \in rectangloid(newRow);$
14:                 **if** $obj == FREE$ **then**

15:                     $rectangloid(newRow).ocup = S_{free}$
16:                 **else**

17:                     $rectangloid(newRow).ocup = S_{occup}$
18:                 **end if**
19:             **end for**
20:         **end if**
21:     **end for**
22:     $nextRow = rows;$
23: **end for**
24: $vertex = rectangloid(:).boundary.center;$
25: $edges = rectangloid(:).neighbour.find;$
26: $\rho = Dijkstra(vertex, edges);$

---

4. Recursive Rewarding Modified Approximate Cell Decomposition (*RR-MACD*)

In this section, a new algorithm for path planning in $3D$ environments is formulated so that using an external planner based on nodes, such as *Dijkstra* [31] or A^* [33], becomes unnecessary. Since it attempts to achieve a final path ρ based on starting conditions, or initial states, each future system state is determined by the present one (each state is a collision-free neighbour voxel).

4.1. Methodology

Let Y denote a work environment as a discrete $3D$ space that contains a finite set of collision-free voxels (S_{free}) and a finite set of busy voxels (S_{occup}).

Let us assume an UAV is included within a collision-free voxel, which is considered as initial state s_k, with the aim of reaching the end point q_f. Let's assume a set formed by voxels of different sizes S_{k+1} as a neighbourhood of s_k. In this context, a state model and a transition matrix can be developed as in Figure 5 to determine the optimal transition from s_k to any state belonging to S_{k+1} based on two transitional measurements (D_1 and D_2). Starting from the current state s_k, the method will try to obtain the optimum neighbour state belonging to S_{k+1} ($s_k \rightarrow S_{k+1} = D_1$). It will then locate which sub-path from each neighbour in S_{k+1} to the final point q_f is best ($S_{k+1} \rightarrow q_f = D_2$).

To solve this problem a discrete deterministic finite automaton (DDFA) [60,61], F, can be defined as $F = (S, G, D, q)^T$ with a set of R_m partial functions, where:

- q are two points in the $3D$ environment space, where

  - q_i is the initial point
  - q_f is the final point.

- S is a finite set of M current states, where

  - S_{free} is the finite set of collision-free voxels. Split in the current voxel $s_k = [s_k(x), s_k(y), s_k(z)]$, and the set of its neighbours $s_{k+1} = [s_{k+1}(x), s_{k+1}(y), s_{k+1}(z)]$.

- S_{occup} is the finite set of occupied voxels.

- $R_m, m = 1 \dots N$ is a set of N partial functions involved in the $3D$ UAV navigation characteristics and determining feasible progress. In this paper, $N = 4$ functions are defined as flight parameters, being:

$$R_1(i,j) = \frac{M_{distance}(s_i \rightarrow s_j)}{M_{trDirect}} \in \mathbb{R} : [0,1]$$
$$M_{distance}(s_i \rightarrow s_j) = \sqrt{(s_i - s_j))^2} \tag{1}$$

where $M_{distance}(s_i \rightarrow s_j)$ is the Euclidean distance between any two states and $M_{trDirect}$ is the distance in a straight line between q_i and q_f.

$$R_2(i,j) = M_{tan}(s_i \rightarrow s_j) \in \mathbb{R} : [-1,1]$$
$$M_{tan}(s_i \rightarrow s_j) = tan^{-1}\theta$$
$$\theta = \left(\frac{\sqrt{[(s_i(y) - s_j(y)]^2 + [s_i(x) - s_j(x)]^2}}{s_i(z) - s_j(z)} \right) \tag{2}$$

where $M_{tan}(s_i \rightarrow s_j)$ is the direction change measurement of the tangent vector to a curve, which shows the inclination angle between any two states.

$$R_3(i,j) = M_{phi}(s_i \rightarrow s_j) \in \mathbb{R} : [-1,1]$$
$$M_{phi}(s_i \rightarrow s_j) = \phi$$
$$\phi = tan^{-1}\left(\frac{s_i(y) - s_j(y)}{s_i(x) - s_j(x)} \right) \tag{3}$$

where $M_{phi}(s_i \rightarrow s_j)$ is the direction change of the bi-normal vector around the tangent vector between any two states.

$R_4(s_i, s_j)$ is associated with the amount of battery and determines the possibility of success on a predefined trajectory:

$$R_4(i,j) = M_{batt}(s_i \rightarrow s_j) \in \mathbb{R} : [0,1]$$
$$M_{batt}(s_i \rightarrow s_j) = \begin{cases} 0, \frac{batt_i \rightarrow batt_j}{Cur_{batt}} > Cur_{batt} \\ 1 - \frac{batt_i \rightarrow batt_j}{Cur_{batt}} \leq Cur_{batt} \end{cases} \tag{4}$$

where $M_{batt}(s_i \rightarrow s_j)$ is the normalised theoretical quantity of battery needed to fly from any state to any other—and the Cur_{batt} is the current amount of battery available for flight.

Further, a Gaussian function $g(R_m)$ is used to determine the reward in executing a possible action and it is defined as:

$$g(R_m) = \frac{\sin(\pi * R_m + \pi/2) + 1}{2} \tag{5}$$

where the transition cost values $(s_k \rightarrow S_{k+1}, S_{k+1} \rightarrow q_f)$ have been normalised within boundaries $[0,1]$. Notice how the greater the effort R_m, the lower the reward $g(R_m)$ and vice-versa. Therefore, the execution of an action from state s_i to different states s_j produces state transitions at different costs—an elevated cost will produce a lower reward on the transition.

All these rewards can be expressed as a vector $G(i,j)$ of flight parameters such as:

$$G(i,j) = [g(R_1(i,j)), g(R_2(i,j)), \ldots, g(R_N(i,j))]^T \quad (6)$$

- $D \in \mathbb{R}^{M-2}$ is the received reward associated with a priority $\mathfrak{p} \in \mathbb{R}^N$ for executing an action on a function $g(R_m)$ and is stated as the sum of two transition priority vectors (D_1 and D_2) defined as:

$$\begin{aligned} D_1 =& (\mathfrak{p} \times G(i,j)) + \xi \\ & i = 1, j = 2 \ldots (M-1)] \end{aligned} \quad (7)$$

$$\begin{aligned} D_2 =& (\mathfrak{p} \times G(i,j)) + \xi \\ & [i = 2 \ldots (M-1), j = M] \end{aligned} \quad (8)$$

$$D = D_1 + D_2 \quad (9)$$

where ξ is a predefined negative reward value in each state belonging to S_{k+1}. Notice that the probability distribution values of the functions $g(R_1), g(R_2), \ldots, g(R_N)$ are independent and the set of answer vectors D which are mappings of $S \times S_{free}$. Therefore, this map is generated with a time-independent probability distribution. Hence, the probability of moving between one instant and the next does not change.

Hence, the best reward value from vector D generates the best x—and the final path, denoted by $\rho_x(F)$, defines a finite labeled graph with vertex $S_x \in S_{free}$.

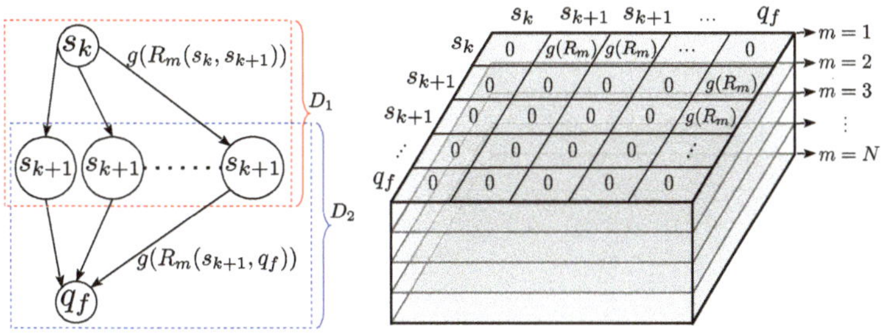

Figure 5. Generic structure state transition. (**a**) Generic state model. (**b**) Generic transition matrix.

4.2. Simple Application

To explain the methodology, and for sake of simplicity, let's state the problem of travelling from the actual state $q_i \in s_k$ or *init* point to the *goal* point q_f in a 2D environment using a standard 2D cell decomposition. Figure 6a shows the initial scenario as well as the different S_{k+1} states (observable and neighbours) that may become a new s_k. In this case, in $t = 0$, it is the same cost to move right or down—this situation appears due to the inherent symmetry of the decomposition methodology. In such a situation, randomness decides the next state.

Since a state cannot point to itself and can be visited only once, when the S_{k+1} states are visited and evaluated, one of them is selected by producing a forwarding movement (see Figure 6b). Hence, the state selected is the new *initial* point s_k, and the process is repeated again (see Figure 6c). The network structure shown indicates a forward movement s_k to the immediately next state s_{k+1}. This movement is independent of any previous state s_k.

Moving towards a 3D environment, a similar scene is represented as a master voxel containing an obstacle inside (Figure 7). This voxel can be split into different levels of voxels with different dimensional characteristics in different spatial locations. To obtain the finite set of collision-free spaces S_{free}, *MACD* is performed recursively.

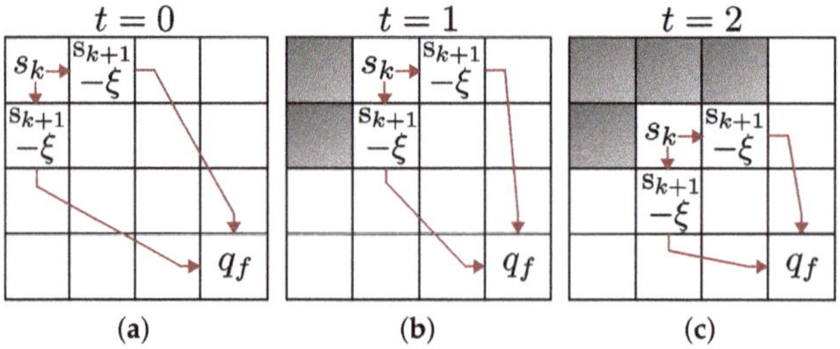

Figure 6. Network structure and state transition. (**a**) state transition $t = 0$. (**b**) state transition $t = 1$. (**c**) state transition $t = 2$.

The main environment boundaries have been defined from the initial coordinates $q_i \equiv (x_i, y_i, z_i)$ to the final one $q_f \equiv (x_f, y_f, z_f)$, resulting in a rectangular shape, being $env = ([x_i, x_f], [y_i, y_f], [z_i, z_f])$. Each obstacle $h_i(x, y, z) \in \mathbb{R}^3 \rightarrow (x, y, z) = \lambda$, is defined as:

$$h_i(\lambda)|_t = h_i(\lambda)|_{t+1} \Rightarrow static \tag{10}$$

$$h_i(\lambda)|_t \neq h_i(\lambda)|_{t+1} \Rightarrow dynamic \tag{11}$$

where $h_i(\lambda) \rightarrow i > 0$ could take two possible states, *static* (Equation (10), the obstacle does not change its position with passing time) or *dynamic* (Equation (11), the obstacle changes its position to another with passing time).

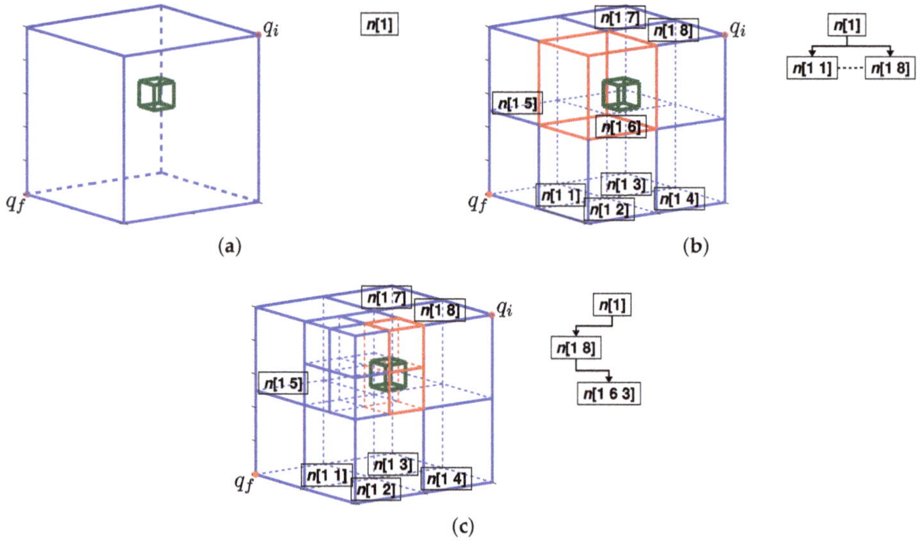

Figure 7. Recursive rewarding *MACD* (*RR-MACD*) start process example. (**a**) Complete environment. (**b**) First *MACD* level $n[1]$. (**c**) second *MACD* Level $n[1\ 8]$.

Figure 7a shows the environment definition (blue lines) as the main node $n[1]$ with an obstacle $h_1(\lambda)$ placed inside (box green lines). Once the first decomposition is performed (Figure 7b), a first octal level $n[[1\ 1], \ldots, [1\ 8]]$ is generated with nodes having different occupancy properties. Each will belong to S_{free} if there is no $h_i(\lambda)$ within its voxel limits ($s_k \bigcap h_i(\lambda) = 0$) (blue lines), or to S_{occup} if the voxel is partial or totally occupied by the obstacle $((s_k \bigcap h_i(\lambda) = 1) \vee (s_k \in h_i(\lambda)))$ (red lines).

The first step is to determine the q_i container voxel (for this example, it is $n[1\ 8]$). In case of obstacle detection in $n[1\ 8]$, a recursive decomposition would be performed on the location. At this level, the node $n[1\ 8]$ is the new starting state s_k and the observable neighbours S_{k+1} are the nodes $n[1\ 4], n[1\ 6]$, and $n[1\ 7]$. At this stage, the state model is depicted in Figure 8 and the transition matrix is detailed in Table 3.

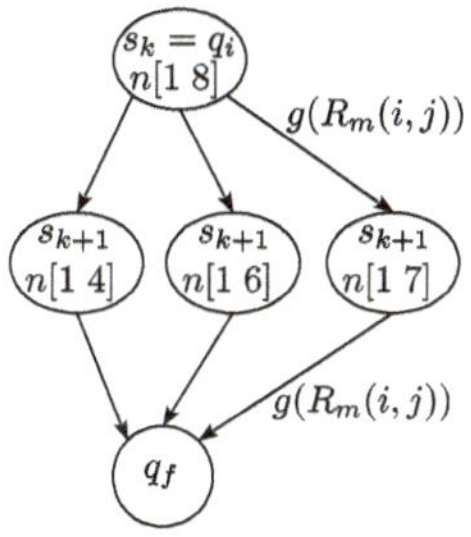

Figure 8. State model $M = 5$ states. List of states: $s_1 \rightarrow n[1\ 8]$, $s_2 \rightarrow n[1\ 4]$, $s_3 \rightarrow n[1\ 6]$, $s_4 \rightarrow n[1\ 7]$, $s_5 \rightarrow [q_f]$.

The probability distribution for the discrete states can be derived from a multidimensional matrix of dimensions $M \times M \times N$ (see Table 3) where, M is the number of states, and N is the number of partial functions $R_{(m=1...N)}$ involved in the 3D UAV navigation.

Table 3. Initial multidimensional transition matrix listed by first column ($i = 1 \ldots M$) and the first row ($j = 1 \ldots M$). The number of levels are given by the number of N functions R_m for the particular navigation characteristics.

m=2 →, m=3 →, m= · · · →, m= N →

		j=1	j=2	j=3	j=4	j=M
	m=1 →	$s_k = q_i$	s_{k+1} $n[1\ 4]$	s_{k+1} $n[1\ 6]$	s_{k+1} $n[1\ 8]$	q_f
i=1	$s_k = q_i$	0	$g(R_m)$	$g(R_m)$	$g(R_m)$	0
i=2	s_{k+1} $n[1\ 4]$	0	0	0	0	$g(R_m)$
i=3	s_{k+1} $n[1\ 6]$	0	0	0	0	$g(R_m)$
i=4	s_{k+1} $n[1\ 8]$	0	0	0	0	$g(R_m)$
i=M	q_f	0	0	0	0	0

This multidimensional matrix equivalent to Figure 5b has been constructed with the information of the state model and the reward values. The aim is find the partial responses coming from the first row and last column, and split in two transition priority vectors $(D_1, D_2) \in \mathbb{R}^{(M-2)}$. Using the priority vector $\mathfrak{p}$ and the offset ξ, vectors D_1 and D_2 deliver partial reward distributions to a possible next state.

The transition priority vector D_1 is built with the set of columns $j = 2 \ldots (M-1)$ and the row $i = 1$ such that

$$D_1 = \left[\mathfrak{p} \times G(i,j)\right] + \xi, i = 1, j = 2 \ldots (M-1)$$
$$G(1,2) = \begin{bmatrix} g(R_1(1,2)) \\ g(R_2(1,2)) \\ \vdots \\ g(R_N(1,2)) \end{bmatrix}$$
$$G(1,3) = \begin{bmatrix} g(R_1(1,3)) \\ g(R_2(1,3)) \\ \vdots \\ g(R_N(1,3)) \end{bmatrix} \tag{12}$$
$$\vdots$$
$$G(1,(M-1)) = \begin{bmatrix} g(R_1(1,(M-1))) \\ g(R_2(1,(M-1))) \\ \vdots \\ g(R_N(1,(M-1))) \end{bmatrix}$$

The second transition priority vector, D_2, is built with the set of rows $i = 2 \ldots (M-1)$ and column $j = M$.

$$D_2 = \left[\mathfrak{p} \times G(i,j)\right] + \xi, i = 2 \ldots (M-1), j = M$$
$$G(2,M) = \begin{bmatrix} g(R_1(2,M)) \\ g(R_2(2,M)) \\ \vdots \\ g(R_N(2,M)) \end{bmatrix}$$
$$G(3,M) = \begin{bmatrix} g(R_1(3,M)) \\ g(R_2(3,M)) \\ \vdots \\ g(R_N(3,M)) \end{bmatrix} \tag{13}$$
$$\vdots$$
$$G((M-1),M) = \begin{bmatrix} g(R_1((M-1),M)) \\ g(R_2((M-1),M)) \\ \vdots \\ g(R_N((M-1),M)) \end{bmatrix}$$

Finally, the final reward vector D expressed as the sum of D_1 and D_2 contains the optimal value which points to the best state (node) for continuing the search of the path ρ_x:

$$D = D_1 + D_2 \tag{14}$$
$$x = best(D) \tag{15}$$

To continue with the example in Figure 7, let us assume that $x = best(D)$ points to $n[1\ 6]$. Nevertheless, $n[1,6]$ is occupied (it belongs S_{occup}), so *MACD* is invoked, creating a new level in the data structure, composed of $n[1\ 6\ (1 \ldots 8)]$ (see Figure 7c). Even though the state s_k remains in $n[1\ 8]$, the new decomposition on $n[1\ 6]$ returns a new set of neighbours, which join with previous ones, and define the new set S_{k+1} defined by $(n[1\ 4], n[1\ 6\ 3], n[1\ 6\ 4], n[1\ 6\ 7], n[1\ 6\ 8], n[1\ 7])$.

So looking for the optimum within S_{k+1} is required. Let us assume the best node from s_{k+1} is $n[1\ 6\ 3]$ (notice that *MACD* is invoked once until now) and so the new state s_k is reassigned to $n[1\ 6\ 3]$ and consequently, the neighbourhood of the new state s_k, is conformed by $S_{k+1} = n[1\ 6\ 1], n[1\ 6\ 4], n[1\ 6\ 7], n[1\ 5], n[1\ 8]$.

The previous actions produce the displacement from a current s_k state to the next best state, and towards the final point. While the container voxel of the resulting better state x does not contain the *goal* point, there is the possibility that x has neighbours in different decomposition levels. Hence, the process continues until the *goal* point is reached.

Once the previous phase has been completed, the optimal path $\rho_x(F)$ is totally determined and the search finishes. The flowchart depicted in Figure 9 and the Algorithm 2 show the pseudo-code for the described actions.

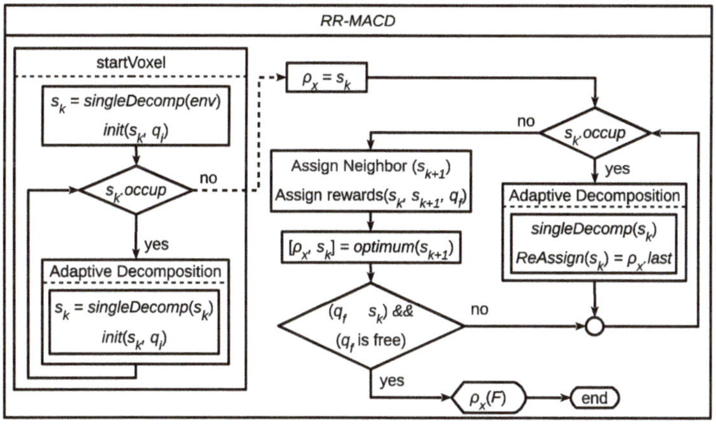

Figure 9. Flowchart of *RR-MACD*. Notice how steps 3 to 8 in Figure 2 are re-used in this chart and renamed as *singleDecomp*.

Algorithm 2 *RR-MACD*

define $targetPoints, Obstacles$
$[s_k] = startVoxel(env)$
while $q_f \in s_k$ **do**
  if $s_k.occup == true$ **then**
    $singleDecomp(s_k);$
    $s_k = \rho_x.last;$
  else
    $[S_{k+1}] = neighbourhood(s_k);$
    $[D_1, D_2] = rewards(s_k, S_{k+1}, q_f);$
    $[\rho_x, s_k] = D.optimum$
  end if
end while
return $\rho_x(F)$

The procedure is split in two stages. Firstly, a start voxel location is defined and, if there is an $h_i(\lambda)$ in the environment, an initial simple decomposition *singleDecomp* is performed (as a result, the collision-free voxel that contains the *init* point will be defined). Once the initial s_k state is assigned—which is also an initial ρ_x—the procedure proceeds depending on its occupation. If s_k is collision-free, the neighbours S_{k+1} are assigned, the rewards are calculated, and the optimal x is located. Therefore, the best state x becomes the new s_k and is added to ρ_x. If this new s_k is occupied, the process requires another decomposition on the current s_k, and s_k will return to its previous state. The procedure is completed when the current s_k contains the *goal* point and s_k is collision-free.

Algorithm 2 summarises the procedure described in Figure 9. In line 2, a search of the first containing $q_i \in s_k$ is performed. The loop continues until the current s_k state contains q_f (line 3). If the current state s_k collides with an obstacle $h_i(\lambda)$, the s_k state is decomposed (line 5: *singleDecomp*) and s_k returns to its previous state (line 6). In line 8, the neighbours of s_k, S_{k+1}, are defined. Line 9 measures the transition rewards for any neighbour in S_{k+1}. Finally, the optimum is added to the path ρ_x and x is assigned as the new s_k in line 10. When the loop finishes, the complete path $\rho_x(F)$ is returned (line 13).

The methodology described presents the following properties:

(a) A stochastic process in discrete time has been defined (it lacks memory), the probability distribution for a future state depends solely on its present values and is independent of the current state history.
(b) The sum of the priorities defined in vector $\mathfrak{p}$ is not equal to 1. $\sum_{i=1}^{N} \mathfrak{p}_i \neq 1$.
(c) The sum of the values of each priority vector, is not equal to 1. $\sum D_1 \neq 1$ and $\sum D_2 \neq 1$.

Finally, it should be noted that the environmental discrete decomposition results in ever smaller voxels of differing sizes. The level of voxel decomposition is variable based on two goals that must be fulfilled: (1) Designer defines the maximum decomposition level (minimum voxel size); however, the algorithm tries to reduce the computational cost and, as a consequence, the minimum voxel size is generally avoided. (2) As soon as a free space meets the defined constraints it is selected by the algorithm regardless of the size of the voxel.

4.3. Dynamic environment approach

The *RR-MACD* can be applied to a 3*D* UAV environment with obstacles for movement. Once q_i and q_f points are defined, the trajectory $\rho_x(F)$ is calculated and the UAV navigates through it. A dynamic environment implies a positional Equation (11). If the obstacles intersect the previously calculated trajectory $(h_i(\lambda)|_t \cap \rho_x(F) = 1)$, a new trajectory must be generated. The described methodology in the previous sections has constant targets q_i and q_f. However, for a dynamic trajectory the path must be updated with a new q_i in the current UAV location.

5. Experiments

In this section, a comparison of computational performance between *MACD* and *RR-MACD* algorithms is made. Three different simulation examples have been carried out with 10 executions of each algorithm over each environment. The 150 set of responses supplies the results to determine the performance. The algorithms have been run in a "8 x Intel(R) Core(TM) i7-4790 CPU @ 3.60 GHz" computer (Manufacturer: Gigabyte Technology Co., Ltd., Model: B85M-D3H) with 8Gb RAM and S.O. Ubuntu Linux 16.04 LTS. The algorithms were programmed in MATLAB version 9.4.0.813654 (R2018a).

For each simulation example, five different 3D environments were defined. Table 4 shows one row per environment, where the column entitled *"UAV target coordinates"* expresses the coordinates in terms of *init* and *goal* points. The environmental dimensions have been defined in distances related to those coordinates (in a scale of *meters m*). The column *"Obstacles Dimensions"* shows the dimensional characteristics of each obstacle and the *"Obstacle Location"* places the obstacles in a specific location. Moreover, the altitude between *init* and *goal* target points are different, guaranteeing that the UAV path will be built in the (x, y, z) axes. It should be highlighted that maps are created with predefined static obstaclesfor the different groups of experiments.

First, an urban environment with several buildings is described. For the construction of this environment, a maximum altitude reference has been taken, such as stated in "The Regulation of Drones in Spain 2019 [62,63]". For environments defined as 2, 3, 4, and 5, larger dimensions have been considered in which a varying number of obstacles in different air spaces are defined.

Table 4. Definition of five different 3D environments for the simulation examples.

#	*UAV Target Coordinates (m)* init			goal			*Obstacles Dimensions (m)*			*Obstacles Ubication (m)*		
	x	*y*	*z*	*x*	*y*	*z*	*x*	*y*	*z*	*x*	*y*	*z*
1	100	100	42	0	0	24	12	12	50	40	30	25
							15	15	30	24	40	15
							30	30	30	70	20	15
							15	15	46	20	70	23
							12	12	54	80	70	27
2	1000	1000	600	0	0	420	200	200	200	333	333	333
							300	300	300	777	777	777
3	1000	1000	300	0	0	700	100	100	100	400	400	400
							150	150	150	400	400	800
4	1200	1200	390	0	0	720	200	300	400	200	800	400
							20	20	20	300	200	700
5	1200	1200	800	0	0	500	10	10	10	600	600	600
							15	15	15	200	800	800
							15	15	15	200	800	200

5.1. Example 1. Static Obstacles and Four Flight Parameters (Constraints)

This example shows the performance of *RR-MACD* when four functions R_m are used ($N = 4$). These functions are the same as those detailed in Section 4. In Figure 10, the specifically results for environments #1 and #2 are shown. Therefore, Figure 10a shows the urban environment. Hence, Figure 10b shows the built path by *RR-MACD* of the environment #1, where black boxes are the obstacles $h_i(\lambda)$, green stars shows the voxel set of vertices in the state s_k and the orange line shows the final path $\rho_x(F)$.

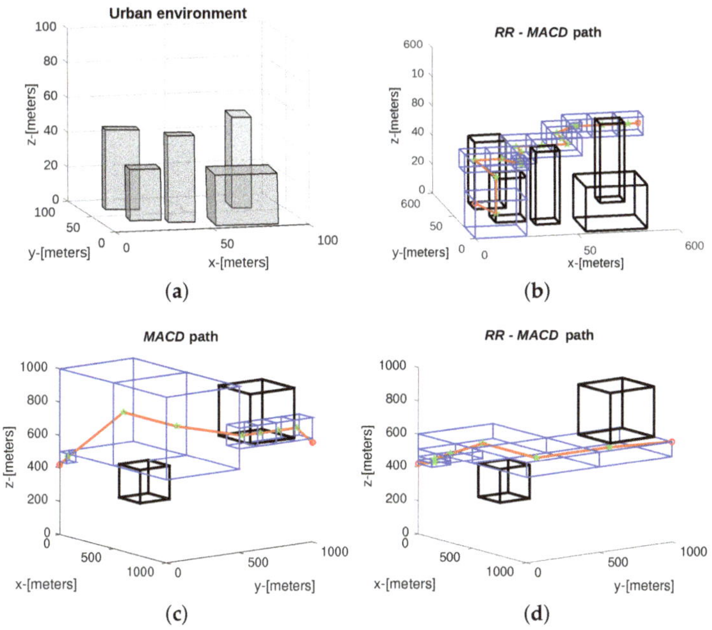

Figure 10. Graphical Results of Example 1. (**a**) Modeled urban environment cluttered with surrounding buildings. (**b**) Results for environment #1 with *RR-MACD*. (**c**) Final path generated using *MACD* for environment #2. (**d**) Final path generated using *RR-MACD* for environment #2.

In Figure 10c, the resulting *MACD* technique on the environment #2 can be appreciated, where black boxes are the static obstacles $h_i(\lambda)$, blue boxes are the set of voxels used for the *Dijkstra's* planner to obtain the optimal final path (green stars are the vertices set of its belonging voxel and the orange line shows the final path ρ. Finally, Figure 10d depicts the final trajectory built by *RR-MACD*, where the orange line shows the final path $\rho_x(F)$.

Comparisons of both algorithms regarding computational time is shown in Table 5. The results show an important advantage in decomposition time S_{free} and $\rho_x(F)$ spaces for *RR-MACD*. Notice that, when an obstacle $h_i(\lambda)$ intersects with a voxel decomposition, *MACD* recursively continues until the predefined level n (for this reason in environment #3 the searching time for *MACD* is considerably greater than other environments).

Table 5. Comparison of both algorithm executions for different environments. Column recursive rewarding modified adaptive cell decomposition (*RR-MACD*) 4 vs. *MACD* (%) shows the average resources (decomposition time (s), number of free voxel decomposition "S_{free}" and number of nodes in the final path "ρ") used for *RR-MACD* in comparison with *MACD*. Column *RR-MACD 10* vs. *RR-MACD 4* (%) shows the average resources (decomposition time (s), number of free voxels decomposition "S_{free}" and number of nodes in the final path) used for *RR-MACD* when the number of flight parameters (constraints) are augmented to 10.

#	*MACD*				*RR-MACD 4 Constraints*			*RR-MACD 4 vs. MACD* (%)			*RR-MACD 10 vs. RR-MACD 4* (%)		
	decom. Time (s)	**Dijks. Time (s)**	**#** S_{free}	**#** ρ	**decom. Time (s).**	**#** S_{free}	**#** $\rho_x(F)$	**decomp. Time**	S_{free}	ρ	**decomp. Time**	S_{free}	ρ
1	0.117	0.038	205	19	0.056	115	18	36.238	54.641	93.684	+51.449	+74.975	+50.000
2	0.104	0.049	496	11	0.012	27	8	8.368	5.443	72.727	+18.710	+26.337	+25.000
3	0.151	0.048	426	13	0.014	19	6	7.105	4.460	46.153	−25.118	−18.723	+1.851
4	3.535	1.021	5201	19	0.003	11	6	0.080	0.211	31.578	+327.894	+363.636	+57.407
5	0.078	0.032	294	23	0.009	19	7	8.470	6.462	30.434	+115.017	+79.532	+33.333

The third column "*RR-MACD 4 vs. MACD* (%)" shows in percentages the differences between *MACD* and *RR-MACD 4* regarding environmental decomposition time, number of S_{free} free-spaces generated during the searching process, and the number of final path nodes ρ. It is important to mention that *MACD* needs an additional time because *Dijks time(s)* shows the seconds needed to find ρ. For example, in the first environment, the *RR-MACD* algorithm just needs 36.238% of the time that *MACD* takes for environment decomposition, and 54.641% of the time that *MACD* takes to generate S_{free}, and 93.684% of the nodes that *MACD* needs to build ρ. Therefore, *RR-MACD* shows a general improvement on the process.

5.2. Example 2. Static Obstacles and 10 Constraints

In example 1, the set of partial functions involved in 3*D* UAV navigation is equal to 4. For this example, an additional set of 6 random values were added as new functions to simulate complex flight characteristics. Therefore, the number of partial functions in 3*D* UAV navigation R_m will be equal to $M = 10$ and the probability distribution multidimensional matrix now has 10 levels. This increment of functions, and its random nature, will provoke inherent changes in the results. These can be observed in the fourth column of Table 5, entitled "*RR-MACD 10 vs. RR-MACD 4 %*", where the relative difference between *RR-MACD 4* and *RR-MACD 10* shows the percentage increase or decrease in decomposition time, S_{free} and $\rho_x(F)$.

For example, let us compare performances between *RR-MACD 10* and *RR-MACD 4* in the environment #1. *RR-MACD 10* needs 51.499% more time to find a final path. It generates 74.975% plus voxels and the number of nodes in the final path ρ is 50% higher.

Table 6 shows a set of additional data corresponding to the results in terms of distances travelled between the *init* point and the *goal* point after the execution of each algorithm. As mentioned in the previous sections, the main objective of planning is not to reach optimality in exclusive terms of distance.

Table 6. Path results in distance metrics.

	Distance Travelled Meters (m)		
Scene	*MACD*	*RR-MACD* 4	*RR-MACD* 10
1	224.060	197.410	241.600
2	1853.000	1592.545	1734.054
3	1768.600	1790.181	1728.300
4	1693.100	1868.463	2221.354
5	1731.800	1829.690	2123.954

5.3. Example 3. Dynamic Obstacles

An additional experiment was performed that considered obstacles in motion. A new environment has been proposed for obstacles intersecting with the calculated $\rho_x(F)$. Hence, a new $\rho_x(F)$ with a new *init* (actual location of the UAV) is built.

Figure 11 represents this new environment with two obstacles in motion (black boxes). The first dynamic obstacle begins its displacement in location (600, 600, 600)m, and performs a continuous motion in an *east* direction with a constant velocity of 15 m/s. The second one begins its flight in location (80, 800, 600)m with a constant velocity of 15 m/s in the same direction. After 11 s, the first obstacle collides with $\rho_x(F)$ (this crash is illustrated with orange line in Figure 11a) and a new $\rho_x(F)$ is calculated, Figure 11b shows the new location of the obstacles (notice that limits in x and y axis have changed from 1000 m to 800 m). At $t = 34$ s, a new collision is detected (Figure 11c), even though the first obstacle is out of the environment, the second one has collided with ρ_x. When the UAV (blue square) has passed this part of $\rho_x(F)$, the new trajectory until *goal* is collision-free.

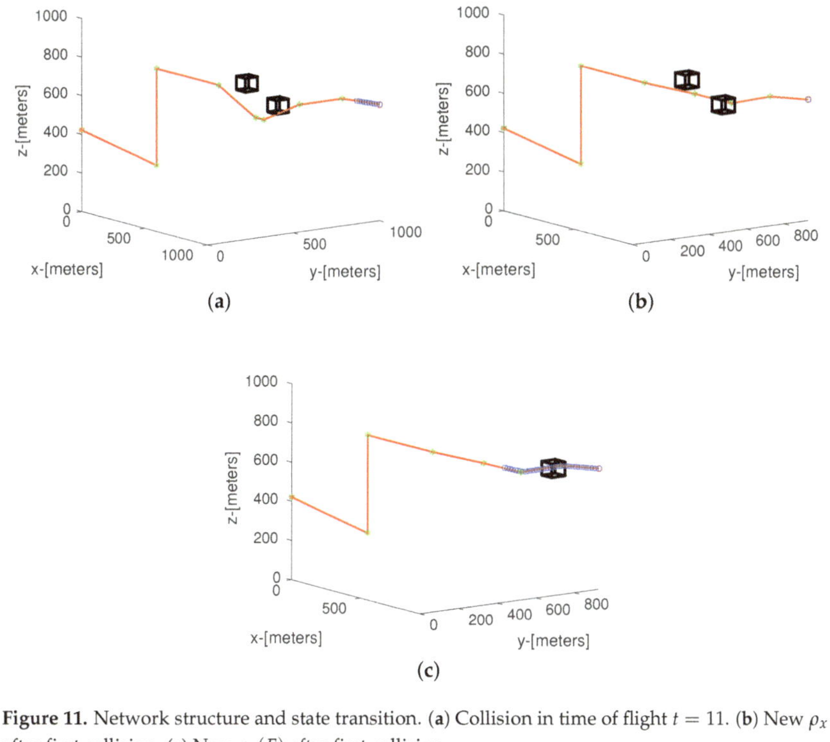

Figure 11. Network structure and state transition. (**a**) Collision in time of flight $t = 11$. (**b**) New $\rho_x(F)$ after first collision. (**c**) New $\rho_x(F)$ after first collision.

6. Conclusions

This paper presents an adaptive grid methodology in 3*D* environments applied to flight path planning. The approach described considers different constraints such as UAV maneuverability and geometry or static and dynamic environment obstacles.

The proposed algorithm, *RR-MACD*, divides the 3*D* environment in a synthesised way and does not need to invoke any additional planner to search (such as *Dijkstra* or A^*) for an optimal path generation. The improvement in the computational effort and the reduction in the number of nodes generated by the *RR-MACD* has been shown. The stochastic process in discrete time involved in the algorithm also shows a future probability distribution that only depends on its present states.

The partition of the 3*D* space into a defined geometric form enables decreasing the number of control points in the generated trajectory. In addition, the issue of computational cost and complexity has been addressed by providing a solution that generates relatively shorter time responses compared to techniques for generating similar trajectories.

In a future work, an additional processing task will be carried out, using the set of nodes, or *control points* $\rho_x(F)$ generated, to create a smoother path—and then the methodology will be tested under real flight conditions on an UAV model as in [64–66].

Author Contributions: Conceptualization, F.S.; Formal analysis, F.S., J.S., S.G.-N. and R.S.; Methodology, F.S.; Supervision, J.S., S.G.-N. and R.S.

Funding: The authors would like to acknowledge the Spanish Ministry of Economy and Competitiveness for providing funding through the project DPI2015-71443-R and the local administration Generalitat Valenciana through the project GV/2017/029. Franklin Samaniego thanks IFTH (Instituto de Fomento al Talento Humano) Ecuador (2015-AR2Q9209), for its sponsorship of this work.

Acknowledgments: The authors would like to thank the editors and the reviewers for their valuable time and constructive comments.

Conflicts of Interest: The authors declare no conflict of interest.

References

1. Valavanis, K.; Vachtsevanos, G. *Handbook of Unmanned Aerial Vehicles*; Springer: Dordrecht, The Netherlands, 2015; pp. 2993–3009. [CrossRef]
2. 20 Great UAV Applications Areas for Drones. Available online: http://air-vid.com/wp/20-great-uav-applications-areas-drones/ (accessed on 28 December 2018).
3. Industry Experts—Microdrones. Available online: https://www.microdrones.com/en/industry-experts/ (accessed on 28 December 2018).
4. Rodríguez, R.; Alarcón, F.; Rubio, D.; Ollero, A. Autonomous Management of an UAV Airfield. In Proceedings of the 3rd International Conference on Application and Theory of Automation in Command and Control Systems, Naples, Italy, 28–30 May 2013; pp. 28–30.
5. Li, J.; Han, Y. Optimal Resource Allocation for Packet Delay Minimization in Multi-Layer UAV Networks. *IEEE Commun. Lett.* **2017**, *21*, 580–583. [CrossRef]
6. Stuchlík, R.; Stachoň, Z.; Láska, K.; Kubíček, P Unmanned Aerial Vehicle—Efficient mapping tool available for recent research in polar regions. *Czech Polar Rep.* **2015**, *5*, 210–221. [CrossRef]
7. Pulver, A.; Wei, R. Optimizing the spatial location of medical drones. *Appl. Geogr.* **2018**, *90*, 9–16. [CrossRef]
8. Claesson, A.; Svensson, L.; Nordberg, P.; Ringh, M.; Rosenqvist, M.; Djarv, T.; Samuelsson, J.; Hernborg, O.; Dahlbom, P.; Jansson, A.; et al. Drones may be used to save lives in out of hospital cardiac arrest due to drowning. *Resuscitation* **2017**, *114*, 152–156. [CrossRef] [PubMed]
9. Reineman, B.; Lenain, L.; Statom, N.; Melville, W. Development and Testing of Instrumentation for UAV-Based Flux Measurements within Terrestrial and Marine Atmospheric Boundary Layers. *J. Atmos. Ocean. Technol.* **2013**, *30*, 1295–1319. [CrossRef]
10. Hernández, E.; Vázquez, M.; Zurro, M. *Álgebra lineal y Geometría*, 3rd ed.; Pearson: Upper Saddle River, NJ, USA, 2012.
11. LaValle, S. *Planning Algorithms*; Cambridge University Press: Cambridge, UK, 2006; pp. 1–826. [CrossRef]

12. Elbanhawi, M.; Simic, M. Sampling-Based Robot Motion Planning: A Review. *IEEE Access* **2014**, *2*, 56–77. [CrossRef]
13. Hernandez, K.; Bacca, B.; Posso, B. Multi-goal path planning autonomous system for picking up and delivery tasks in mobile robotics, *IEEE Lat. Am. Trans.* **2017**, *15*, 232–238. [CrossRef]
14. Kohlbrecher, S.; Von Stryk, O.; Meyer, J.; Klingauf, U. A flexible and scalable SLAM system with full 3D motion estimation. In Proceeding of the 9th IEEE International Symposium on Safety, Security, and Rescue Robotics SSRR 2011, Tokyo, Japan, 22–24 August 2011; pp. 155–160. [CrossRef]
15. Abbadi, A.; Matousek, R.; Jancik, S.; Roupec, J. Rapidly-exploring random trees: 3D planning. In Proceedings of the 18th International Conference on Soft Computing MENDEL, Brno, Czech Republic, 27–29 June 2012; pp. 594–599
16. Aguilar, W.; Morales, S. 3D Environment Mapping Using the Kinect V2 and Path Planning Based on RRT Algorithms. *Electronics* **2016**, *5*, 70. [CrossRef]
17. Aguilar, W.; Morales, S.; Ruiz, H.; Abad, V. RRT* GL Based Optimal Path Planning for Real-Time Navigation of UAVs. *Soft Comput.* **2017**, *17*, 585–595._50. [CrossRef]
18. Yao, P.; Wang, H.; Su, Z. Hybrid UAV path planning based on interfered fluid dynamical system and improved RRT. In Proceedings of the IECON 2015—41st Annual Conference of the IEEE Industrial Electronics Society, Yokohama, Japan, 9–12 November 2015; pp. 829–834. [CrossRef]
19. Yan, F.; Liu, Y.; Xiao, J. Path Planning in Complex 3D Environments Using a Probabilistic Roadmap Method. *Int. J. Autom. Comput.* **2013**, *10*, 525–533. [CrossRef]
20. Yeh, H.; Thomas, S.; Eppstein, D.; Amato, N. UOBPRM: A uniformly distributed obstacle-based PRM. In Proceedings of the 2012 IEEE/RSJ International Conference on Intelligent Robots and Systems, Vilamoura, Algarve, Portugal, 7–12 October 2012; pp. 2655–2662. [CrossRef]
21. Denny, J.; Amatoo, N. Toggle PRM: A Coordinated Mapping of C-Free and C-Obstacle in Arbitrary Dimension. *Algorithm. Found. Robot. X* **2013**, *86*, 297–312 doi:10.1007/978-3-642-36279-8_18. [CrossRef]
22. Li, Q.; Wei, C.; Wu, J.; Zhu, X. Improved PRM method of low altitude penetration trajectory planning for UAVs. In Proceedings of the 2014 IEEE Chinese Guidance, Navigation and Control Conference, Yantai, China, 8–10 August 2014; pp. 2651–5656. [CrossRef]
23. Ortiz-Arroyo, D. A hybrid 3D path planning method for UAVs. In Proceedings of the 2015 Workshop on Research, Education and Development of Unmanned Aerial Systems (RED-UAS), Cancún, México, 23–25 November 2015; pp. 123–132. [CrossRef]
24. Thanou, M.; Tzes, A. Distributed visibility-based coverage using a swarm of UAVs in known 3D-terrains. In Proceedings of the 2014 6th International Symposium on Communications, Control and Signal Processing (ISCCSP), Athens, Greece, 21–23 May 2014; pp. 425–428. [CrossRef]
25. Qu, Y.; Zhang, Y.; Zhang, Y. Optimal flight path planning for UAVs in 3-D threat environment. In Proceedings of the 2014 International Conference on Unmanned Aircraft Systems (ICUAS), Orlando, FL, USA, 27–30 May 2014; pp. 149–155. [CrossRef]
26. Fang, Z.; Luan, C.; Sun, Z. A 2D Voronoi-Based Random Tree for Path Planning in Complicated 3D Environments. *Intell. Auton. Syst.* **2017**, *531*, 433–445._31. [CrossRef]
27. Khuswendi, T.; Hindersah, H.; Adiprawita, W. UAV path planning using potential field and modified receding horizon A* 3D algorithm. In Proceedings of the 2011 International Conference on Electrical Engineering and Informatics, Bandung, Indonesia. 17–19 July 2011; pp. 1–6. [CrossRef]
28. Chen, X.; Zhang, J. The Three-Dimension Path Planning of UAV Based on Improved Artificial Potential Field in Dynamic Environment. In Proceedings of the 2013 5th International Conference on Intelligent Human-Machine Systems and Cybernetics, Hangzhou, Zhejiang, China, 26–27 August 2013; Volume 2, pp. 144–147. [CrossRef]
29. Rivera, D.; Prieto, F.; Ramirez, R. Trajectory Planning for UAVs in 3D Environments Using a Moving Band in Potential Sigmoid Fields. In Proceedings of the 2012 Brazilian Robotics Symposium and Latin American Robotics Symposium, Fortaleza, Cear a, Brazil, 16–19 October 2012; pp. 115–119. [CrossRef]
30. Liu, L.; Shi, R.; Li, S.; Wu, J. Path planning for UAVS based on improved artificial potential field method through changing the repulsive potential function. In Proceedings of the 2016 IEEE Chinese Guidance, Navigation and Control Conference (CGNCC), Nanjing, China, 12–14 August 2016; pp. 2011–2015. [CrossRef]

31. Dijkstra, E. A note on two problems in connexion with graphs. *Numer. Math.* **1959**, *1*, 269–271. [CrossRef]
32. Verscheure, L.; Peyrodie, L.; Makni, N.; Betrouni, N.; Maouche, S.; Vermandel, M. Dijkstra's algorithm applied to 3D skeletonization of the brain vascular tree: Evaluation and application to symbolic. In Proceedings of the 2010 Annual International Conference of the IEEE Engineering in Medicine and Biology, Buenos Aires, Argentina, 31 August–4 September 2010; pp. 3081–3084. [CrossRef]
33. Hart, P.E.; Nils, J. A formal basis for the Heuristic Determination of minimum cost paths. *IEEE Trans. Syst. Sci. Cybern.* **1968**, *4*, 100–107. [CrossRef]
34. Niu, L.; Zhuo, G. An Improved Real 3D a* Algorithm for Difficult Path Finding Situation. In Proceeding of the International Archives of the Photogrammetry, Remote Sensing and Spatial Information Sciences, Beijing, China, 3–11 July 2008; Volume 37, pp. 927–930.
35. Stentz, A. Optimal and Efficient Path Planning for Partially-Known Environments. *ICRA* **1994**, *94*, 3310–3317.
36. Ferguson, D.; Stentz, A. Field D*: An Interpolation-Based Path Planner and Replanner. *Robot. Res.* **2007**, *28*, 239–253._22. [CrossRef]
37. De Filippis, L.; Guglieri, G.; Quagliotti, F. Path Planning Strategies for UAVS in 3D Environments. *J. Intell. Robotic Syst.* **2012**, *65*, 247–264. [CrossRef]
38. Nosrati, M.; Karimi R.; Hasanvand, H. Investigation of the * (Star) Search Algorithms: Characteristics, Methods and Approaches. *World Appl. Program.* **2012**, *2*, 251–256.
39. Kroumov, V.; Yu, J.; Shibayama, K. 3D path planning for mobile robots using simulated annealing neural network. In Proceeding of the 2009 International Conference on Networking, Sensing and Control, Okayama, Japan, 29 March 2009; pp. 130–135.
40. Gautam, S.; Verma, N. Path planning for unmanned aerial vehicle based on genetic algorithm & artificial neural network in 3D. In Proceeding of the 2014 International Conference on Data Mining and Intelligent Computing (ICDMIC), Odisha, India, 20–21 December 2014; pp. 1–5. [CrossRef]
41. Maturana, D.; Scherer, S. 3D Convolutional Neural Networks for landing zone detection from LiDAR. In Proceeding of the 2015 IEEE International Conference on Robotics and Automation (ICRA), Beijing, China, 2–5 August 2015; Volume 2015, pp. 3471–3478. [CrossRef]
42. Iswanto, I.; Wahyunggoro, O.; Cahyadi, A. Quadrotor Path Planning Based on Modified Fuzzy Cell Decomposition Algorithm. *TELKOMNIKA Telecommun. Comput. Electron. Control* **2016**, *14*, 655–664. [CrossRef]
43. Liu, S.; Wei, Y.; Gao, Y. 3D path planning for AUV using fuzzy logic. In Proceeding of the Computer Science and Information Processing (CSIP)), Xi'an, Shaanxi, China, 24–26 August 2012; pp. 599–603.
44. Duan, H.; Yu, Y.; Zhang, X.; Shao, S. Three-dimension path planning for UCAV using hybrid meta-heuristic ACO-DE algorithm. *Simul. Model. Pract. Theory* **2010**, *18*, 1104–1115. [CrossRef]
45. He, Y.; Zeng, Q.; Liu, J.; Xu, G.; Deng, X. Path planning for indoor UAV based on Ant Colony Optimization. In Proceeding of the 2013 25th Chinese Control and Decision Conference (CCDC), Guiyang, China, 25–27 May 2013; pp. 2919–2923. [CrossRef]
46. Zhang, Y.; Wu, L.; Wang, S. UCAV Path Planning by Fitness-Scaling Adaptive Chaotic Particle Swarm Optimization. *Math. Probl. Eng.* **2013**, 1–9. [CrossRef]
47. Goel, U.; Varshney, S.; Jain, A.; Maheshwari, S.; Shukla, A. Three Dimensional Path Planning for UAVs in Dynamic Environment using Glow-worm Swarm Optimization. *Procedia Comput. Sci.* **2013**, *133*, 230–239. [CrossRef]
48. Chen, Y.; Mei, Y.; Yu, J.; Su, X.; Xu, N. Three-dimensional unmanned aerial vehicle path planning using modified wolf pack search algorithm. *Neurocomputing* **2017**, *226*, 4445–4457. [CrossRef]
49. Wang, G.; Chu, H.E.; Mirjalili, S. Three-dimensional path planning for UCAV using an improved bat algorithm. *Aerosp. Sci. Technol.* **2016**, *49*, 231–238. [CrossRef]
50. Aghababa, M. 3D path planning for underwater vehicles using five evolutionary optimization algorithms avoiding static and energetic obstacles. *Appl. Ocean Res.* **2012**, *38*, 48–62. [CrossRef]
51. Mac, T.; Copot, C.; Tran, D.; De Keyser, R. Heuristic approaches in robot path planning: A survey. *Rob. Auton. Syst.* **2016**, *86*, 13–28. [CrossRef]
52. Szirmay-Kalos, L.; Márton, G. Worst-case versus average case complexity of ray-shooting. *Computing* **1998**, *61*, 103–131. [CrossRef]
53. Berger, M.J.; Oliger, J. Adaptive mesh refinement for hyperbolic partial differential equations. *Comput. Phys.* **1984**, *53*, 484–512. [CrossRef]

54. Min, C.; Gibou, F. A second order accurate projection method for the incompressible Navier-Stokes equations on non-graded adaptive grids. *J. Comput. Phys.* **2006**, *219*, 912–929. [CrossRef]
55. Hasbestan, J.J.; Senocak, I. Binarized-octree generation for Cartesian adaptive mesh refinement around immersed geometries, *J. Comput. Phys.* **2018**, *368*, 179–195. [CrossRef]
56. Pantano, C.; Deiterding, R.; Hill, D.J.; Pullin, D.I. A low numerical dissipation patch-based adaptive mesh refinement method for large-eddy simulation of compressible flows. *J. Comput. Phys.* **2007**, *221*, 3–87. [CrossRef]
57. Ryde, J.; Hu, H. 3D mapping with multi-resolution occupied voxel lists. *Auton. Robots* **2010**, *28*, 169–185. [CrossRef]
58. Samet, H.; Kochut, A. Octree approximation an compression methods. In Proceeding of the First International Symposium on 3D Data Processing Visualization and Transmission, Padova, Italy, 19–21 June 2002; pp. 460–469. [CrossRef]
59. Samaniego, F.; Sanchis, J.; García-Nieto, S.; Simarro, R. UAV motion planning and obstacle avoidance based on adaptive 3D cell decomposition: Continuous space vs discrete space. In Proceeding of the 2017 IEEE Ecuador Technical Chapters Meeting (ETCM), Salinas, Ecuador, 6–20 October 2017; pp. 1–6. [CrossRef]
60. Markus, S.; Akesson, K.; Martin, F. Modeling of discrete event systems using finite automata with variables. In Proceeding of the 2007 46th IEEE Conference on Decision and Control, New Orleans, LA, USA, 12–14 December 2007; pp. 3387–3392. [CrossRef]
61. Yang, Y.; Prasanna, V. Space-time tradeoff in regular expression matching with semi-deterministic finite automata. In Proceeding of the 2011 IEEE INFOCOM, Shanghai, China, 17 August 2007; pp. 1853–1861. [CrossRef]
62. Normativa Sobre Drones en España [2019]—Aerial Insights. Available online: http://www.aerial-insights.co/blog/normativa-drones-espana/ (accessed on 28 December 2018).
63. Disposici6n 15721 del BOE núm. 316 de 2017 - BOE.es. Available online: https://www.boe.es/boe/dias/2017/12/29/pdfs/BOE-A-2017-15721.pdf (accessed on 28 December 2018).
64. Velasco-Carrau, J.; García-Nieto, S.; Salcedo, J.; Bishop, R. Multi-Objective Optimization for Wind Estimation and Aircraft Model Identification. *J. Guid. Control Dyn.* **2016**, *39*, 372–389. [CrossRef]
65. Vanegas, G.; Samaniego, F.; Girbes, V.; Armesto, L.; Garcia-Nieto, S. Smooth 3D path planning for non-holonomic UAVs. In Proceeding of the 2018 7th International Conference on Systems and Control (ICSC), Universitat Politècnica de València, Spain, 24–26 October 2018; pp. 1–6. [CrossRef]
66. Samaniego, F.; Sanchís, J.; García-Nieto, S.; Simarro, R. Comparative Study of 3-Dimensional Path Planning Methods Constrained by the Maneuverability of Unmanned Aerial Vehicles. In Proceeding of the 2018 7th International Conference on Systems and Control (ICSC), Universitat Politècnica de València, Spain, 24–26 October 2018; pp. 13–20. [CrossRef]

electronics

MDPI

Article

Preliminary Design of an Unmanned Aircraft System for Aircraft General Visual Inspection

Umberto Papa [1,*] and Salvatore Ponte [2]

1 Department of Science and Technology, University of Naples "Parthenope", 80143 Naples, Italy
2 Department of Engineering, University of Campania "L. Vanvitelli", 81031 Aversa (CE), Italy; salvatore.ponte@unicampania.it
* Correspondence: umberto.papa@uniparthenope.it

Received: 26 October 2018; Accepted: 13 December 2018; Published: 14 December 2018

Abstract: Among non-destructive inspection (NDI) techniques, General Visual Inspection (GVI), global or zonal, is the most widely used, being quick and relatively less expensive. In the aeronautic industry, GVI is a basic procedure for monitoring aircraft performance and ensuring safety and serviceability, and over 80% of the inspections on large transport category aircrafts are based on visual testing, both directly and remotely, either unaided or aided via mirrors, lenses, endoscopes or optic fiber devices coupled to cameras. This paper develops the idea of a global and/or zonal GVI procedure implemented by means of an autonomous unmanned aircraft system (UAS), equipped with a low-cost, high-definition (HD) camera for carrying out damage detection of panels, and a series of distance and trajectory sensors for obstacle avoidance and inspection path planning. An ultrasonic distance keeper system (UDKS), useful to guarantee a fixed distance between the UAS and the aircraft, was developed, and several ultrasonic sensors (HC-SR-04) together with an HD camera and a microcontroller were installed on the selected platform, a small commercial quad-rotor (micro-UAV). The overall system concept design and some laboratory experimental tests are presented to show the effectiveness of entrusting aircraft inspection procedures to a small UAS and a PC-based ground station for data collection and processing.

Keywords: UAS; aircraft maintenance; General Visual Inspection; sensor fusion; image processing; flight mechanics

1. Introduction

General Visual Inspection (GVI) is a common method of quality control, data acquisition and data analysis, involving raw human senses such as vision, hearing, touch, smell, and/or any non-specialized inspection equipment. Unmanned aerial systems (UAS), i.e., unmanned aerial vehicles (UAV) equipped with the appropriate payload and sensors for specific tasks, are being developed for automated visual inspection and monitoring in many industrial applications. Maintenance Steering Group-3 (MSG-3), a decision logic process widely used in the airline industry [1], defines GVI as a visual examination of an interior or exterior area, installation or assembly to detect obvious damage, failure or irregularity, made from within touching distance and under normally available lighting condition such as daylight, hangar lighting, flashlight or drop-light [2]. As stated by the Federal Aviation Administration (FAA) Advisory Circular 43-204 [3], visual inspection of aircrafts is used to:

- Provide an overall assessment of the condition of a structure, component, or system;
- Provide early detection of typical airframe defects (e.g., cracks, corrosion, engine defects, missing rivets, dents, lightning scratches, delamination, and disbonding) before they reach critical size;
- Detect errors in the manufacturing process;
- Obtain more information about the condition of a component showing evidence of a defect.

As far as aircraft maintenance, safety and serviceability are concerned, entrusting GVI tasks to a UAS with hovering capabilities may be a good alternative, allowing more accurate inspection procedures and extensive data collection of damaged structures [4]. Traditional aircraft inspection processes, performed in hangars from the ground or using telescopic platforms, can typically take up to 1 day, whereas drone-based aircraft GVI could significantly reduce the inspection time. Moreover, a recent study [5] quantified aircraft visual inspection error rates up to 68%, classifying them as omissions (missing a defect) rather than commissive errors (false alarms, i.e., flagging a good item as defective). The use of a reliable automatic GVI procedure with a UAS could substantially lower the incidence of omissions, which could be catastrophic for aircraft serviceability, whereas false positives, or commissive errors, are not a major concern. From this standpoint, the development of a drone-based GVI approach is justified by the following aspects:

- Reduced aircraft permanence in the hangar and reduced cost of conventional visual inspection procedures;
- accelerated and/or facilitated visual checks in hard-to-reach areas, increased operator safety;
- possibility of designing specific and reproduceable inspection paths around the aircraft, capturing images at a safe distance from the structures and at different viewpoints, and transmitting data via dedicated links to a ground station;
- accurate defect assessment by comparing acquired images with 3D structural models of the airplane;
- ease of use, no pilot qualification needed;
- possibility of gathering different information by installing on the UAV cost-effective sensors (thermal cameras, non-destructive testing sensors, etc.);
- increased quality of inspection reports, real-time identification of maintenance issues, damage and anomalies comparison with previous inspections;
- possibility of producing automatic inspection reports and performing accurate inspections after every flight.

Ease of access to the inspection area is important in obtaining reliable GVI procedures. Access consists of the act of getting into an inspection position (primary access) and performing the visual inspection (secondary access). Unusual and unreachable positions (i.e., crouching, lying on back, etc.) are examples of difficult primary access [3].

This paper describes the preliminary design of a GVI system onboard a commercial VTOL (Vertical Take-Off and Landing) UAV (a quadrotor). The experimental setup is composed of a ground station (PC-based), an embedded control system installed on the airframe, and sensors devoted to GVI. To ensure a minimum safety distance from the aircraft is inspected, an ultrasonic distance keeper system (UDKS) has been designed. A high-definition (HD) camera will detect visual damage caused by hail strikes or lightning strikes, which are among the most dangerous threats for the airframe. The incidence of lightning strikes on aircrafts in civil operation is of the order of one strike per aircraft per year, whereas hail strikes are much more common and dangerous. The hovering UAS will be pushed along the fuselage and wings, and the vehicle automatically maintains a safe standoff distance from the aircraft due to ultrasonic sensors which are part of the devised UDKS.

The paper is structured as follows: After justifying in the Introduction the use of a UAV-based aircraft inspection system, a Background (Section 2) focuses on literature review of related work. In Section 3, the chosen UAV, the GVI equipment and the UDKS are presented. Section 4 shows preliminary results obtained from initial test flights, evaluating the performance of the UDKS and of the image acquisition/processing module for damage detection. Conclusions and further developments as outlined in Section 5.

2. Background

The potential for UAV-based generation of high-quality images of structural details to detect structural damage and support condition assessment of civil structures is very high, particularly in difficult-to-access areas [6]. Drone-based visual inspection is used for monitoring oil and gas platforms [7], drilling rigs, pipelines [8], transmission networks [9], infrastructures like bridges [10] and buildings [11], assessing road conditions [12], 3D mapping of pavement distresses [13], inspecting power lines [14] and equipment [15], photovoltaic (PV) plants [16], cracks in concrete civil infrastructures [17], construction sites [18], elevators [19], as well as viaducts, subways, tunnels, level crossings, dams, reservoirs, etc. [20]. Integration of Unmanned Ground Vehicles (UGV) and UAVs is also being exploited for industrial inspection tasks [21].

Nowadays, a broad range of UAVs exist, from small and lightweight fixed-wing aircraft to rotor helicopters, large-wingspan airplanes and quadcopters, each one for a specific task, generally providing persistence beyond the capabilities of manned vehicles [22]. UAVs can be categorized with respect to mass, range, flight altitude, and endurance (Table 1).

Table 1. UAV categorization (adapted from References [23,24]).

	Mass (kg)	Range (km)	Altitude (m)	Endurance (h)
Micro UAV (MAV)	<5	<10	Up to 250	≤1
Mini UAV	<20 or 25	<10	Up to 300	<2
Low Altitude, Long Endurance (LALE)	15–25	>500	3000	>24
Low Altitude, Deep Penetration (LADP)	250–2500	>250	50–9000	0.5–1
Medium Altitude, Long Endurance (MALE)	1000–1500	>500	3000	24–48
High Altitude, Long Endurance (HALE)	2500–5000	>2000	20000	24–48
Tactical UAV (TUAV), Close Range (CR)	25–150	10–30	3000	2–4
TUAV, Medium Range (MR)	150–500	>500	8000	10–18

VTOL aircrafts offer many advantages over Conventional Take-Off and Landing (CTOL) vehicles, mainly due to the small area required for take-off and landing, and the ability to hover in place and fly at very low altitudes and in narrow, confined spaces. Among VTOL aircrafts, such as conventional helicopters and crafts with rotors like the tiltrotor and fixed-wing aircraft with directed jet thrust capability, the quadrotor, or quadcopter, is very frequently chosen, especially in academic research on mini or micro-size UAVs, as an effective alternative to the high cost and complexity of conventional rotorcrafts, due to its ability to hover and move without the complex system of linkages and blade elements present in single-rotor vehicles [25–27]. Quadrotors are widely used in close-range photogrammetry, search and rescue operations, environment monitoring, industrial component inspection, etc., due to their swift maneuverability in small areas. The quadrotor has good ranking among VTOL vehicles, yet it has some drawbacks. The craft size is comparatively larger, energy consumption is greater, implying lower flight time, and the control algorithms are very complicated due to the fact that only four actuators are used to control all six degrees of freedom (DOF) of the quadrotor (which is classified as an underactuated system), and that the changing aerodynamic interference patterns between the rotors have to be taken into account [26,28–30].

3. UAV and GVI Equipment

3.1. Unmanned Aerial Vehicle

The unmanned aircraft chosen for this application is a micro-UAV (MAV, see Table 1) quadrotor, model "RC EYE One Xtreme" (Figure 1), produced by RC Logger [31]. It spans 23 cm from the tip of one rotor to the tip of the opposite rotor; has a height of 8 cm, a total mass of 260 g; with a 100-g, 7.4-V LiPo battery; and a payload capacity of 150 g. Typical flight time is of the order of 10 min, with a 1150-mAh battery and the payload. This vehicle consolidates rigid design, modern aeronautic technology as well as easy maintainability into a versatile multi-rotor platform, durable enough to survive most crashes. Six-axis gyro stabilization technology and a robust brushless motor-driven flight control are embedded within a rugged yet stylish frame design. The One Xtreme is an ideal platform for flight applications (the maximum transmitter range is approximately 150 m in free field) ranging from aerial surveillance, imaging or simply unleashing acrobatic fun flight excitement. A low-level built-in flight control system can balance out small undesired changes to the flight altitude, and allows one to select three flight modes: beginner, sport and expert. In this work, all experimental data sessions were performed flying the UAV in the beginner's mode.

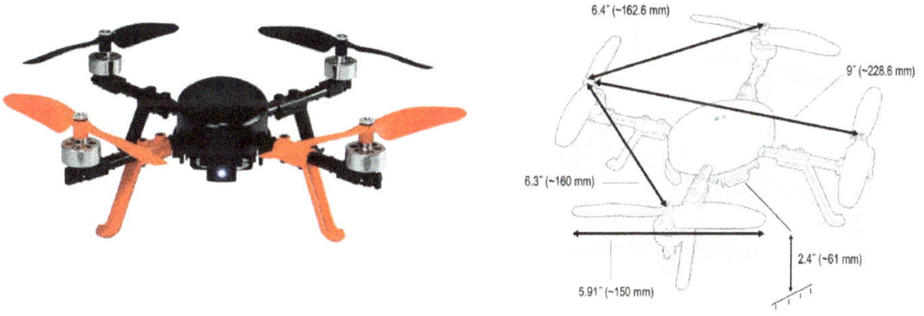

(a) (b)

Figure 1. Aspect (**a**) and dimensions (**b**) of the RC EYE One Xtreme (Mode 2).

The One Xtreme can be operated both indoors and outdoors during calm weather conditions, since the empty weight of the UAV (260 g) makes it react sensitively to wind or draughts. In order to control and manage all the phases of flight during a GVI session (in particular, the landing procedure and the cruise at a fixed distance from the fuselage of the aircraft to be inspected), a distance acquisition system using sonic ranging (SR) sensors was built and tested.

3.2. GVI and Image Processing Equipment

The image acquisition subsystem devised for the GVI equipment consists of a Raspberry Pi Camera Module v2, released in 2016 [32], mounted on a small single-board computer, the Raspberry Pi 2 Model B [33], with a 900-MHz quad-core ARM Cortex-A7 CPU, 1-GB RAM, two USB ports, and a camera interface (CSI port, located behind the Internet port) with dedicated camera software. A Wi-Fi dongle was installed on the Raspberry board for real-time transmission of images to the PC-based control station. The total weight of the image acquisition hardware (close-range camera, computer module and case) is 50 g.

Table 2 reports some technical specifications of the Pi Camera Module v2. The module allows manual control of the focal length, in order to set up specific and constant values. It can acquire images with a maximum resolution of about 6 Megapixels at a maximum rate of 90 frames per second. To avoid buffer overload and excessive computational cost, the resolution was set to 1920×1080 (2 Megapixel), corresponding to HD video. Figures 2 and 3 show the standalone assembled image acquisition system and the installation on the quadrotor, respectively.

Table 2. Raspberry Pi Camera Module (v2) hardware specification.

Features	Properties
Size, weight	25 × 24 × 9 mm, 3 g
Still resolution	8 Megapixels
Video modes	1080p30, 720p60 and 640 × 480p60/90
Sensor resolution	3280 × 2464 pixel (Sony IMX219)
Focal length	3.04 mm
Pixel size	1.12 × 1.12 µm
Sensor size	Width: 6.004 ± 0.006 mm Height: 3.375 ± 0.005 mm
Fixed focus	1m to infinity
Frame rate	max 90 fps
Horizontal/Vertical FOV (Field Of View)	62.2/48.8 degrees

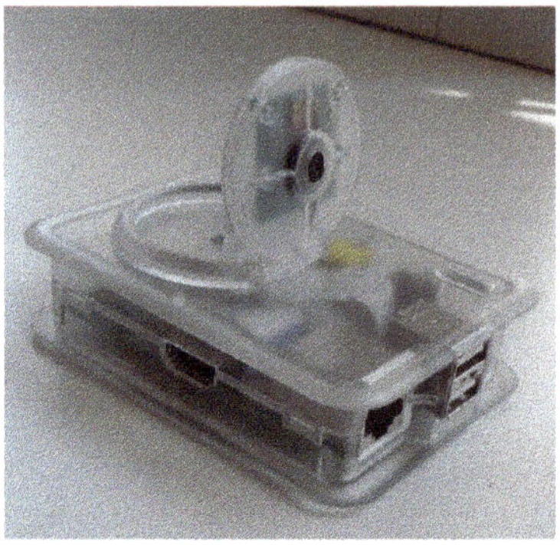

Figure 2. Image acquisition subsystem (Raspberry Pi 2 Model B and Pi camera), standalone configuration.

Figure 3. Image acquisition subsystem mounted on RC EYE One Xtreme.

Close-range camera calibration, i.e., the process that allows determining the interior orientation of the camera and the distortion coefficients [34], has been performed by means of a dedicated toolbox developed in the Matlab environment and using some coded targets [35]. The principal point or image center, the effective focal length and the radial distortion coefficients were estimated to account for the variation of lens distortion within the field of view, and to determine the location of the camera in the scene.

The basic image-processing task for the GVI system is object recognition, i.e., the identification of a specific object in an image. Several methodologies exist for automatic detection of circular targets in a frame. They can be divided in two categories, namely, no-initial-approximation-required, which allows to obtain the coordinate center of a circular target on an image in a completely autonomous way, and initial-approximation-required, which needs an approximated position of the target center and can detect the center with high accuracy.

The damages (modeled as circular anomalies) can be detected automatically through a Circle Hough Transform (CHT), based on the Hough Transform applied to an edge map [36] and by means of edge detection techniques such as the Canny edge detector [37,38]. The CHT has been chosen due to its effectiveness for automatic recognition in real time and because it is already developed in the OpenCV library and customizable in the Python environment.

The visual inspector can use CHT-processed images to highlight probable damage, delimited by a circular or rectangular area. The image processing algorithm, developed by the authors [35] and transferred to the embedded computer board of the quadrotor, essentially loads and blurs the image to reduce the noise, applies the CHT to the blurred image, and displays the detected circles (probable damaged areas) in a window.

3.3. Ultrasonic Distance Keeper System (UDKS) and Data Filtering

Detecting and avoiding obstacles in the inspection area is a fundamental task for a safe, automatic GVI procedure. The system developed is based on a commercial off-the-shelf ultrasonic sensor, the HC-SR04, as shown in Figure 4 [39]. It provides distance measurements in the range of 2–400 cm with a target accuracy of 3 mm. These features guarantee a correct distance from the aircraft and other obstacles in the inspection area.

Figure 4. HC-SR04 sonic ranging sensor.

The module includes the ultrasonic transmitter and receiver and control circuitry. The shield has four pins (VCC and Ground for power supply, Echo and Trigger for initialization and distance acquisition). Technical specifications of the HC-SR04 are reported in Table 3.

Table 3. HC-SR04 technical specifications.

HC-SR04	
Supply Voltage	+5 V DC
Working Current	15 mA
Ranging distance	2–400 cm
Range resolution	0.3 cm
Input Trigger	10-μs TTL pulse
Echo pulse	Pos. TTL pulse
Burst Frequency	40 kHz
Measuring Angle	30 degrees
Weight	20 g
Dimensions	45 × 20 × 15 mm

To start measurements, the trigger pin of the sensor needs a high pulse (5V) for at least 10 μs, which initiates the sensor. Eight cycles of ultrasonic burst at 40 kHz are transmitted, and when the sensor detects ultrasonic echo returns, the Echo pin is set to High. The delay between the Trig pulse and the Echo pulse is proportional to the distance from the reflecting object. Figure 5 summarizes the ranging measurement sequence. If no obstacles are detected, the output pin will give a 38-ms-wide, high-level signal. For optimal performance, the surface of the object to be detected should be at least 0.5 square meters. Assuming a value of 340 m/s for the speed of sound, the distance (in centimeters) is given by the delay time divided by 58.

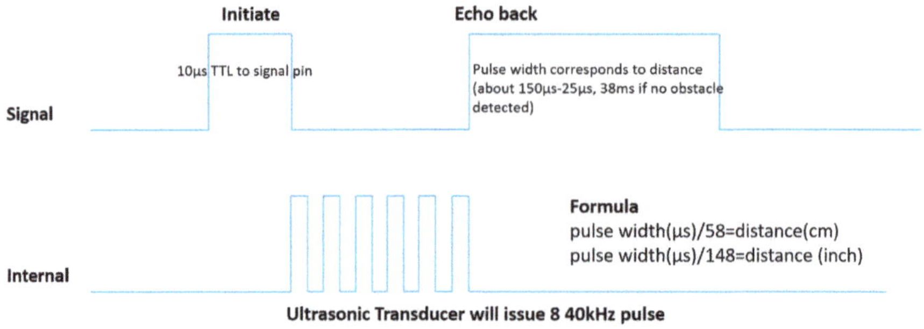

Figure 5. Sonic Ranger sequence chart.

To improve measurement accuracy, the relationship between the speed of sound a and temperature and relative humidity variation has been considered:

$$a = \sqrt{\gamma R T} \tag{1}$$

where γ is the heat capacity ratio (1.44 for air), R is the universal gas constant (287 J/kg K for air) and T is the temperature (K). A low-cost digital temperature and humidity sensor (DHT11) has been integrated in the sensor suite. The sensor provides a digital signal with 1-Hz sampling rate. The humidity readings are in the range of 20–80% with 5% accuracy, and the temperature readings are in the range 0–50 °C with ±2 °C accuracy [40]. The adjustment is performed during the acquisition of distance data by means of Arduino sketches and libraries for the management of sensors (DHT11 and HC-SR04) and for data collection. Experimental results and an analysis on the improved performance of the sonic ranging sensor were presented in previous works [41–43].

To obtain high reliability, a wider scanning area and accurate distance measurements, the designed UDKS employs four ultrasonic sensors [44] in a cross configuration (Figures 6 and 7). The frame has been designed and realized by the authors with a 3D printer. The total weight (frame and sensors) is 78 g.

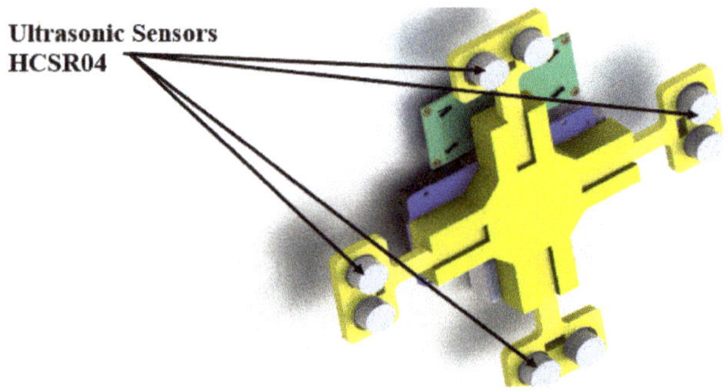

Figure 6. Rendering of the UDKS frame.

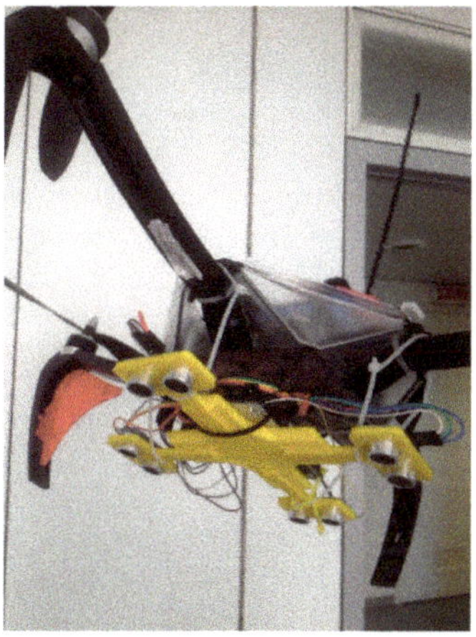

Figure 7. UDKS mounted on the UAV.

The microcontroller which acquires and manages measurements from the four-sensor structure is an Arduino Mega2560 board, powered by the LiPo battery of the MAV. The board acquires distance data from the sensors at 20-Hz sampling rate and sends the measurements to the PC-based ground station via a Wi-Fi link. The distance readings are corrected for temperature and humidity variations (see Equation. (1)) and are successively smoothed with a simple 1D Kalman filter.

The data acquisition system and the observation model are described by the following equations:

$$\begin{cases} x_{k+1} = Ax_k + w_k \\ y_k = Cx_k + v_k \end{cases} \tag{2}$$

where x_k is the k-th distance; w_k is the model noise, assumed as Gaussian with zero mean and variance Q, y_k is the k-th measurement; v_k is the measurement noise, assumed as Gaussian with zero mean and variance R (in our case, 0.09 cm^2, as declared by the manufacturer, see Table 3). In our case, A and C are scalar quantities equal to 1. The Kalman filter algorithm is based on a predictor-corrector iterative sequence, as follows:

$$\hat{x}_{k+1|k} = \hat{x}_{k|k} \text{ (predicted distance value, before measurement)} \tag{3}$$

$$P_{k+1|k} = P_{k|k} + Q \text{ (\textit{a-priori} estimation error)} \tag{4}$$

$$K_{k+1} = \frac{P_{k+1|k}}{P_{k+1|k} + R} \text{ (Kalman gain)} \tag{5}$$

$$\hat{x}_{k+1|k+1} = \hat{x}_{k+1|k} + K_{k+1}(y_{k+1} - \hat{x}_{k+1|k}) \text{ (new estimate from current measurement)} \tag{6}$$

$$P_{k+1|k+1} = P_{k+1|k}(1 - K_{k+1}) \text{ (\textit{a-priori} estimation error)} \tag{7}$$

The implementation of Equations (4) to (7) in the Arduino IDE (Integrated Development Environment) is reported in Figure 8 as a function (`kalmanSmooth`) called whenever a new distance measurement (the variable `Reading`) is available from the sensors.

All data (raw distance measurements, smoothed distances, raw and processed images) received via the Wi-Fi link are stored in the PC-based ground station for analysis and further processing with Matlab. As an alternative, a LabVIEW VI (Virtual Instrument), developed by the authors, performs offline data processing (noise removal, calibration).

```
float kalmanSmooth(float Reading)
{
P += Q; //Arduino compound addition (+=) for the a-priori estimation error, Eq. (4)
K = P/(P+R); //Kalman gain, Eq. (5)
X += K*(Reading - X); //Compound addition for the smoothed value, Eq. (6)
P = (1-K)*P; //A-posteriori smoothing (estimation) error, Eq. (7)
return X;
}
```

Figure 8. Arduino function `kalmanSmooth` for Kalman filtering of distance data acquired by the SR sensor. P, Q and R are the variances of the smoothing error, of the measurement model and of the acquired measurement, respectively.

During the first experimental sessions, the UDKS has been positioned below the quadrotor. A complete UDKS will use a second cross configuration, like the one depicted in Figure 7, coupled to the lateral side of the UAV, to insure range detection when the quadrotor is flying over the fuselage and along the sides of the aircraft to be inspected. Figure 9 shows the assembly. The total payload weight slightly exceeds the limits imposed by the manufacturer, but by using wider blades (or, obviously, a greater quadrotor) the lift-off weight will increase. The complete configuration has not been tested and will be the object of future experimental campaigns.

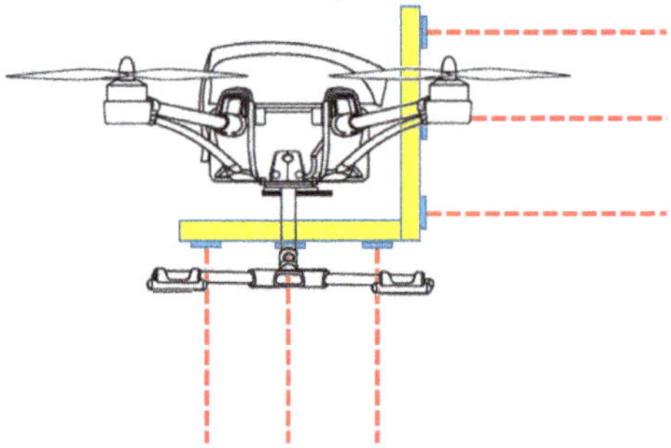

Figure 9. UDKS installation, configuration in front of and below the quad rotor.

Another idea to be developed is a real-time obstacle mapping system, obtained by coupling the SR sensor with a servomotor in a rotary configuration, as shown in Figure 10—a sort of ultrasonic radar which sends data to a PPI (Plan Position Indicator)-like screen provided by the Arduino IDE and allows the inspector to visually detect the presence of obstacles in the surroundings of the drone (Figure 11).

Figure 10. Rotary/radar configuration.

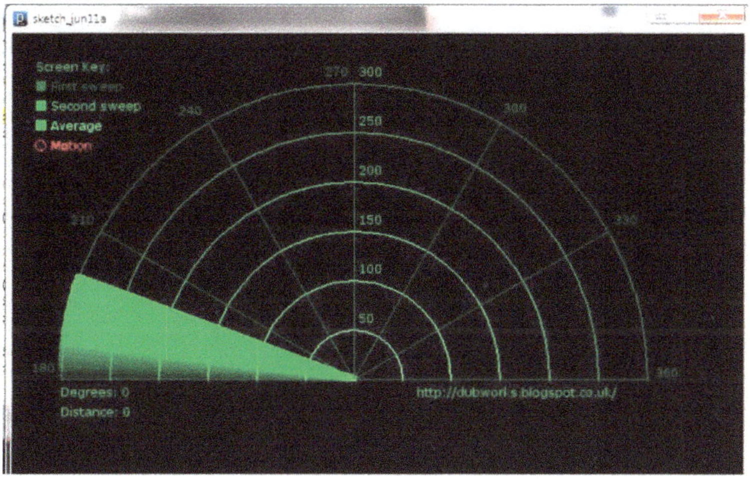

Figure 11. Real-time distance mapping and manual obstacle detection through the embedded Arduino IDE.

4. Experimental Results

This section shows some preliminary results of a test flight (in manual mode) of the MAV equipped with the GVI hardware (camera, UDKS, microcontroller) previously illustrated. The main objectives of this first flight test are to verify correct distance acquisition by the sensors and to evaluate the performance of the image-processing software. The inspected targets are different aircraft test panels damaged by hail and lightning strikes. The minimum distance considered for the GVI procedure is about 1.5–2 m from the aircraft and other obstacles in the area. The UAS flew in a cylindrical "safety area", with the axis corresponding to the nominal inspection path to avoid collisions with the aircraft. The flight time, due to the high current absorption of the payload, was of the order of 5 min, about 50% of the nominal endurance as specified by the manufacturer.

According to Illuminating Engineering Society (IES), direct, focused lighting is the recommended general lighting for aircraft hangars [45]. Generally, most maintenance tasks require between 75 and 100 fc (foot-candles), i.e., 800-1100 lux (or lumen per square meter). The room in which the flight tests were performed is $10 \times 6\text{-m}^2$, 4-m high, with high-reflectance walls and floor, helping in reflecting light and distributing it uniformly. The average measured brightness was 80 fc (860 lux). We arranged the experimental setup in order to obtain direct, focused lighting and minimize specular reflection from the test panels, avoiding glare. Compatibility with particular lighting conditions such as flashlight or drop-light, has not been tested.

As an example of the UDKS measurements, Figure 12 shows an acquisition at 1.20-m horizontal distance from the test panel and 1.60-m hovering height from the floor. Kalman filtering has been applied in real-time by the microcontroller to reduce measurement noise. A small drift (about 10 cm) can be noted during the movement of the craft along the direction of the test panel, due to the manual mode: In this initial test, no automatic path planning was implemented, and the UAS was piloted by a human operator located on the ground station a few meters away from the quadrotor.

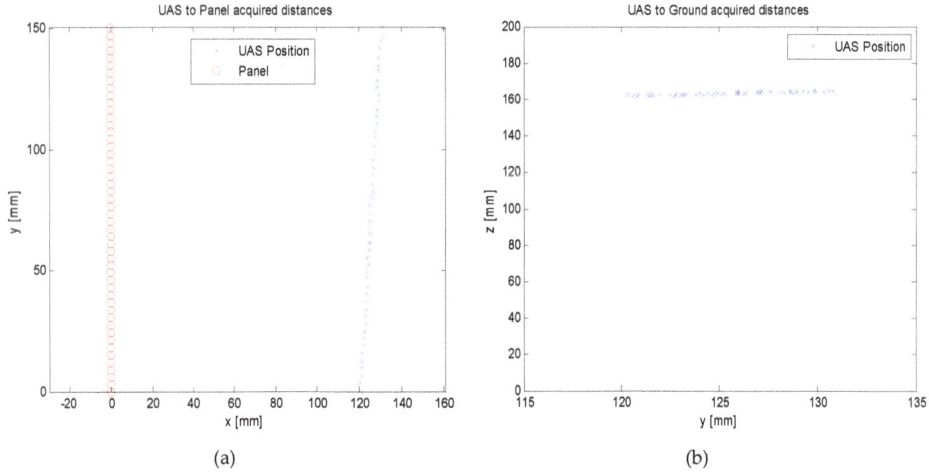

Figure 12. (**a**) Acquired distance from the damaged panel; (**b**) distance from the ground.

Figure 13a shows the first test panel, with damage caused by hail. The HD camera acquired the image of the panel, and the embedded image-processing software module, run by the Raspberry Pi 2 computer board, sent the image to the ground control station highlighting the damaged areas (Figure 13b).

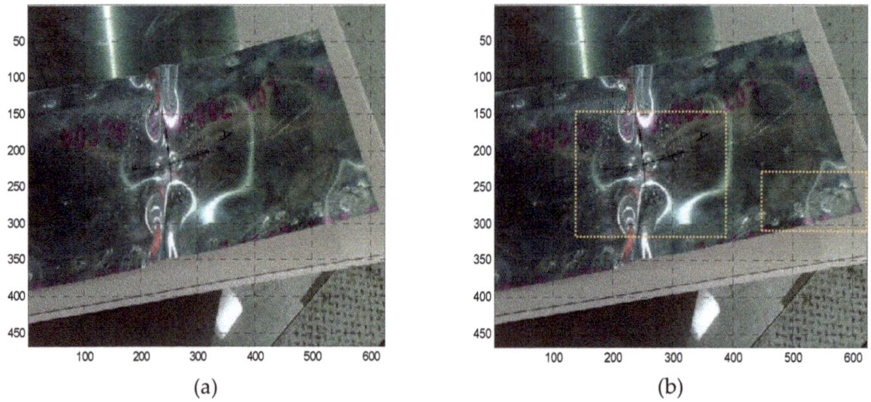

Figure 13. (**a**) Test panel (aluminum) damaged by hail. (**b**) Image of the test panel with the hail-strike damage highlighted. The image processing module has also identified another defective area (small rectangle on the right) not damaged by hail.

Figure 14a shows the second test item damaged by a lightning strike, which was simulated in a laboratory by inducing a current on the panel. The damaged area was correctly identified by the image-processing algorithms and is highlighted in Figure 14b.

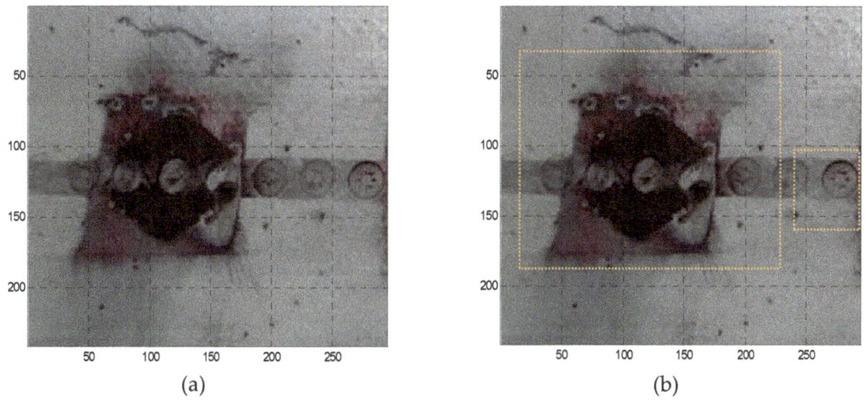

Figure 14. (**a**) Lightning-strike damage on an aluminum panel. (**b**) Image of the damaged area highlighted. A circular pattern has been correctly identified by the CHT routine and is enclosed in the small rectangle on the right side of the image.

From these preliminary tests, it can be deduced that automated damage detection, even in areas not easily reachable by a human inspector, allows us to significantly reduce the inspection time. Moreover, the remote operator can identify structural damages in real time and verify the correctness of the UAS flight path by analyzing the distance measurements provided by the UKDS.

5. Conclusions and Further Work

This paper has presented the conceptual design of an embedded system, useful for GVI in aircraft maintenance, installed on a commercial micro unmanned aircraft vehicle (MAV) and coupled to a remotely-operated PC-based ground station. The architecture of the ultrasonic distance keeper system (UDKS) and of the image acquisition subsystem have been presented and analyzed. The tools and processing used for this research were:

- MAV (quadrotor);
- Microprocessor and embedded HD camera (Raspberry);
- Open-source image processing libraries;
- Wi-Fi link for data transmission to a PC-based ground station;
- SR sensors for distance measurements;
- Microcontroller (Arduino) and IDE;
- Matlab/LabVIEW for post-processing and data presentation.

The first tests have shown the correctness of the distance acquisition system (UDKS) and the capability of the HD camera, managed by a computer board with embedded image processing software, to identify damaged areas and send in real time images of the critical (damaged) zones to a remote operator in a ground control station, equipped with a laptop computer.

Future developments of the research activity are mainly focused on the full-field trajectory planning. During a GVI procedure, the UAS must be able to plan and follow a specific path, which mainly depends on the type and size of the aircraft to be inspected. Collecting top views of the aircraft will allow us to define a safe inspection trajectory, and to perform fully automatic inspection flights around the structure of the craft. Work is currently underway in defining algorithms for generating optimal paths that maximize the coverage of the aircraft structure [46]. Different path-planning algorithms will be tested, and the most promising, from preliminary studies, seems to be the Rapidly-exploring Random Tree (RRT) methodology [47].

Furthermore, different sensors are being evaluated and tested to increase the accuracy of distance measurements. Currently, experiments with a Lidar (Laser Detection and Ranging) are giving good results in terms of increased accuracy and reduced power consumption. Another issue to be dealt with in future work is the design of an effective automated Detect-And-Avoid (DAA) strategy, to ensure safe flight of the UAS and to eliminate possible collisions with obstacles in the flight path [48,49]. Other issues, such as increasing the endurance (flight time) with high-capacity LiPo batteries, or using a larger platform to accommodate different sensors, still remaining in the MAV category, are also under examination.

Author Contributions: Conceptualization, Software, U.P.; Methodology, Resources, Data Curation, U.P. and S.P.; Writing-Original Draft preparation, U.P.; Writing-Review& Editing, S.P.

Funding: This Research received no external fundings.

Conflicts of Interest: The authors declare no conflict of interest.

References

1. Air Transport Association of America Inc. *ATA MSG-3, Operator/Manufacturer Scheduled Maintenance Development, Vol. 1- Fixed Wing Aircraft*; ATA: Washington, DC, USA, 2015.
2. Intergraph Corp. *Maintenance Steering Group-3 (MSG-3)-Based Maintenance and Performance-Based Planning and Logistic (PBP&L) Programs—A White Paper*; Intergraph Corporation: Madison, AL, USA, 2006.
3. US Department of Transportation—FAA. *Visual Inspection for Aircraft*; Advisory Circular (AC) No. 43-204; Federal Aviation Administration: Washington, DC, USA, 1997.
4. Airbus. Innovation Takes Aircraft Visual Inspection to New Heights. 2018. Available online: https://www.airbus.com/newsroom/news/en/2018/04/innovation-takes-aircraft-visual-inspections-to-new-heights.html (accessed on 13 October 2018).
5. Office of the Secretary of Defense. *Unmanned Aircraft Systems Roadmap: 2005–2030*; Office of the Secretary of Defense: Washington, DC, USA, 2005.
6. Morgenthal, G.; Hallermann, N. Quality Assessment of Unmanned Aerial Vehicle (UAV) Based Inspection of Structures. *Adv. Struct. Eng.* **2014**, *17*, 289–302. [CrossRef]
7. Marinho, C.A.; de Souza, C.; Motomura, T.; Gonçalves da Silva, A. In-Service Flares Inspection by Unmanned Aerial Vehicles (UAVs). In Proceedings of the 18th World Conference on Nondestructive Testing, Durban, Africa, 16–20 April 2012.
8. Sadovnychiy, S. Unmanned Aerial Vehicle System for Pipeline Inspection. In Proceedings of the 8th WSEAS International Conference on Systems, Athens, Greece, 12–14 July 2004.
9. Tatum, M.C.; Liu, J. Unmanned Aerial Vehicles in the Construction Indstry. In Proceedings of the 53rd ASC Annual International Conference, Seattle, WA, USA, 5–8 April 2017; pp. 383–393.
10. Metni, N.; Hamel, T. A UAV for bridge inspection: Visual servoing control law with orientation limits. *Autom. Constr.* **2007**, *17*, 3–10. [CrossRef]
11. Eschmann, C.; Kuo, C.-M.; Kuo, C.-H.; Boller, C. Unmanned Aircraft Systems for Remote Building Inspection and Monitoring. In Proceedings of the 6th European Workshop on Structural Health Monitoring (EWSHM 2012), Dresden, Germany, 3–6 July 2012.
12. Zhang, C. An UAV-based photogrammetric mapping system for road condition assessment. *Remote. Sens. Spat. Inf. Sci.* **2008**, *37*, 627–632.
13. Leonardi, G.; Barrile, V.; Palamara, R.; Suraci, F.; Candela, G. *3D Mapping of Pavement Distresses Using an Unmanned Aerial Vehicle (UAV) System*; Springer: Berlin, Germany, 2018; Volume 101, pp. 164–171.
14. Montambault, S.; Beaudry, J.; Touissant, K.; Pouliot, N. On the application of VTOL UAVs to the inspection of Power Utility Assets. In Proceedings of the 1st International Conference on Applied Robotics for the Power Industry (CARPI 2010), Montreal, QC, Canada, 5–7 October 2010; pp. 1–7.
15. Deng, C.; Wang, S.; Huang, Z.; Tam, Z.; Liu, J. Unmanned Aerial Vehicles for Power Line Inspection: A Cooperative Way in Platforms and Communications. *J. Commun.* **2014**, *9*, 687–692. [CrossRef]
16. Bellezza Quarter, P.; Grimaccia, F.; Leva, S.; Mussetta, M.; Aghaei, M. Light Unmanned Aerial Vehicles (UAVs) for Cooperative Inspection of PV Plants. *IEEE J. Photovolt.* **2014**, *4*, 1107–1113.

17. Kim, H.; Sim, S.H.; Cho, S. Unmanned Aerial Vehicle (UAV)-powered Concrete Crack Detection based on Digital Image Processing. In Proceedings of the 6th International Conference on Advances in Experimental Structural Engineering, Urbana-Champaign, IL, USA, 1–2 August 2015.
18. Rodrigues Santos de Melo, R.; Bastos Costa, D.; Sampaio Alvares, J.; Irizarri, J. Applicability of unmanned aerial system (UAS) for safety inspection on construction sites. *Safety Sci.* **2015**, *98*, 174–185. [CrossRef]
19. Kit, H.T.; Chen, H. Autonomous Elevator Inspection with Unmanned Aerial Vehicle. In Proceedings of the 3rd Asia-Pacific World Congress on Computer Science and Engineering (APWC on CSE), Nadi, Fiji, 5–6 December 2016.
20. Ellenberg, A.; Branco, L.; Krick, A.; Baroli, I.; Kontsos, A. Use of Unmanned Aerial Vehicle for Quantitative Infrastructure Evaluation. *J. Infrastruct. Syst.* **2014**, *21*. [CrossRef]
21. Blumenthal, S.; Holz, D.; Linder, T.; Molitor, P.; Surmann, H.; Tretyakov, V. Teleoperated Visual Inspection and Surveillance with Unmanned Ground and Aerial Vehicles. In Proceedings of the REV2008—Remote Engineering & Virtual Instrumentation, Düsseldorf, Germany, 23–25 June 2008.
22. See, J.E. *Visual Inspection: A Review of the Literature*; SANDIA Report SAND2012-8590; Sandia National Laboratories: Albuquerque, NM, USA, 2012; p. 77.
23. Eisenbeiss, H. A Mini Unmanned Aerial Vehicle (UAV): System Overview and Image Acquisition. *Remote Sens. Spat. Inf. Sci.* **2004**, *36.5/W1*, 1–7.
24. Dalamagkidis, K. Classification of UAVs. In *Handbook of Unmanned Aerial Vehicles*; Springer Science+Business Media: Dordrecht, The Netherlands, 2015; pp. 83–91.
25. Valavanis, K.P. Introduction. In *Advances in Unmanned Aerial Vehicles. State of the Art and the Road to Autonomy*; Springer: Dordrecht, The Netherlands, 2007; pp. 3–13.
26. Bouabdallah, S.; Siegwart, R. Design and Control of a Miniature Quadrotor. In *Advances in Unmanned Aerial Vehicles. State of the Art and the Road to Autonomy*; Valavanis, Springer: Dordrecht, The Netherlands, 2007; pp. 171–210.
27. Schmidt, M.D. Simulation and Control od a Quadrotor Unmanned Aerial Vehicle. Ph.D. Thesis, College of Engineering, University of Kentucky, Lexington, Kentucky, 2011.
28. Nonami, K. *Autonomous Flying Robots: Unmanned Aerial Vehicles and Micro Aerial Vehicles*; Springer: Heidelberg, Germany, 2010.
29. Dief, T.N.; Yoshida, S. Review: Modeling and Classical Controller of Quad-rotor. *IRACST Int. J. Comput. Sci. Inf. Tecnol. Secur.* **2015**, *5*, 314–319.
30. Hoffmann, G.; Huang, H.; Waslander, S.L.; Tomlin, C.J. Quadrtor Helicopter Flight Dynamics and Control: Theory and Experiment. In Proceedings of the AIAA Guidance, Navigation and Control Conference and Exhibit, AIAA 2007-6461, Hilton Head, CA, USA, 20–23 August 2007.
31. CEI (Conrad Electronic International). *RC Logger®EYE One Xtreme—Operating Instructions (88008RC—Mode 2)*; CEI Ltd.: Hong Kong, China, 2015; p. 60.
32. Raspberry Pi Foundation. Raspberry Pi Camera Module v2. Hardware Specification. 2015. Available online: https://www.raspberrypi.org/documentation/hardware/camera/ (accessed on 13 October 2018).
33. Raspberry Pi Foundation. Raspberry Pi 2 Model B Specifications. 2015. Available online: https://www.raspberrypi.org/products/raspberry-pi-2-model-b/ (accessed on 13 October 2018).
34. Brown, D.C. Close-range camera calibration. *Photogramm. Eng.* **1971**, *37*, 855–866.
35. Del Pizzo, S.; Papa, U.; Gaglione, S.; Troisi, S.; Del Core, G. A Vision-based navigation system for landing procedure. *Acta IMEKO* **2018**, *7*, 102–109. [CrossRef]
36. Hough, P.V.C. Method and Means for Recognizing Complex Patterns. US Patent Office No. US 3069654, 18 December 1962.
37. Canny, J. A computational approach to edge detection. *IEEE Trans. Pattern Anal. Mach. Intell.* **1986**, *6*, 679–698. [CrossRef]
38. Luhman, T.; Robson, S.; Kyle, S.; Harley, I. *Close Range Photogrammetry: Principles, Techniques and Applications*; Whittles Publishing/John Wiley & Sons, Inc.: Hoboken, NJ, USA, 2006.
39. Cytron Technologies. *HC-SR04 Ultrasonic Sensor—Product User's Manual, V 1.0*; Cytron Technologies Sdn. Bhd.: Penang, Malaysia, 2013.
40. Adafruit Industries. Adafruit Learning System—DHT11, DHT12 and AM2302 Sensors. Available online: https://cdn-learn.adafruit.com/downloads/pdf/dht.pdf?timestamp=1543644105 (accessed on 13 October 2018).

41. Papa, U.; Picariello, F.; Del Core, G. Atmosphere Effect on Sonar Sensor System. *Aerosp. Electron. Syst. Mag.* **2016**, *31*, 34–40. [CrossRef]
42. Papa, U.; Del Core, G. Design of Sonar Sensor Model for Safe Landing of an UAV. In Proceedings of the 2nd IEEE Workshop on Metrology for Aerospace, Benevento, Italy, 3–5 June 2015; pp. 361–365.
43. Papa, U.; Ponte, S.; Del Core, G.; Giordano, G. Obstacle Detection and Ranging Sensor Integration for a Small Unmanned Aircraft System. In Proceedings of the 4th International Workshop on Metrology for Aerospace (MetroAeroSpace), Padova, Italy, 21–23 June 2017.
44. Hamel, T.; Mahony, R. Visual servoing of under-actuated dynamic rigid body system: An image space approach. *IEEE Trans. Robot. Autom.* **2002**, *18*, 187–198. [CrossRef]
45. DiLaura, D.L.; Houser, K.W.; Mistrick, R.G.; Stelly, G.R. *The Lighting Handbook*, 10th ed.; Illuminating Engineering Society (IES): New York, NY, USA, 2011.
46. Almadhound, R.; Taha, T.; Seneviratne, L.; Dias, J.; Cai, G. Aircraft Inspection Using Unmanned Aerial Vehicles. In Proceedings of the International Micro Air Vehicle Conference and Competition 2016 (IMAV 2016), Beijng, China, 17–21 October 2016.
47. Bry, A.; Roy, N. Rapidly-exploring Random Belief Trees for Motion Planning Under Uncertainty. In Proceedings of the IEEE Intenational Conference on Robotics and Automation (ICRA 2011), Shangai, China, 9–13 May 2011.
48. Griffiths, S.; Saunders, J.; Curtis, A.; Barber, B.; McLain, T.; Beard, R. Obstacle and Terrain Avoidance for Miniature Aerial Vehicles. In *Advances in Unmanned Aerial Vehicles. State of the Art and the Road to Autonomy*; Springer: Dordrecht, The Netherlands, 2007; pp. 213–244.
49. Papa, U. *Embedded Platforms for UAS Landing Path and Obstacle Detection*; Book Series of Studies in Systems, Decision and Control; Springer International Publishing: Dordrecht, The Netherlands, 2018; Volume 136.

electronics

MDPI

Article

Real-Time Ground Vehicle Detection in Aerial Infrared Imagery Based on Convolutional Neural Network

Xiaofei Liu [1], Tao Yang [1,2,*] and Jing Li [3,*]

1 SAIIP, School of Computer Science, Northwestern Polytechnical University, Xi'an 710072, China; xiaofei@mail.nwpu.edu.cn
2 Research & Development Institute of Northwestern Polytechnical University in Shenzhen, Shenzhen 518057, China
3 School of Telecommunications Engineering, Xidian University, Xi'an 710071, China
* Correspondence: tyang@nwpu.edu.cn (T.Y.); jinglixd@mail.xidian.edu.cn (J.L.); Tel.: +86-150-0291-9079 (T.Y.); +86-139-9132-0168 (J.L.)

Received: 3 April 2018; Accepted: 19 May 2018; Published: 23 May 2018

Abstract: An infrared sensor is a commonly used imaging device. Unmanned aerial vehicles, the most promising moving platform, each play a vital role in their own field, respectively. However, the two devices are seldom combined in automatic ground vehicle detection tasks. Therefore, how to make full use of them—especially in ground vehicle detection based on aerial imagery–has aroused wide academic concern. However, due to the aerial imagery's low-resolution and the vehicle detection's complexity, how to extract remarkable features and handle pose variations, view changes as well as surrounding radiation remains a challenge. In fact, these typical abstract features extracted by convolutional neural networks are more recognizable than the engineering features, and those complex conditions involved can be learned and memorized before. In this paper, a novel approach towards ground vehicle detection in aerial infrared images based on a convolutional neural network is proposed. The UAV and the infrared sensor used in this application are firstly introduced. Then, a novel aerial moving platform is built and an aerial infrared vehicle dataset is unprecedentedly constructed. We publicly release this dataset (NPU_CS_UAV_IR_DATA), which can be used for the following research in this field. Next, an end-to-end convolutional neural network is built. With large amounts of recognized features being iteratively learned, a real-time ground vehicle model is constructed. It has the unique ability to detect both the stationary vehicles and moving vehicles in real urban environments. We evaluate the proposed algorithm on some low–resolution aerial infrared images. Experiments on the NPU_CS_UAV_IR_DATA dataset demonstrate that the proposed method is effective and efficient to recognize the ground vehicles. Moreover it can accomplish the task in real-time while achieving superior performances in leak and false alarm ratio.

Keywords: aerial infrared imagery; real-time ground vehicle detection ; convolutional neural network; unmanned aerial vehicle

---

1. Introduction

Vehicle detection is an essential and pivotal role in several applications like intelligent video surveillance [1–4], car crash analysis [5], autonomous vehicle driving [6]. Most traditional approaches adopt the way that the camera is installed on a low-altitude pole or mounted on the vehicle itself. For instance, Sun and Zehang [7] present a method which jointly uses Gabor filters and Support Vector Machines for on-road vehicle detection. The Gabor filters are for feature extraction and these extracted features are used to train a classifier for detection. The authors in [8] propose a method to detect vehicles from stationary images using colors and edges. Zhou, Jie and Gao [9] propose a moving vehicle detection method based on example-learning. With regard to these approaches, the coverage of camera is limited despite rotating in multiple directions, they only detect vehicles on a small scale.

Given the installed camera's limited coverage, researchers turn to other moving platforms. Satellites [10,11], aircrafts, helicopters and unmanned aerial vehicles have been used to solve the bottleneck. The cost of images collected by the satellites, aircrafts and helicopters is remarkably high, and this equipment isn't able to make a quick response according to the time and weather. With the rapid development of the unmanned aerial vehicles industry, the price of small drones has dropped in recent years. The UAVs(unmanned aerial vehicles) have been the research focus. It seems easily accessible for the general public to obtain it, so all kinds of cameras including optical and infrared, have started to be installed on it. In this way, the unmanned aerial vehicle can be used as a height–adjustable moving camera platform on a large scale. Many researchers have made great efforts in this field. For example, Luo and Liu [12] propose an efficient static vehicle detection framework on aerial range data provided by the unmanned aerial vehicle, which is composed of three modules–moving vehicle detection, road area detection and post processing. The authors in [13] put forward a vehicle detection method from UAVs, which is integrated with Scalar Invariant Feature Transform and Implicit Shape Model. Another work [14], a hybrid vehicle detection method that integrates the Viola–Jones (V–J) and linear SVM classifier with HOG(Histogram of Oriented Gradient) features, is proposed for vehicle detection for aerial vehicle images obtained in low-altitude. It is not able to choose a robust feature to aim at the small size vehicles. Besides, some researchers in [15–17] have adopted the other sensors (e.g., depth sensors, RGB-D imagery) into the object detection area.

Another vehicle detection methods are accomplished by background modeling or foreground segmentation. The authors in [18] put forward with a method of moving object detection on non–stationary cameras and bring it to vehicle detection on mobile device. They model the background through dual-mode single Gaussian Model with life–cycle model and compensate the motion of the camera via mixing neighboring models. The authors in [19,20] propose a method of detecting and locating moving object under realistic condition based on the motion history images representation, which incorporates the timed–MHI for motion trajectory representation. Afterwards, a spatio–temporal segmentation procedure is employed to label motion regions by estimating density gradient. However these methods might cause a great number false alarms and fail to detect the stationary vehicles.

All these adaptive detection methods above are same in essentials while differing in minor points. They employ the similar strategy: manual designed features (e.g., SIFT, SURF, HOG, Edge, Color or their combinations) [21–23], background modeling or foreground segmentation, common classifiers (e.g., SVM, Adaboost) and sliding window search. These manual features might not hold the diversity of vehicles' shapes, illumination variations and background changes. The sliding window is an exhaustive traversal, which is time–consuming, not purposeful. This might cause too many redundant bounding boxes and has a bad influence on the following extraction and classification's speed and efficiency.

However, deep learning [24–29] establishes convolutional neural network that could automatically extract abundant representative features from the vast training samples. It has an outstanding performance on diverse data. Lars et al. [25] propose a network for vehicle detection in aerial images, which has overcome the shortcoming of original approach in case of handling small instances. Deng et al. [26] propose a fast and accurate vehicle detection framework, they develop an accurate–vehicle– proposal–network based on hyper feature map and put forward with a coupled R-CNN(convolutional neural network) method. A novel double focal loss convolutional neural network is proposed in [27]. In this paper, the skip connection is used in the CNN structure to enhance the feature learning and the focal loss function is used to substitute for conventional cross entropy loss function in both the region proposed network and the final classifier. They [27,30,31] all adopt the same framework, Region Proposal plus Convolutional Neural Network. By virtue of the CNN's strong feature extraction capacity, it achieves a higher detection ratio. Inspired by these work above, the authors in [29,32] introduce the elementary framework on aerial vehicle detection and recognition.

As described in [29], a deep convolutional neural network is adopted to mine highly descriptive features from candidate bounding boxes, then a linear support vector machine is employed to classify the region into "car" or "no–car" labels. The authors in [32] propose a hyper region proposal network to extract potential vehicles with a combination of hierarchical feature maps, then a cascade of boosted classifiers are employed to verify the candidate regions, false alarm ratio is further reduced.

All these work above have achieved tremendous advances in vehicle detection [33,34]. For object detection, images matching plays a vital role in searching part. The authors in [33] propose a novel visible-infrared image matching algorithm, and they construct a co–occuring feature by cross-domain image database and feature extraction. Jing et al. [34] extend the visible–infrared matching to photo-to-sketch matching by constructing visual vocabulary translator. The authors in [15] extract object silhouettes from the noisy background by a sequence of depth maps captured by a RGB-D sensor and track it using temporal motion information from each frame. The authors in [17] present a novel framework of 3D object detection, tracking and recognition from depth video sequences using spatiotemporal features and modified HMM. They use spatial depth shape features and temporal joints features to improve object classification performance. However, those approaches are not suitable for aerial infrared vehicle detection. The vehicle detections based on deep neural network and classification can't reach the real-time demands. The vehicle detections based on these manual designed features' matching have poor performances on the detection ratio measurement, for the aerial infrared images are low resolution and fuzzy and the manually extracted features are rare.

Considering the trade-off between the real-time demand and quantified index–*Precision*, *Recall* and *F1-score*, we adopt the convolutional neural network (the number of layers is not deep) to extract abundant features in the aerial infrared images, treat the vehicle detection as a typical regressive problem to accelerate the bounding boxes generations. Some detection results are illustrated in Figure 1. The majority of vehicles are detected, and these bounding boxes approximately cover the vehicles. The proposed method unexpectedly runs at a sampling frequency of 10 fps. The real-time vehicle detection is demanding. In the detection system, we might not demand an extremely accurate vehicle position, but an approximate position obtained in time is more necessary. Once detection speed falls behind the sample frequency, the information provided is lagged. This might mislead surveillance system.

The main contributions of this paper can be summarized as follows:

- We propose a method of detecting ground vehicles in aerial imagery based on convolutional neural network. Firstly, we combine the UAV and infrared sensor to the real-time system. There exist some great challenges like scale, view changes and scene's complexity in ground vehicle detection. In addition, the aerial imagery is always low-resolution, fuzzy and low-contrast, which adds difficulties to this problem. However, the proposed method adopts a convolutional neural network instead of traditional feature extraction, and uses the more recognized abstract features to search the vehicle, which have the unique ability to detect both the stationary and moving vehicles. It can work in real urban environments at 10 fps, which has a better real-time performance. Compared to the mainstream background model methods, it gets double performances in the *Precision* and *Recall* index.
- We construct a real-time ground vehicle detection system in aerial imagery, which includes the DJI M-100 UAV (Shenzhen, China), the FLIR TAU2 infrared camera (Beijing, China.), the remote controls and iPad (Apple, California, US). The system is built to collect large amounts of training samples and test images. These images are captured on different scenes includes road and multi-scenes. Additionally, this dataset is more complex and diversified in vehicle number, shape and surroundings. The aerial infrared vehicle dataset (The dataset (NPU_CS_UAV_IR_DATA) is online at [35],) which is convenient for the future research in this field.

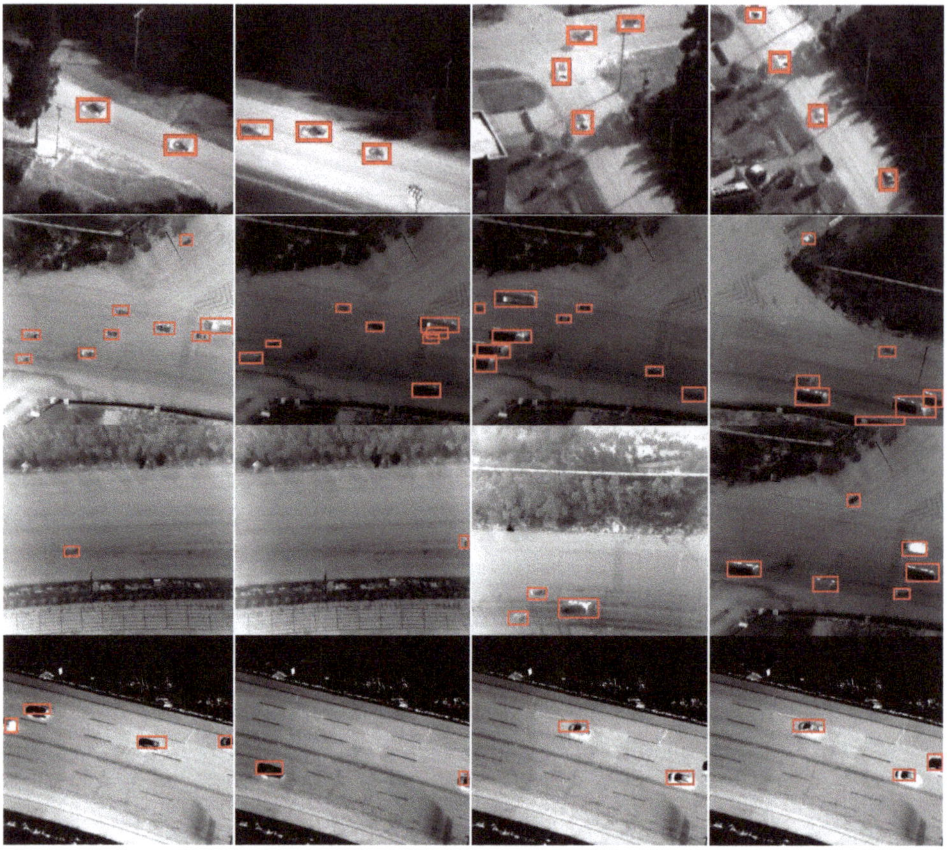

Figure 1. The detection method's performances on four tests, from top to bottom: VIVID_pktest1 [36], NPU_DJM100_1, NPU_DJM100_2, Scenes Change [35].

2. Aerial Infrared Ground Vehicle Detection

The proposed method is illustrated in Figure 2. It can be mainly divided into three steps. First, we manually segment vehicles by the help of a *labelimg* toolbox [37]. The labeled results are shown in Figure 3. This labeling step is pivotal to training [38]. The second step is devoted to sample region feature extraction in a convolutional neural network. We use data augmentations like rotation, crops, exposure shifts and more to expand samples. For training, we adopt a pre-trained classification network on ImageNet [39], and then fine-tune this. The pre-trained model on the ImageNet has many optimization parameters. On the basis of this, the loss function can be convergent rapidly in the training process. We add a region proposal layer to predict vehicles' coordinates (x, y, width, height) and corresponding confidence. These outputs contain many false alarms and redundant bounding boxes. We remove false alarms by confidence threshold. Finally, non-maximum suppression is adopted to eliminate redundant bounding boxes.

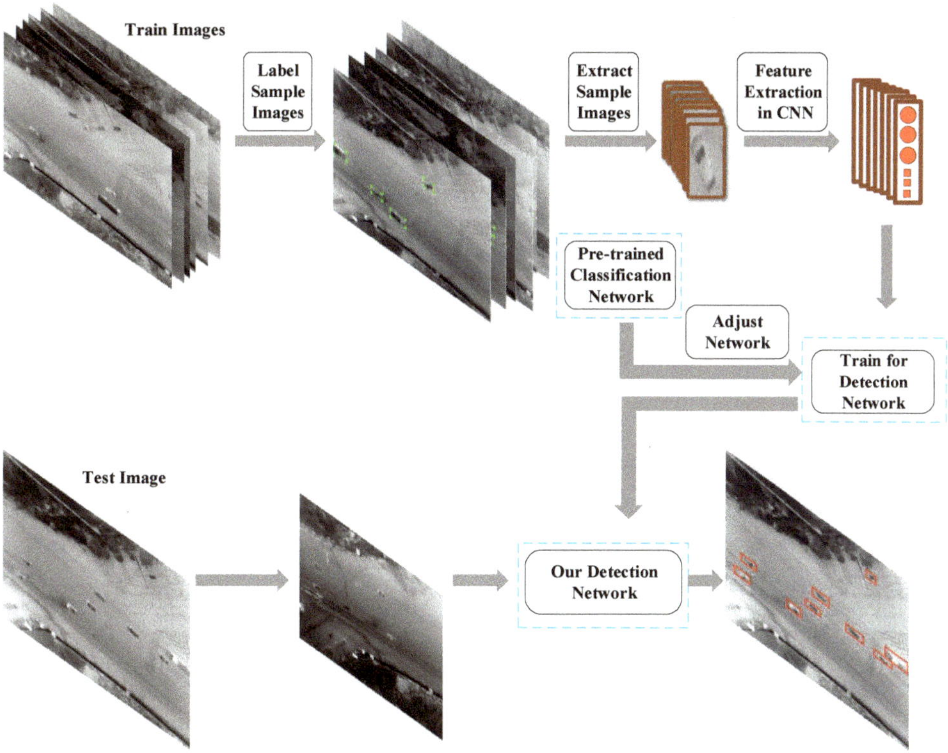

Figure 2. Flowchart of the proposed vehicle detection method in infrared images. Before training, we manually label vehicles in images via *labelimg* tool. In the label, we rectangle the vehicle by locating the top left corner and down right corner, and then keep these location and label information as a xml file. Afterwards, it is necessary to expand the sample by the operations—rotation, crop and shift. We load a pre-trained classification network for training. The Pre-Trained Network: classification network pre-trained on ImageNet [39].

2.1. Label Train Samples

Before labeling, it is necessary to construct an aerial infrared system to capture images for training samples. The aerial infrared system is mainly composed of the DJI Matrice 100 and the FLIR TAU2 camera. The DJI Matrice 100's major components are made of carbon fiber, which makes it light and solid, in order to guarantee it flies smoothly. The infrared sensor possesses the ability of temperature measurement and various color models' conversion, which meets the rigorous demands in several environments. In the capture, an intersection filled with a large volume of traffic is chosen as a flight place. Aerial images are captured at five different times alone. Images chosen from the first four times are train samples. Furthermore, aerial infrared vehicle samples from the public data VIVID_pktest are added.

Before training, it is necessary to label large amounts of training samples. These green rectangular regions rectangled in Figure 3 are some labeled samples. Partial vehicles appear in the image due to the limited view of infrared sensors, especially when it turns a corner, passes through the road or starts to enter into view, so we may catch the front or rear of some vehicles. These pieces of information are helpful because the vehicles often pass through an intersection or make a turn. This information mentioned above can ensure sample integrity. The information captured is crucial for vehicle detection.

In the label process, we obtain some vehicle patches in the infrared images. This guarantees that more training samples are captured and more situations are collected as much as possible. Although this operation is time-consuming and implemented offline, it insures vehicle samples' integrity [38]. This can avoid rough sample segmentation. We could observe some part vehicles in the left in 1–3 (row-col) in Figure 3. This is because the vehicle starts to come into view. There are some moving vehicles close to each other in (3–4) and (2–3). Once roughly segmented, the neighborhoods could be mistaken for just one, but there are two or more vehicles in practice.

All the vehicles are labeled in the training sample, then each image and vehicle position are made into a xml format as the voc [40].

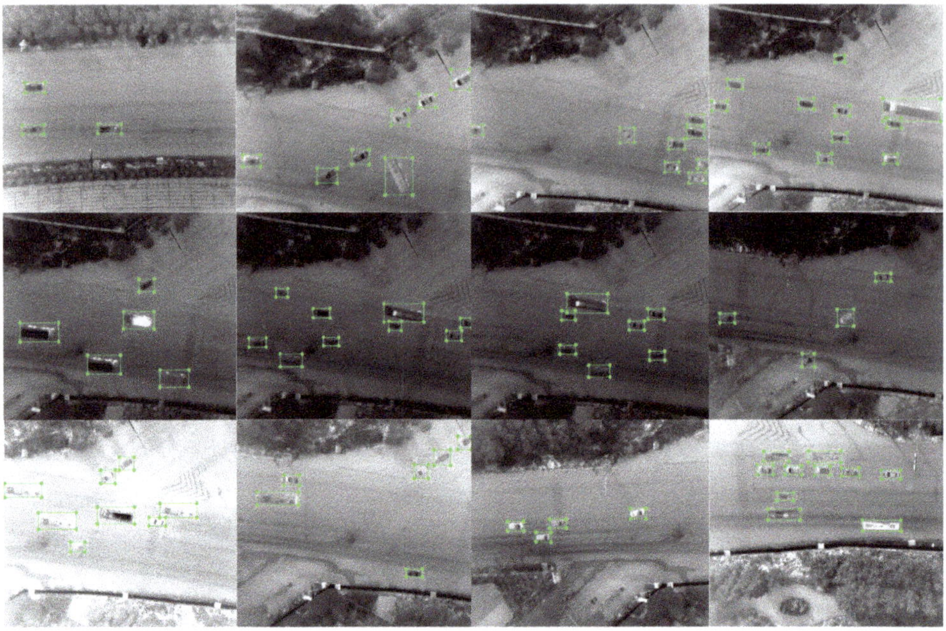

Figure 3. The labeled vehicle samples captured by the unmanned aerial vehicle DJI MATRICE-100. The manual label process is accomplished by the *labelimg* toolbox, which is a widely used tool in the sample label. Firstly, the label "vehicle" is written in this tool before label process. Then, we put green rectangles around the vehicles in images by searching the two locations: the left-top corner and the right-bottom corner. Finally, we keep this position and label information in a xml format like the voc.

2.2. Convolutional Neural Network

With respect to vehicle detection in aerial infrared images, we apply a convolutional neural network to the full image. It is based on the regressive idea to accomplish the object (vehicles) detection, rather than a typical classification problem. The network designed extracts features and trains on the full images, not the local positive and negative samples. The neural network's architecture is shown in Figure 4. Firstly, we resize the input image into 416 × 416, and utilize the convolutional layer and pooling layer by turning to an extract feature. Inspired by the fact that the Faster R-CNN [31] predicts offset and confidence for bounding boxes using the region proposal, we adopt a region proposal layer to predict bounding boxes. A lot of bounding boxes are obtained this way. We remove some false bounding boxes with low confidence by a threshold filter, and then eliminate redundant boxes using the non maximum suppression.

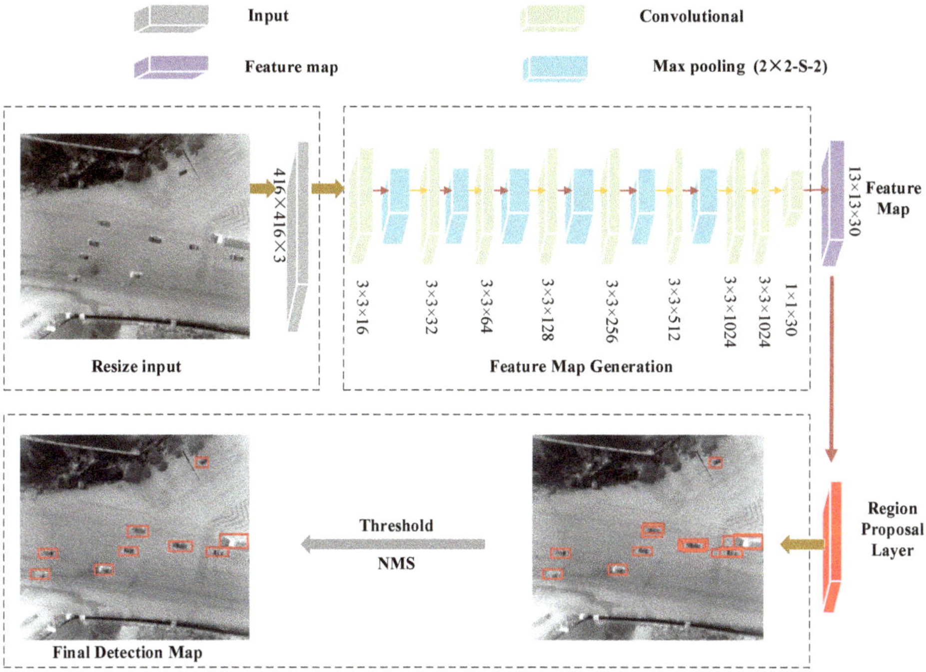

Figure 4. The architecture: The detection network is composed of nine convolutional layers, six pooling layers and a region proposal layer. Input image : 416 × 416 × 3. Output feature map: 13 × 13 × 30. The convolutional layers: 3 × 3 filters; The pooling layers: max pooling (2 × 2 with 2 stride). After each pool, filter channels double. We add a region proposal layer following the feature map, which is designed to generate bounding boxes, and then carry out threshold and NMS (Non Maximum Suppression) disposal.

Feature Map Generation

The detection framework can be mainly divided into two parts: **feature map generation** and **candidate bounding boxes generation**. The details of feature map are illustrated in Table 1. The process is composed of 15 layers: nine convolutional layers and six max pooling layers. Table 1 illustrates the filters channel, size, input and output of each layer. The original image is resized into 416 × 416 as the input. The convolutional layers downsample it by a factor 32, and the output size is 13 × 13. After this, there exists a single center cell in the feature map. The location prediction is based on the center location mechanism.

We carry out 16 filters (3 × 3) convolution operation on the input (416 × 416 × 3), followed by a 2 × 2 with two strides. Subsequently, the number of filers doubles, but the number of strides for pooling layer remains unchanged. Executing the above operations until the number of filters increases to 512, then the channel of stride on pooling layer is set as 1. This disposal wouldn't change the channel of the input (13 × 13 × 512). Based on this, we add two 3 × 3 convolutional layers with 1024 filters, following a 1 × 1 convolutional layer with 30 filters. Finally, the original image is turned into 13 × 13 × 30.

Table 1. Feature Map Generation includes the input and output of each layer.

Number	Layer	Filters	Size/Stride	Input	Output
0	convolutional	16	3×3/1	416 × 416 × 3	416 × 416 × 16
1	max pooling		2 × 2/2	416 × 416 × 16	208 × 208 × 16
2	convolutional	32	3 × 3/1	208 × 208 × 16	208 × 208 × 32
3	max pooling		2 × 2/2	208 × 208 × 32	104 × 104 × 32
4	convolutional	64	3 × 3/1	104 × 104 × 32	104 × 104 × 64
5	max pooling		2 × 2/2	104 × 104 × 64	52 × 52 × 64
6	convolutional	128	3 × 3/1	52 × 52 × 64	52 × 52 × 128
7	max pooling		2 × 2/2	52 × 52 × 128	26 × 26 × 128
8	convolutional	256	3 × 3/1	26 × 26 × 128	26 × 26 × 256
9	max pooling		2 × 2/2	26 × 26 × 256	13 × 13 × 256
10	convolutional	512	3 × 3/1	13 × 13 × 256	13 × 13 × 512
11	max pooling		2 × 2/1	13 × 13 × 512	13 × 13 × 512
12	convolutional	1024	3 × 3/1	13 × 13 × 512	13 × 13 × 1024
13	convolutional	1024	3 × 3/1	13 × 13 × 1024	13 × 13 × 1024
14	convolutional	30	1 × 1/1	13 × 13 × 1024	13 × 13 × 30

2.3. Bounding Boxes Generation

After convolutional and pooling operations, the final output is a 13 × 13 × 30 feature map. We add a region proposal layer following the feature map to predict the vehicle's location. Inspired by the RPN (region proposal network) of Faster-RCNN [31], we adopt a region proposal layer to service for vehicle border regression. The core purpose of the region proposal layer is to directly generate region proposals by the convolutional neural network. To generate region proposals, we slide a small network over the feature map output by the last shared convolutional layer. The small network takes a 3 × 3 spatial window as input on the feature map. The sliding window is mapped to a 30-dimensional feature vector. The feature is fed into a box-regression layer this way.

At each sliding-window, we simultaneously obtain a great deal of region proposals. Supposing the number of the proposals for each location is R, the output of region proposal layer is $4R$ coordinates, which are the R boxes' parametric expressions. The five classes about the proposals' percentages of width and height are (1.08, 1.09), (3.42, 4.41) , (6.63, 11.38) , (9.42, 5.11), (16.62, 10.52).

2.3.1. Vehicle Prediction on Bounding Boxes

The detection network is an end-to-end neural network. The vehicle's bounding boxes are accomplished directly by the network, The bounding boxes are achieved in the **bounding boxes generation** section. The confidence is computed as Equation (1):

$$C = P_{vehicle} * I_{pred}^{truth}, \tag{1}$$

where $P_{vehicle}$ indicates whether there exists a vehicle in the current prediction box, the I_{pred}^{truth} is the intersection over union between the predicted box and the ground truth. If no vehicle, the $P_{vehicle}$ is 0, 1, otherwise.

The confidence reflects the confidence level if the box contains a vehicle. A new parameter: $confidence$ is added, and each bounding box can be parameterized by $x, y, w, h, confidence$.

During the practical evaluating process, these above values are normalized to the range of $[0, 1]$. The $confidence$ reflects the probability of predicted boxes belonging to the vehicle. The $P_{vehicle}$ is defined as follows:

$$P_{vehicle} = \begin{cases} 1, & \text{vehicle,} \\ 0, & \text{no vehicle.} \end{cases} \tag{2}$$

2.3.2. Non Maximum Suppression

In order to eliminate redundant bounding boxes, we use non maximum suppression to find the best bounding box for each object. It is used to suppress non-maxima elements and search the local maxima value. The NMS [41,42] is for selecting high score detections and skipping windows covered by a previously selected detection.

The left-top corner (X_{min},Y_{min}), right-bottom corner (X_{max},Y_{max}), and *confidence* of detection boxes are the inputs in the NMS. The (X_{min},Y_{min}) and (X_{max},Y_{max}) are calculated by the following equations:

$$X_{min} = x - w, \tag{3}$$
$$X_{max} = x + w, \tag{4}$$
$$Y_{min} = y - h, \tag{5}$$
$$Y_{max} = y + h. \tag{6}$$

The area of each bounding box is calculated by Equation (7):

$$area = (X_{max} - X_{min} + 1) * (Y_{max} - Y_{min} + 1). \tag{7}$$

Then, the bounding boxes are sorted by confidence, and the overlap area of box i and j is computed by Equation (12):

$$X_{cross1} = max(X_{min}(i), X_{min}(j)), \tag{8}$$
$$Y_{cross1} = max(Y_{min}(i), Y_{min}(j)), \tag{9}$$
$$X_{cross2} = min(X_{max}(i), X_{max}(j)), \tag{10}$$
$$Y_{cross2} = min(Y_{max}(i), Y_{max}(j)), \tag{11}$$

$$cover_{(i,j)} = \frac{(X_{cross2} - X_{cross1} + 1) * (Y_{cross2} - Y_{cross1} + 1)}{min(area(i), area(j))}. \tag{12}$$

Once the $cover_{(i,j)}$ is over the suppression threshold, the bounding box with lower confidence would be discarded and the bounding box with highest confidence would be finally kept.

We unconditionally retain the box with higher confidence in each iteration, then calculate the overlap percentage between the box with the highest confidence and the other boxes. If the overlap percentage is bigger than 0.3, the current iteration terminates. The best box is determined until all the regions have been traversed.

3. Aerial Infrared System and Dataset

How we obtain the aerial infrared images (equipments and flight height) and prepare training samples and test images will be demonstrated in this section. In the test, we verify the method on the **VIVID_pktest1** [36], which has a pretty outstanding performance. However, these images in VIVID_pktest1 can not represent the aerial infrared images in the actual flight.

We capture the actual aerial infrared images (five different times) at an intersection. Experiments are implemented based on the Darknet [43] framework and run on a graphics mobile workstation with Intel core i7-3770 CPU (Santa Clara, California, US), a Quard K1100M of 2 GB video memory, and 8 GB of memory. The operating system is Ubuntu 14.04 (Canonical company, London, UK).

3.1. Aerial Infrared System

To evaluate the proposed vehicle detection approach, we have constructed an aerial infrared system, which is composed of the DJI Matrice 100 and the FLIR TAU2 camera.

Experiments are conducted by using aerial infrared images with 640 × 512 resolution, which are captured by a camera mounted on a quad rotor with a flight altitude of about 120 m above the ground. Figure 5 shows the basic components of the system, its referenced parameters are illustrated in Table 2. The dataset is online at [35].

Table 2. The equipment and parameters.

Camera and UAV	Specification	Parameter
	aircraft	DJI-MATRICE 100
	infrared sensor	FLIR TAU2
	capture solution	640 × 512
	capture frame rate	10 fps
	focal length	19 mm
	head rotation	32° × 26°

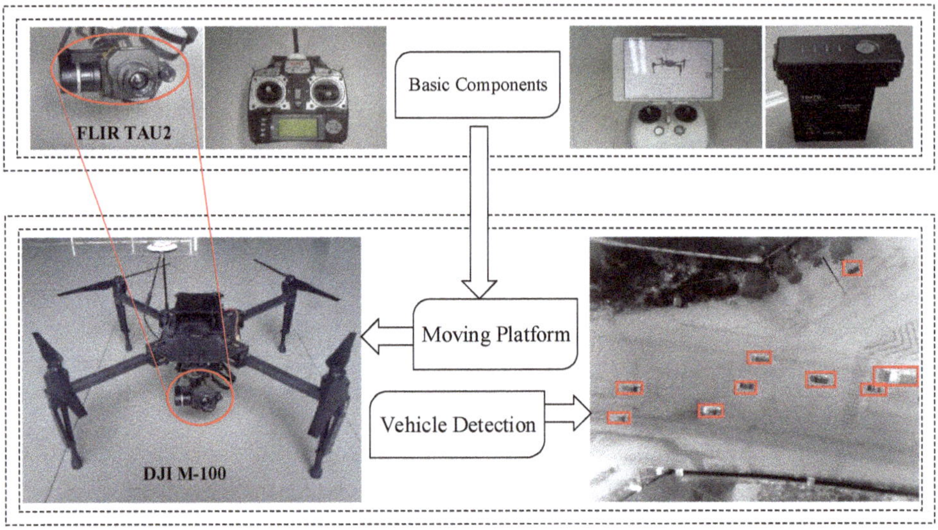

Figure 5. The real-time detection system is mainly composed of the DJI M-100, the FLIR TAU2 camera, the color model remote control, the flight remote control unit and iPad. The sensor installed on the UAV can capture the ground vehicles in real time, then transmit this information to the processor, and finally the processor shows the real-time detector on the screen.

3.2. Dataset

3.2.1. Training Samples

VIVID_pktest Sample: As we all know, the VIVID is a public data set for object tracking, which is composed of three subparts. The second part(VIVID_pktest2) [44] is a training sample. The image sequences are continuous in time, and the adjacent frames are similar to each other. If all images are put into training, the samples are filled with redundancy, so we only choose a set of 151, but which cover all vehicles appearing in VIVID_pktest2.

The authentical infrared sample: For the actual aerial infrared images, an intersection filled with large traffic volume is chosen as a flight place. We capture vehicle samples at five different times alone and choose sample images from the the first four times. The sampling frequency is 10 fps. Finally, we select 368 images, each of which is different in vehicle number, shape and color, as training samples considering redundant samples.

3.2.2. Test Images

As for evaluating the proposed method, we prepare four aerial infrared test image groups. The NPU_DJM100_1, NPU_DJM100_2 and Scenes Change are all captured over Xi'an, China. Four scenes with different backgrounds, flying altitudes, recording times and outside temperatures are used for testing (seen Table 3).

Table 3. Basic information of four test image groups.

Test	Size	Flying Altitude	Scenario	Date/Time	Temperature
VIVID_pktest1	320 × 240	80 m	Multi-scene	Unknown	Unknown
NPU_DJM100_1	640 × 512	120 m	Road	18 May 2017/16:00 pm	29° C
NPU_DJM100_1	640 × 512	120 m	Road	18 May 2017/16:30 pm	29° C
Scenes Change	640 × 512	80 m	Road	14 April 2017/10:30 pm	18° C

- **VIVID_pktest1**: The VIVID_pktest1 [36] is the first test image group, which is used for testing the detection network trained by the sample from the VIVID_pktest2 [44]. The VIVID_pktest1 is captured at an 80 m high altitude, which contains 100 images and 446 vehicles. The size is 320 × 240.
- **NPU_DJM100_1**: The sample chosen and their adjacent images from the aerial infrared images captured in the first four times is removed, then the remaining is used as the second test image group.
- **NPU_DJM100_2**: The images captured at the fifth time are the third test image group. There are few connections with images belonging to the previous four times.
- **Scenes Change**: The images are captured at earlier times and 80 m flight height. There is not a lot of traffic. This scene is totally different from all of the above. It is used to eliminate scenario training possibilities.

3.3. *Training*

In training, we use a batch size of 32, a max batch of 5000, a momentum of 0.9 and a decay of 0.0005. Through the training, the learning rate is set as 0.01. In each convolutional layer, we implement a batch normalized disposal except for the final layer before the feature map. With respect to the exquisitely prepared sample images, we divide them at a ratio of 7:3. Seventy percent were used for training, the remaining is for validation.

Loss function: In the objective module, we use the Mean Squared Error (MSE) for training.

$$Loss = \sum_{i=0}^{S^2} coordError + iouError + classError, \tag{13}$$

where S is the dimension's number of the network's output, $coordError$ is the error of coordinates between the predicted and the labeled, $iouError$ is the overlap's error, and $classError$ is the category of error (vehicle or non-vehicle). In the experiment, we amend Equation (13) by the following:

(1) The coordinates and the IOU (intersection over union) have different contribution degrees to the Loss, so we set the $\lambda_{coord} = 5$ to amend the $coordError$.

(2) For the IOU's error, the gridding includes the vehicle and the gridding having no vehicle should make various contributions to $Loss$. We use the $\lambda_{noobj} = 0.5$ to amend the $iouError$.

(3) As for the equal error, these large objects' impacts should be lower than small ones on vehicle detection because the percentage of error belonging to large objects is far less than those belonging to small ones. We square the w, h to improve it. The final Loss is as Equation (14):

$$Loss = \lambda_{coord} \sum_{i=0}^{S^2} \sum_{j=0}^{B} \prod_{ij}^{obj} [(x_i - \hat{x}_i)^2 + (y_i - \hat{y}_i)^2 + (\sqrt{w_i} - \sqrt{\hat{w}_i})^2 + (\sqrt{h_i} - \sqrt{\hat{h}_i})^2]$$
$$+ \sum_{i=0}^{S^2} \sum_{j=0}^{B} \prod_{ij}^{obj} [(C_i - \hat{C}_i)^2] + \lambda_{noobj} \sum_{i=0}^{S^2} \sum_{j=0}^{B} \prod_{ij}^{noobj} [(C_i - \hat{C}_i)^2]$$
$$+ \sum_{i=0}^{S^2} \prod_{ij}^{obj} \sum_{c \in classes} [(p_i(c) - \hat{p}_i(c))^2], \quad (14)$$

where the x, y, w, h, C, p are the predicted, and the $\hat{x}$, $\hat{y}, \hat{w}, \hat{h}, \hat{C}, \hat{p}$ are the labeled. The $\prod_{ij}^{obj}$ and the $\prod_{ij}^{noobj}$ reflect that the probability of that object is in, and not in, the j bounding boxes, respectively.

4. Experimental Results and Discussion

The method's performances on four test image groups are respectively shown in Figures 6–9. We rectangle the vehicles with red color. Some representative detection results will be demonstrated in **achievement exposition** section. The concert efficiency is given in the **statistical information** section.

4.1. Achievement Exposition

As seen in Figure 6, almost all of the vehicles have been detected by the proposed method. There exist many shadows of trees in the 2-3 (row-col) of Figure 6. This would cause great disturbances to vehicle detection tasks. This problem can be solved by storing and learning the similar cases before. In (2-2) of Figure 6, when the vehicle abruptly turns in the intersection, it might escape from surveillance. However, the method could catch the tendency and locate it in time. The test images are of good quality in the VIVID_pktest1 [36] and they haven't yet involved more complicated conditions like large illuminations. The test images of VIVID_pktest1 [36] are much simpler than the real aerial images in both the number of vehicles and the conditions' complexity. The performances of the VIVID_pktest1 [36] were not sufficiently convincing, hence the actual aerial infrared images obtained by the DJI Matrice 100 are used to test the method.

Figure 6. The performances of the method on part of the images of **VIVID_pktest1** [36].

As Figure 7 shows, the NPU_DJM100_1 is more challenging in vehicle's quantity and environmental complexity. There exist distinct illumination variations in the (1-1) and (2-1) of Figure 7. The traffic flow is very large in the intersection and the vehicles shuttling back and forth is very common. There are two detection boxes on the same vehicle in (2-2). There are two rectangles put around the same vehicle on the right but little overlap. The suppression threshold can not be set as a very small value, as this would have a bad influence on eliminating redundant rectangles. There still exist several false alarms in (2-4) of Figure 7.

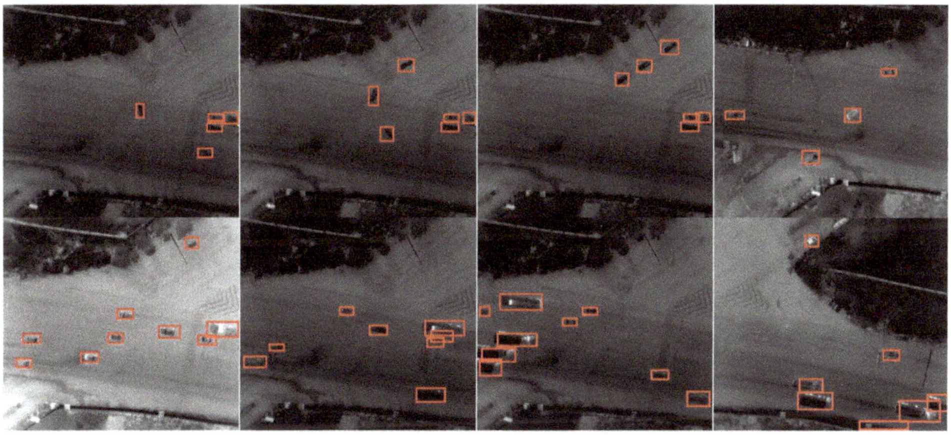

Figure 7. The performances of the method on part images of **NPU_DJM100_1** [35].

According to the analysis above, the method based on the neural network is able to accomplish vehicle detection in aerial infrared images, even in some harsh environments. The NPU_DJM100_1 and all the training samples are captured at the same time. Although we choose the NPU_DJM100_1, which is far from the the training sample in time, someone may suspect that the method may be achieved by scenario training. To remove this suspicion, we prepare additional test images (NPU_DJM100_2) captured in a scene, which are different from the scenes of the fourth times. The partial detection results are shown in Figure 8.

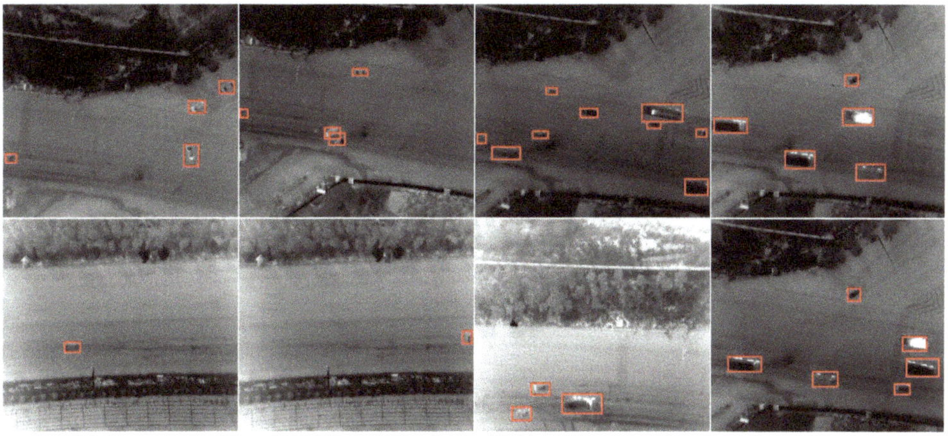

Figure 8. The performances of the method on part images of **NPU_DJM100_2** [35].

At first glance, the scenes of NPU_DJM100_2 are similar to NPU_DJM100_1 when comparing Figures 7 and 8. However, they are partly different from each other. A vehicle in the center of Figure 8 (1-4) is much brighter than all the vehicles in Figure 7 because of the blazing sunlight. The method locates the vehicle from being partly in the camera's view (1-3) to being completely in the camera's view. Figure 7 concentrates on a crossing road, but Figure 8 focuses on the one-way traffic, especially in (2-1, 2-2, 2-3) of Figure 8. These scenarios of (2-1, 2-2, 2-3) are different from Figure 7, but the proposed method locates these vehicles appearing in those images.

Scenes Change: To further validate the expansibility of the method, we test it on some infrared images captured 80 m above the ground. The road of Scenes Change is two-way traffic. The lamp post can be clearly seen on the ground. The guard bars on the two sides of the road are exposed to high temperatures for the long term, which are similar to vehicles in brightness. This causes great disturbances for detection in the aerial infrared images.

As can be seen from Figure 9, the proposed method is capable of detecting the vehicles that run in the same or opposite directions, locating the vehicle partially when it starts to come into or escape from the view in changed scenes. This evidence above proves that the method is feasible and dependable for aerial infrared vehicle detection.

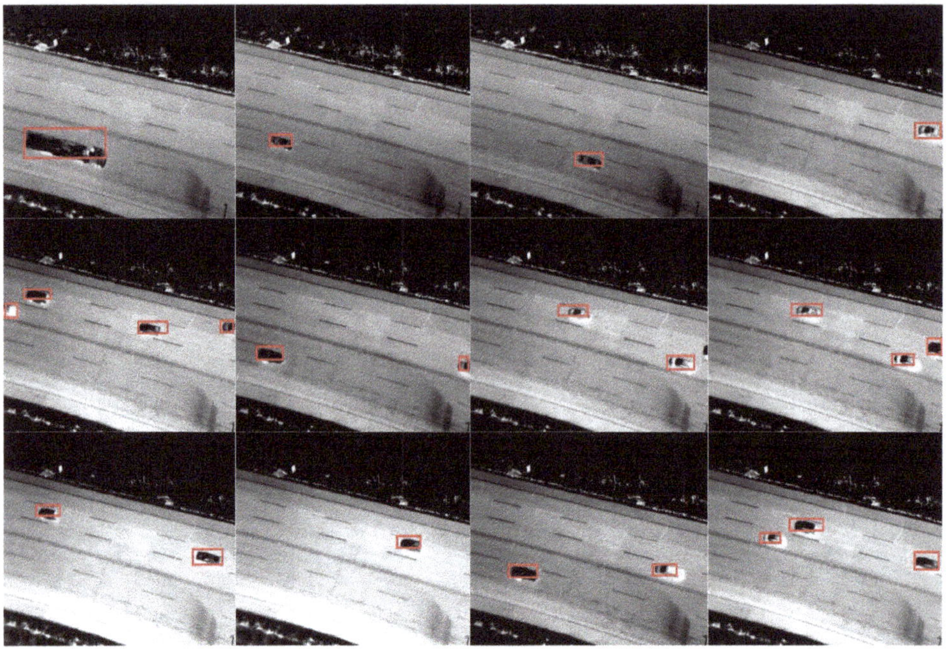

Figure 9. The performances of the method on partial images of **Scenes Change** [35].

4.2. Assessment Method

To evaluate the capability of the methodology on vehicle detection in aerial infrared images, we adopt these measurements: *Precision*, *Recall* and *F1-Score* defined as follows:

$$Precision = \frac{TP}{TP + FP}. \tag{15}$$

The *Precision* is the percentage of the correctly-detected vehicles' number over the total detected vehicles:

$$Recall = \frac{TP}{TP + FN}. \tag{16}$$

The *Recall* is the percentage of the correctly-detected vehicles number over the total true vehicles:

$$F1 - score = \frac{2 \times Recall \times Precision}{Recall + Precision}. \tag{17}$$

The *F1-Score* is a trade-off between *Recall* and *Precision*, where *TP* is the true positives (i.e., the number of vehicles correctly detected), *FP* is the false positives (i.e., the number of vehicles incorrectly detected), and *FN* is the false negatives (i.e., the number of other objects are wrongly regarded as vehicles).

4.3. Statistical Information

Figures 6–9 show some detection results, and then we conduct a statistical analysis about the method's performance. The details (quantitative results) are shown in Table 4.

Table 4. The performances on the test images. FP: false positive; TP: true positive; FN: false negative.

Test	Images	Vehicles	TP	FP	FN	Precision	Recall	F1-Score	Time(s)
VIVID_pktest1	100	446	388	58	7	87.00%	98.23%	92.27%	4.50
NPU_DJM100_1	189	642	612	30	22	95.33%	96.53%	95.93%	16.07
NPU_DJM100_2	190	922	903	19	30	97.94%	96.78%	97.36%	17.48
Scenes Change	100	85	79	6	0	92.94%	100%	96.34%	9.20
	Total	2095	1982	113	59	94.61%	97.11%	95.84%	

On the whole, the majority of the vehicles have been detected by the method. In total, the mean of *Precision* is 94.61%. The average of *Recall* is 97.11%. The *F1-Score* is basically flat . This measured information sufficiently demonstrates that the method is available for the ground vehicle detection in aerial infrared images.

4.4. Discussion

Comparison with State-of-the-Art

Figure 10 and Table 5 display a comparison about the method to a state-of-the-art method in [18]. This is a method for detecting moving objects with non-stationary cameras. It models the background through a dual-modal single Gaussian model (SGM) with age, which prevents the background model from being contaminated by the foreground pixels while still allowing the model to adapt to changes in the background, and compensates the motion of the camera by mixing neighboring models, which reduces the errors arising from motion compensation, in order to achieve rapid vehicle detection.

Table 5 illustrates that the proposed method achieves absolute advantages in *Precision*, *Recall*, and *F1-score* measurements. The performances of the **Scenes Change** group of [18] are equal to the proposed method, but there were very few vehicles in this group. With the vehicle's number increasing, the performance generally degrades, but the performance of the proposed is smooth and steady, maintaining a high level.

As can be seen from Figure 10, the method of [18] locates the running vehicles incorrectly, which only puts a rectangle around part of the running vehicle part even when the vehicle is completely in the image. There exist many false alarms (#77,#78) and residual errors (#97,#98,#89,#90). The proposed method is superior in the location accuracy and detection rate, which is able to detect almost all the vehicles. The red rectangles are the proposed method's detection results, the green rectangles belong to [18]. It is obvious that the detection results of [18] fluctuate drastically, especially in the # 71, #72, #75,#76 of Scenes Change.

Table 5. Comparison with the state-of-the art method, the bold value of each row is the best performance.

Test	Images Number	Vehicles	The Proposed Method			Method in [18]		
			Precision	Recall	F1-Score	Precision	Recall	F1-Score
VIVID_pktest1	1–100	446	87.00%	**98.23%**	92.27%	42.82%	77.45%	55.15%
NPU_DJM100_1	1–60	38	**100%**	**100%**	**100%**	52.63%	31.75%	39.61%
NPU_DJM100_2	1–100	501	**98.20%**	97.62%	97.91%	34.35%	40.78%	37.29%
Scenes Change	70–90	20	**100%**	**100%**	**100%**	**100%**	**100%**	**100%**
Total		1005	91.34%	92.08%	91.71%	37.98%	54.44%	44.74%

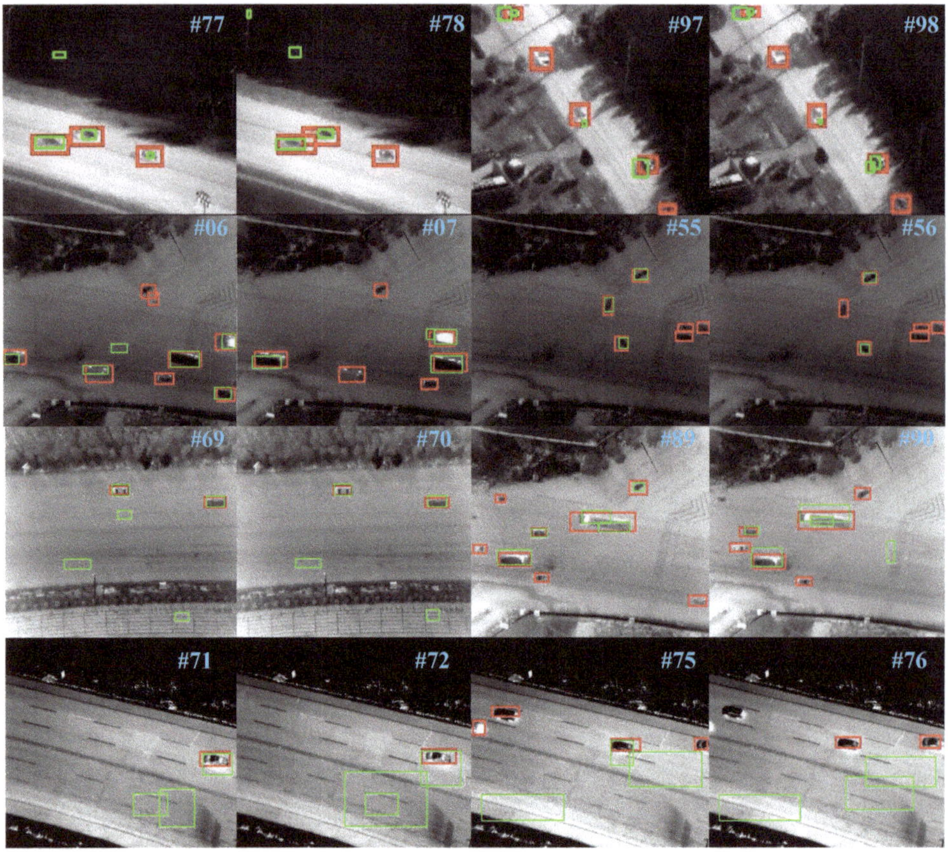

Figure 10. Comparison with a state-of-the-art method in [18]. The red rectangle is the proposed method's detection results, the green rectangles belong to [18]. From top to bottom: VIVID_pktest1, NPU_DJM100_1, NPU_DJM100_2, Scene Changes [35].

5. Conclusions

This paper proposes an efficient method for real-time ground vehicle detection in infrared imagery based on a convolutional neural network. In the proposed approach, we exploit a convolutional neural network to mine the abundant abstract features among the aerial infrared imagery. These features are more distinguished in ground vehicle detection. For ground vehicle detection, we firstly build a real-time ground vehicle detection system to capture real scene aerial images. All of the manually labeled training samples and test images are publicly posted. Then, we construct the convolutional and pooling layers and region proposal layer to achieve feature extraction. The convolutional and pooling layers are adopted to explore the vehicle's inherent features, and the rear region proposal layer is exploited to generate candidate vehicle boxes. Finally, on the basis of a labeled sample's feature, the method iteratively learns and memorizes these features to generate a real-time ground vehicle model. It has the unique ability to detect both the stationary vehicles and moving vehicles in real urban environments. Experiments on the four different scenes demonstrate that the proposed method is effective and efficient to recognize the ground vehicles. In addition, it can accomplish the task in real time while achieving superior performances in leak and false alarm ratio. Furthermore, the current work shows great potential for ground vehicle detection in aerial imagery.

In the real world, the real-time ground vehicle detection can be applied to intelligent surveillance, traffic safety, wildlife conservation and so on. In the intelligent surveillance, the system can rapidly give the vehicle's location in imagery under day and night, which is helpful for traffic monitoring and traffic flow statistics. Traffic crashes might occur in our daily lives all the time, but it is difficult to confirm the responsibility for the accident in complex backgrounds. The system can be used to identify the principal responsible party for its real-time detection capacity. As for the wildlife conservation, most of the protected animals are caught and killed during the night. The system can locate the hunter's vehicle at night, and this helps some regulatory agencies to take countermeasures in time.

Author Contributions: X.L., T.Y. and J.L. constructed the overall system and developed the neural network. In addition, they wrote and revised the paper. X.L. and J.L. participated in the research data collection, analysis and interpretation. T.Y. guided the experiments and the statistical analysis. Additionally, they jointly designed and performed the experiments.

Acknowledgments: This research was funded by the National Natural Science Foundation of China (No. 61672429), and the ShenZhen Science and Technology Foundation (JCYJ20160229172932237).

Conflicts of Interest: The authors declare no conflict of interest.

Abbreviations

The following abbreviations are used in this manuscript:

SVM	Support Vector Machines
UAV	Unmanned Aerial Vehicle
SIFT	Scalar Invariant Feature Transform
HOG	Histogram of Oriented Gradient
ISM	Implicit Shape Model
NMS	Non Maximum Suppression
CNN	Convolutional Neural Network
MHI	Motion History Image
HMM	Hidden Markov Model

References

1. Diamantopoulos, G.; Spann, M. Event detection for intelligent car park video surveillance. *Real-Time Imaging* **2005**, *11*, 233–243. [CrossRef]
2. Cheng, H.; Weng, C.; Chen, Y. Vehicle detection in aerial surveillance using dynamic bayesian networks. *IEEE Trans. Image Process. A Publ. IEEE Signal Process. Soc.* **2012**, *21*, 2152–2159. [CrossRef] [PubMed]
3. Chen, H.; Zhou, Y.; Deng, C. Study and implementation of car video surveillance system. *Commun. Technol.* **2012**, *45*, 55–56.
4. Zhao, G.; Hong-Bing, M.A.; Chen, C. Design of a wireless in-car video surveillance dedicated file system. *Electron. Des. Eng.* **2016**, *24*, 10–13.
5. Bohn, B.; Garcke, J.; Iza-Teran, R.; Paprotny, A.; Peherstorfer, B.; Schepsmeier, U.; Thole, C.A. Analysis of car crash simulation data with nonlinear machine learning methods. *Procedia Comput. Sci.* **2013**, *18*, 621–630. [CrossRef]
6. Saust, F.; Wille, J.M.; Maurer, M. Energy-optimized driving with an autonomous vehicle in urban environments. In Proceedings of the Vehicular Technology Conference, Yokohama, Japan, 6–9 May 2012; pp. 1–5.

7. Sun, Z.; Bebis, G.; Miller, R. On-road vehicle detection using gabor filters and support vector machines. *Int. Conf. Digit. Signal Process.* **2002**, *2*, 1019–1022.
8. Tsai, L.; Hsieh, J.; Fan, K. Vehicle detection using normalized color and edge map. *IEEE Int. Conf. Image Process.* **2007**, *16*, 850–864. [CrossRef]
9. Zhou, J.; Gao, D.; Zhang, D. Moving vehicle detection for automatic traffic monitoring. *IEEE Trans. Veh. Technol.* **2007**, *56*, 51–59. [CrossRef]
10. Elmikaty, M.; Stathaki, T. Car detection in high-resolution urban scenes using multiple image descriptors. In Proceedings of the International Conference on Pattern Recognition, Stockholm, Sweden, 24–28 August 2014; pp. 4299–4304.
11. Yang, T.; Wang, X.; Yao, B.and Li, J.; Zhang, Y.; He, Z.; Duan, W. Small moving vehicle detection in a satellite video of an urban area. *Sensors* **2016**, *16*, 1528. [CrossRef] [PubMed]
12. Luo, P.; Liu, F.; Liu, X.; Yang, Y. Stationary Vehicle Detection in Aerial Surveillance With a UAV. In Proceedings of the 8th International Conference on Information Science and Digital Content Technology (ICIDT), Jeju, Korea, 26–28 June 2012; pp. 567–570.
13. Chen, X.; Meng, Q. Vehicle detection from UAVs by using SIFT with implicit shape model. In Proceedings of the IEEE International Conference on Systems, Man, and Cybernetics, Manchester, UK, 13–16 October 2013; pp. 3139–3144.
14. Xu, Y.; Yu, G.; Wang, Y.; Wu, X.; Ma, Y. A hybrid vehicle detection method based on Viola-Jones and HOG + SVM from UAV images. *Sensors* **2016**, *16*, 1325. [CrossRef] [PubMed]
15. Kamal, S.; Jalal, A. A hybrid feature extraction approach for human detection, tracking and activity recognition using depth sensors. *Arab. J. Sci. Eng.* **2016**, *41*, 1043–1051. [CrossRef]
16. Farooq, A.; Jalal, A.; Kamal, S. Dense RGB-D map-based human tracking and activity recognition using skin joints features and self-organizing map. *Ksii Trans. Internet Inf. Syst.* **2015**, *9*, doi:10.3837/tiis.2015.05.017. [CrossRef]
17. Kamal, S.; Jalal, A.; Kim, D. Depth images-based human detection, tracking and activity recognition using spatiotemporal features and modified HMM. *J. Electr. Eng. Technol.* **2016**, *11*, 1857–1862. [CrossRef]
18. Yi, K.M.; Yun, K.; Kim, S.W.; Chang, H.J.; Jin, Y.C. Detection of moving objects with non-stationary cameras in 5.8 ms: Bringing motion detection to your mobile device. In Proceedings of the IEEE Conference on Computer Vision and Pattern Recognition Workshops, Portland, OR, USA, 23–28 June 2013; pp. 27–34.
19. Li, L.; Zeng, Q.; Jiang, Y.; Xia, H. Spatio-temporal motion segmentation and tracking under realistic condition. In Proceedings of the International Symposium on Systems and Control in Aerospace and Astronautics, Harbin, China, 19–21 January 2006; pp. 229–232.
20. Yin, Z.; Collins, R. Moving object localization in thermal imagery by forward-backward MHI. In Proceedings of the Conference on Computer Vision and Pattern Recognition Workshop, New York, NY, USA, 17–22 June 2006; p. 133.
21. Shao, W.; Yang, W.; Liu, G.; Liu, J. Car detection from high-resolution aerial imagery using multiple features. In Proceedings of the Geoscience and Remote Sensing Symposium, Munich, Germany, 22–27 July 2012; pp. 4379–4382.
22. Tuermer, S.; Kurz, F.; Reinartz, P.; Stilla, U. Airborne vehicle detection in dense urban areas using HOG features and disparity maps. *IEEE J. Sel. Top. Appl. Earth Observ. Remote Sens.* **2013**, *6*, 2327–2337. [CrossRef]
23. Chen, Z.; Wang, C.; Luo, H.; Wang, H.; Chen, Y.; Wen, C.; Yu, Y.; Cao, L.; Li, J. Vehicle detection in high-resolution aerial images based on fast sparse representation classification and multiorder feature. *IEEE Trans. Intell. Transp. Syst.* **2016**, *17*, 2296–2309. [CrossRef]
24. Yu, S.L.; Westfechtel, T.; Hamada, R.; Ohno, K.; Tadokoro, S. Vehicle detection and localization on bird's eye view elevation images using convolutional neural network. In Proceedings of the IEEE International Symposium on Safety, Security and Rescue Robotics, Shanghai, China, 11–13 October 2017; pp. 102–109.
25. Sommer, L.; Schuchert, T.; Beyerer, J. Fast deep vehicle detection in aerial images. In Proceedings of the Applications of Computer Vision, Santa Rosa, CA, USA, 24–31 March 2017; pp. 311–319.
26. Deng, Z.; Sun, H.; Zhou, S.; Zhao, J.; Zou, H. Toward fast and accurate vehicle detection in aerial images using coupled region-based convolutional neural networks. *IEEE J. Sel. Top. Appl. Earth Observ. Remote Sens.* **2017**, *10*, 3652–3664. [CrossRef]
27. Yang, M.; Liao, W.; Li, X.; Rosenhahn, B. Vehicle detection in aerial images. *arXiv* **2018**, arXiv:1801.07339. [CrossRef]

28. Konoplich, G.V.; Putin, E.; Filchenkov, A. Application of deep learning to the problem of vehicle detection in UAV images. In Proceedings of the IEEE International Conference on Soft Computing and Measurements, St. Petersburg, Russia, 25–27 May 2016; pp. 4–6.
29. Ammour, N.; Alhichri, H.; Bazi, Y.; Benjdira, B.; Alajlan, N.; Zuair, M. Deep learning approach for car detection in UAV imagery. *Remote Sens.* **2017**, *9*, 312. [CrossRef]
30. Girshick, R.; Donahue, J.; Darrell, T.; Malik, J. Region-based convolutional networks for accurate object detection and segmentation. *IEEE Trans. Pattern Anal. Mach. Intell.* **2016**, *38*, 142–158. [CrossRef] [PubMed]
31. Ren, S.; He, K.; Girshick, R.; Sun, J. Faster R-CNN: Towards real-time object detection with region proposal networks. *IEEE Trans. Pattern Anal. Mach. Intell.* **2017**, *39*, 1137–1149. [CrossRef] [PubMed]
32. Tang, T.; Zhou, S.; Deng, Z.; Zou, H.; Lei, L. Vehicle detection in aerial images based on region convolutional neural networks and hard negative example mining. *Sensors* **2017**, *17*, 336. [CrossRef] [PubMed]
33. Li, J.; Li, T.; Yang, T.; Lu, Z. Cross-domain co-occurring feature for visible-infrared image matching. *IEEE Access* **2018**, *6*, 17681–17698. [CrossRef]
34. Li, J.; Li, C.; Yang, T.; Lu, Z. A novel visual vocabulary translator based cross-domain image matching. *IEEE Access* **2017**, *5*, 23190–23203. [CrossRef]
35. Test Images and Train Samples. Available online: https://shanxiliuxiaofei.github.io/ (accessed on 2 April 2018).
36. Pktest01. Available online: http://vision.cse.psu.edu/data/vividEval/datasets/PETS2005/PkTest01/index.html (accessed on 2 April 2018).
37. LabelImg Tool. Available online: https://github.com/tzutalin/labelImg (accessed on 2 April 2018).
38. Sande, K.E.A.V.D.; Uijlings, J.R.R.; Gevers, T.; Smeulders, A.W.M. Segmentation as selective search for object recognition. In Proceedings of the IEEE International Conference on Computer Vision, Barcelona, Spain, 6–13 November 2012; pp. 1879–1886.
39. Russakovsky, O.; Deng, J.; Su, H.; Krause, J.; Satheesh, S.; Ma, S.; Huang, Z.; Karpathy, A.; Khosla, A.; Bernstein, M. ImageNet large scale visual recognition challenge. *Int. J. Comput. Vis.* **2015**, *115*, 211–252. [CrossRef]
40. Everingham, M.; Gool, L.; Williams, C.; Winn, J.; Zisserman, A. The pascal visual object classes (VOC) challenge. *Int. J. Comput. Vis.* **2010**, *88*, 303–338. [CrossRef]
41. Felzenszwalb, P.F.; Girshick, R.B.; Mcallester, D.; Ramanan, D. Object detection with discriminatively trained part-based models. *IEEE Trans. Pattern Anal. Mach. Intell.* **2010**, *32*, 1627–1645. [CrossRef] [PubMed]
42. Girshick, R.; Donahue, J.; Darrell, T.; Malik, J. Rich feature hierarchies for accurate object detection and semantic segmentation. In Proceedings of the IEEE Conference on Computer Vision and Pattern Recognition, Washington, DC, USA, 23–28 June 2014; pp. 580–587.
43. Redmon, J. Darknet: Open Source Neural Networks in C. 2013–2016. Available online: http://pjreddie.com/darknet/ (accessed on 2 April 2018).
44. Pktest02. Available online: http://vision.cse.psu.edu/data/vividEval/datasets/PETS2005/PkTest02/index.html (accessed on 2 April 2018).

electronics

MDPI

Article

Research on Air Confrontation Maneuver Decision-Making Method Based on Reinforcement Learning

Xianbing Zhang, Guoqing Liu, Chaojie Yang and Jiang Wu *

School of Automation Science and Electrical Engineering, Beihang University, Beijing 100191, China; zhangxianbing@buaa.edu.cn (X.Z.); liuguoqing@buaa.edu.cn (G.L.); yangchaojie@buaa.edu.cn (C.Y.)

* Correspondence: wujiang@buaa.edu.cn; Tel.: +86-139-1006-9931

Received: 26 September 2018; Accepted: 22 October 2018; Published: 27 October 2018

Abstract: With the development of information technology, the degree of intelligence in air confrontation is increasing, and the demand for automated intelligent decision-making systems is becoming more intense. Based on the characteristics of over-the-horizon air confrontation, this paper constructs a super-horizon air confrontation training environment, which includes aircraft model modeling, air confrontation scene design, enemy aircraft strategy design, and reward and punishment signal design. In order to improve the efficiency of the reinforcement learning algorithm for the exploration of strategy space, this paper proposes a heuristic Q-Network method that integrates expert experience, and uses expert experience as a heuristic signal to guide the search process. At the same time, heuristic exploration and random exploration are combined. Aiming at the over-the-horizon air confrontation maneuver decision problem, the heuristic Q-Network method is adopted to train the neural network model in the over-the-horizon air confrontation training environment. Through continuous interaction with the environment, self-learning of the air confrontation maneuver strategy is realized. The efficiency of the heuristic Q-Network method and effectiveness of the air confrontation maneuver strategy are verified by simulation experiments.

Keywords: over-the-horizon air confrontation; maneuver decision; Q-Network; heuristic exploration; reinforcement learning

1. Introduction

The intelligent air confrontation decision-making system can be effectively applied to automatic/autonomous simulated air confrontation, maneuver confrontation, anti-interception and various auxiliary decision-making systems of manned/unmanned aerial vehicles. The world's major military powers are conducting in-depth research in this field. The intelligent decision-making system will, thus, become an important part of future decision on air confrontation.

In the process of over-the-horizon air confrontation, reasonable maneuver decision-making is the premise of making weapons attack, sensor use, electronic countermeasures, and other decisions. It is accompanied by the entire air confrontation process and is an extremely important part. This paper mainly studies the intelligent maneuver decision-making method in this environment, based on the single-to-single air confrontation in super-horizon air confrontation.

The current air confrontation decision-making methods can be divided into two main categories: non-learning strategies and self-learning strategies. Among them, the non-learning strategy mainly adopts the optimization theory or the game method. There is no data-based training process in the strategy solving process, and there is no process of updating and optimizing the strategy by interacting with the environment. The methods adopted by non-learning strategies mainly include: differential countermeasure [1,2], matrix game [3], expert system [4], and impact map [5] among others.

The self-learning strategy refers to the information generated by the interaction between the historical data and the environment; and strategy learning is carried out, and finally a better strategy is solved. The self-learning strategy has characteristics of offline and online learning training, and has strong adaptability and can cope with complex and changeable environments. The main methods used in self-learning strategies include: genetic algorithm [6,7], artificial immune system [8,9], supervised learning [10], reinforcement learning [11], etc.

Reinforcement learning is a self-learning method which, through constant trial and error, interacts with the environment, gradually acquires knowledge, and improves action plans to adapt to the environment. Reinforcement learning has good application in decision-making fields such as robot control and automatic driving.

2. Air Confrontation Learning Training Environment Design

2.1. Aircraft Modelling

In the decision-making process of over-the-horizon air confrontation, the main focus is on real-time position and speed information of the two sides, but there is no requirement for the attitude information of the enemy aircraft. Therefore, the model of the aircraft is modeled by a three-degree-of-freedom model.

In order to facilitate the study, the paper made multiple assumptions [12,13]:

- The aircraft does not have a side-slip motion, that is, the side-slip angle is 0.
- Air speed is not considered when the aircraft is moving.
- The mass of the aircraft is constant, and the acceleration of gravity and atmospheric density do not change with changes in flight altitude.
- The Earth is regarded as an inertial system, that is, it regards the Earth as stationary, ignoring the effects of the Earth's rotation and revolution.

Based on the above assumptions, the force diagram of the aircraft is shown in Figure 1:

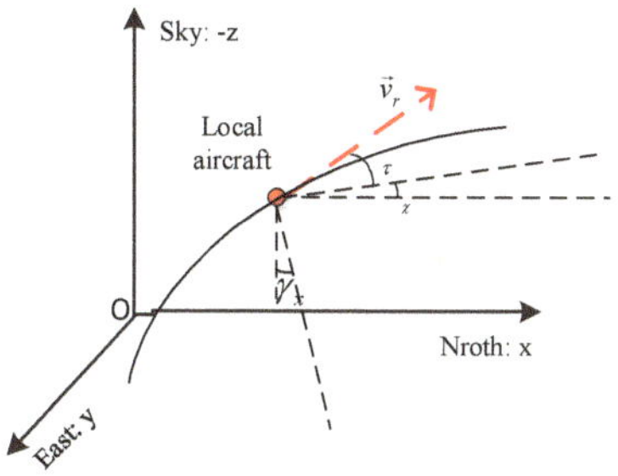

Figure 1. The geometric situation of confrontation.

where v is the speed of the aircraft, g represents the acceleration of gravity, n_x and n_f indicate the tangential overload and the normal overload, τ, χ and γ_x respectively indicate the aircraft's track inclination angle, track azimuth, and track roll angle. The following formula can be obtained by analyzing the force of the aircraft.

$$\begin{aligned} &mgn_x - mg\sin\tau = m\dot{v} \\ &mgn_f\cos\gamma_x - mg\cos\tau = mv\dot{\tau} \\ &mgn_f\sin\gamma_x = mv\cos\tau\dot{\chi}_x \end{aligned} \quad (1)$$

Transforming the above formula, we can get the dynamic equation of the aircraft as follows:

$$\begin{aligned} \dot{v} &= g(n_x - \sin\tau) \\ \dot{\tau} &= \tfrac{g}{v}(n_f \cos\gamma_x - \cos\tau) \\ \dot{\chi} &= \tfrac{g}{v\cos\tau} n_f \sin\gamma_x \end{aligned} \tag{2}$$

In this paper, the movement of the aircraft can be controlled by three quantities of n_x, n_f and γ_x. n_x can control the speed of flight, n_f and γ_x can control the track tilt angle and track azimuth to control flight speed direction.

Based on the above symbol representations, the kinematic equation of the aircraft can be expressed as:

$$\begin{aligned} \dot{x} &= v\cos\tau\cos\chi \\ \dot{y} &= v\cos\tau\sin\chi \\ \dot{z} &= -v\sin\tau \end{aligned} \tag{3}$$

where x, y, and z represent the coordinates of the aircraft in the ground coordinate system (using the North East coordinate system).

2.2. Learning Training Scene Design

Over-the-horizon air confrontation, unlike short-range air confrontation, has powerful missiles, radars, and support for various ground-to-air equipment information, which allows air confrontation to occur at a greater distance. Both parties can speculate through various information support. The opponent's position is then attacked by the precise guidance of the missile. This paper only studies the maneuvering strategy of over-the-horizon air confrontation, and air confrontation in close range is not considered.

The airspace in which the over-the-horizon air battle is located is assumed as follows (Table 1): The initial distance between the two sides is 65~100 km; when the distance between the two sides is less than 20 km, it is considered to have entered the close range, and the air battle is over. The height of both sides is 5~7 km.

Table 1. Air confrontation airspace.

Initial Distance/km	End Distance/km	Height/km
65~100	<20	6

This paper assumes that the local aircraft has a perception of enemy aircraft during the over-the-horizon air battle. When the enemy aircraft falls within the radar detection range of the local aircraft, enemy information can be obtained more accurately; when the enemy aircraft is not in the radar detection area of the local aircraft, it is assumed that the aircraft can obtain enemy aircraft information through other sources of information in the confrontation system (e.g., ground station radar, airborne early warning aircraft, etc.), but the information obtained by this method has a large error. This assumption is also to ensure that both sides have effective decision-making factors in the one-to-one over-the-horizon air confrontation decision-making process. Otherwise, if the other party's information is unknown, it is difficult to obtain an effective strategy through the learning algorithm of this paper. This is an area of incomplete information game, which is beyond the scope of this paper.

In order to maintain the balance of the two fighters, the performance of both sides is different: the enemy's missile attack capability is dominant, and the aircraft is dominant in the radar detection range. The specific configuration of the fighter parameters of both parties is shown in Tables 2 and 3.

Table 2. Local aircraft performance.

Parameter	Range
Aircraft speed	200 m/s~300 m/s
Radar detection distance	80 km
Radar detection angle	−60°~60°
Missile off-axis launch angle	30°
Missile inescapable cone angle	20°
Missile maximum launch distance	50 km
Missile maximum escape distance	35 km
Missile minimum escape distance	20 km

Table 3. Enemy aircraft performance.

Parameter	Range
Aircraft speed	200 m/s~300 m/s
Radar detection distance	70 km
Radar detection angle	−60°~60°
Missile off-axis launch angle	30°
Missile inescapable cone angle	20°
Missile maximum launch distance	55 km
Missile maximum escape distance	40 km
Missile minimum escape distance	25 km

According to the configuration in the table, the radar detection area of the unit and the enemy aircraft can be represented by the Figure 2:

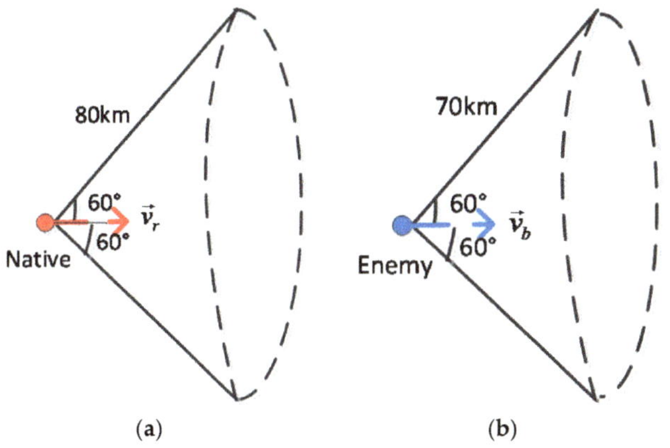

Figure 2. The radar detection area of both sides. (**a**) Native radar; (**b**) Enemy radar.

The missile attack zone of both fighters can be expressed in Figure 3:

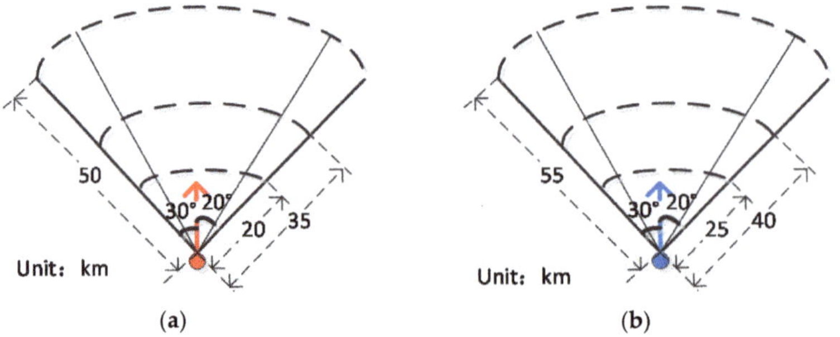

Figure 3. The missile attack area of both sides. (**a**) Native radar; (**b**) Enemy radar.

2.3. Enemy Strategy Design

The enemy aircraft strategy is a very important part of the air confrontation training environment. It determines the fidelity of the over-the-horizon air confrontation environment and also has a great influence on the strategy learned by the algorithm. This paper focuses on the study of air confrontation maneuver strategies with reinforcement learning methods, focusing on the design and improvement of methods, and does not put too much energy into the study of enemy aircraft strategy. Because the air confrontation maneuver strategy learning method studied in this paper is a general method, it is also applicable to training on change in strategy design of the enemy aircraft.

Therefore, this paper identifies enemy strategy as a relatively simple one, which is shown in Figure 4. First, the battlefield situation of the over-the-horizon air confrontation is evaluated based on expert experience. Then, assume that the other party maintains the current state of motion, adopts a method similar to the matrix strategy, and selects the optimal action from the action set as the decision result.

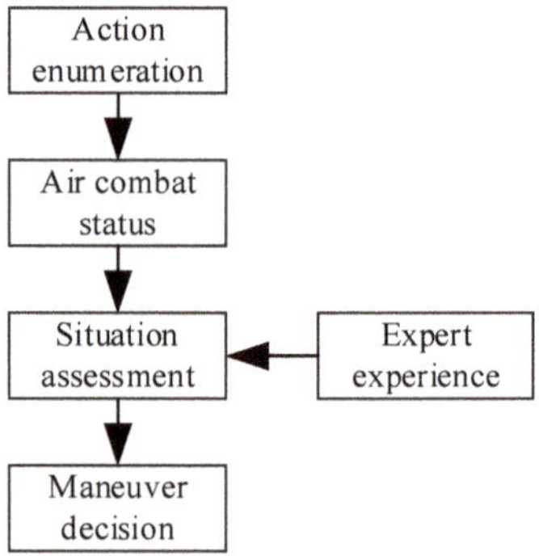

Figure 4. Strategy design of enemy aircraft.

2.4. Reward and Punishment Signal Design

When using the reinforcement learning algorithm to solve practical problems, it is necessary to adjust and optimize the strategy according to the reward and punishment signals fed back by the environment. In the process of constructing the over-the-horizon air confrontation training environment, the reward and punishment signals are mainly considered from two aspects: the detection ability of the aircraft against the enemy aircraft and the threat of the attack on the enemy aircraft.

In the process of over-the-horizon air confrontation, the geometric situation of the battlefield is shown in Figure 5.

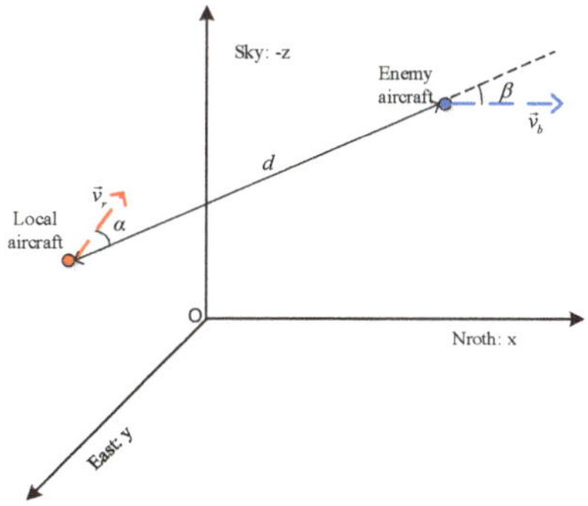

Figure 5. The geometric situation of confrontation.

In Figure 5, $\vec{v}_r$ and $\vec{v}_b$ respectively represent the speed vector of the local aircraft and the speed vector of the enemy aircraft, α indicating the azimuth angle of the enemy aircraft with respect to the local aircraft, β indicating the enemy's entry angle with respect to the local aircraft, and d indicates the distance between the two sides.

2.4.1. Detection Capability

The detection capability of the aircraft to the enemy aircraft is mainly affected by three factors: azimuth α, entry angle β, and distance d between the two sides.

1. Azimuth factor

When the enemy aircraft is located within the maximum detection angle range of the local radar, the aircraft has the ability to detect the enemy aircraft, and thus constructs the azimuth detection advantage:

$$T_{\text{det\_}\alpha} = \begin{cases} 0, & |\alpha| > \alpha_{FR\max} \\ e^{-\frac{|\alpha|}{\alpha_{FR\max}}}, & |\alpha| \leq \alpha_{FR\max} \end{cases} \tag{4}$$

$\alpha_{FR\max}$ is the maximum detection angle of the local fight radar, and *FR* is the abbreviation of the fight radar.

2. Entry angle factor

This paper assumes that the aircraft airborne radar is a pulse Doppler radar. The characteristics of the radar are: when the target and the local aircraft are head-on, they have strong detection capability, and when the target is on the positive side, the detection capability is poor. The ability to detect a trailing target is less than the ability to detect at the head. Based on this, the advantage of entering the angle detection is constructed as:

$$T_{\text{det\_}\beta} = \begin{cases} \cos(180 - |\beta|) * e^{-\frac{\pi*(180-|\beta|)}{180}} \\ (90^\circ \leq |\beta| < 180^\circ) \\ 0.5\cos(|\beta|) * e^{-\frac{\pi*|\beta|}{180}} \\ (0^\circ \leq |\beta| < 90^\circ) \end{cases} \tag{5}$$

3. Distance factor

When the enemy aircraft is located within the maximum detection distance of the local radar, the aircraft has the ability to detect the enemy aircraft. Based on this, build a distance detection advantage as:

$$T_{\text{det\_}d} = \begin{cases} 0, & d > D_{FR\text{max}} \\ e^{-\frac{3*d}{D_{FR\text{max}}}}, & d \leq D_{FR\text{max}} \end{cases} \tag{6}$$

$D_{FR\text{max}}$ is the maximum detection distance of the local radar.

4. Total detection advantage

In an actual air confrontation scenario, the azimuth detection advantage $T_{\text{det\_}\beta}$ and the entry angle detection advantage $T_{\text{det\_}d}$ have a certain coupling, and the overall angle detection advantage is constructed as follows:

$$T_{\text{det\_}ag} = (T_{\text{det\_}\alpha})^{\gamma_1} * (T_{\text{det\_}\beta})^{\gamma_2} \tag{7}$$

γ_1 and γ_2 are the two parameters that can control the proportion of $T_{\text{det\_}\beta}$ and $T_{\text{det\_}d}$ in the total angular detection advantage. They meet the following conditions: $0 \leq \gamma_1, \gamma_2 \leq 1$ and $\gamma_1 + \gamma_2 = 1$.

In addition, considering the distance d and the coupling relationship between these angles, the overall detection advantages of constructing the local aircraft to the enemy aircraft are:

$$T_{\text{det}} = (T_{\text{det\_}ag})^{u_1} * (T_{\text{det\_}d})^{u_2} \tag{8}$$

The role of u_1 and u_2 is similar to γ_1 and γ_2, and they meet the following conditions: $0 \leq u_1, u_2 \leq 1$ and $u_1 + u_2 = 1$.

2.4.2. Attack Threat

The attack threat of the aircraft to the enemy aircraft is mainly affected by three factors: azimuth α, energy E and distance d.

1. Azimuth factor

Based on the target azimuth and the performance of the local radar and missile, build an angle threat factor:

$$T_{thr\_\alpha} = \begin{cases} 0 & \alpha > \alpha_R \\ 0.3(1 - \frac{|\alpha| - \alpha_M}{\alpha_R - \alpha_M}) & \alpha_M \leq |\alpha| \leq \alpha_R \\ 0.8 - \frac{|\alpha| - \alpha_{Mk}}{2(\alpha_M - \alpha_{Mk})} & \alpha_{Mk} \leq |\alpha| < \alpha_M \\ 1 - \frac{|\alpha|}{5\alpha_{Mk}} & 0 \leq |\alpha| < \alpha_{Mk} \end{cases} \tag{9}$$

α_R is the maximum search angle of the local radar, α_M is the maximum attack angle of the local missile, α_{Mk} is the maximum angle of the non-escape zone of the local missile.

2. Energy factor

In the air confrontation process, the higher the energy of the fighter, the stronger the attacking ability of the launched missile, and the greater the threat to the enemy aircraft. The energy here is mainly composed of kinetic energy, according to the kinetic energy formula, which is simplified as follows:

$$E = \frac{v^2}{2g} \tag{10}$$

v is the speed of the local aircraft, g is the gravitational acceleration, and weight can be ignored considering the particle model.

Based on this, build the native energy threat factor:

$$T_{thr\_E} = \begin{cases} 1, & \frac{E}{E_T} \geq 2 \\ 0.5^{2-\frac{E}{E_T}}, & 0.5 \leq \frac{E}{E_T} < 2 \\ \frac{E}{2E_T}, & \frac{E}{E_T} < 0.5 \end{cases} \tag{11}$$

E is the energy of the aircraft and E_T is the enemy aircraft's energy.

3. Distance factor

The distance threat factor is constructed based on the distance between the enemy and the enemy and the performance of the local radar and missile:

$$T_{thr\_d} = \begin{cases} 0 & d \geq D_R \\ 0.5e^{-\frac{d-D_{Mmax}}{D_R-D_{Mmax}}} & D_{Mmax} \leq d < D_R \\ 2^{-\frac{d-D_{Mkmax}}{D_{Mmax}-D_{Mkmax}}} & D_{Mkmax} \leq d < D_{Mmax} \\ 1 & D_{Mkmin} \leq d < D_{Mkmax} \\ 2^{-\frac{d-D_{Mkmin}}{10-D_{Mkmin}}} & 10 \leq d < D_{Mkmin} \\ 0 & d < 10 \end{cases} \tag{12}$$

D_R is the maximum search distance of the local radar, D_{Mmax} is the maximum attack distance of the local missile, D_{Mkmax} is the maximum inescapable distance of the local missile, and D_{Mkmin} is the minimum inescapable distance of the local missile.

4. Total attack threat

Considering that the distance factor and the angle factor have a certain coupling relationship, the total attack threat of the aircraft to the enemy aircraft is:

$$T_{thr} = k_1 * (T_{thr\_\alpha})^{\eta_1} * (T_{thr\_d})^{\eta_2} + k_2 * T_{thr\_E} \tag{13}$$

k_1, k_2, η_1 and η_2 are control parameters, and they meet the following conditions:
$0 \leq \eta_1, \eta_2 \leq 1, \eta_1 + \eta_2 = 1, 0 \leq k_1, k_2 \leq 1$ and $k_1 + k_2 = 1$.

2.4.3. Reward and Punishment Signal Synthesis

According to the above-mentioned advantages of the detection capability of the enemy aircraft and the threat of attack, the total threat of constructing the local aircraft is:

$$T = (T_{\text{det}})^{\gamma_1} * (T_{thr})^{\gamma_2} \tag{14}$$

The two parameters γ_1, γ_2 are the index of the local aircraft detection capability and the attack threat, which determine the importance ratio of the two in the reward and punishment function. These two values can be obtained empirically.

In the same way, the enemy's threat to the local aircraft can be found, which is defined as T_t, and the reward and punishment signals are designed accordingly:

$$R = T - T_t \tag{15}$$

R is the relative threat value of the enemy aircraft to the enemy aircraft. When the local threat is greater than the enemy aircraft threat, the reward is positive, otherwise it is negative.

3. Markov Decision Process Modeling

The Markov decision process [14] can be represented by a six-tuple $\langle S, A, P, R, \gamma, V \rangle$. The aircraft constructed in this paper is a model; there is no random item. The element P can be omitted here. At the same time, the reward and punishment function R has also been designed before. Therefore, in this section, only state space S, action space A, discount factor γ, and objective function V of MDP [15] need to be determined.

3.1. Air Confrontation State Space

According to the battlefield geometry map, the battlefield situation can be expressed in 9 quantities: α, β, d, v_r, v_b, τ_r, τ_b, γ_r, γ_b. They respectively indicate the azimuth angle of the enemy aircraft relative to the aircraft, the angle of entry of the enemy aircraft with respect to the aircraft, the distance between the two sides, the speed of the aircraft, the speed of the enemy aircraft, the inclination angle of the local aircraft, the inclination angle of the enemy aircraft track, and the present aircraft track roll angle and enemy aircraft track roll angle. Considering whether the enemy aircraft is located in the local radar detection range, the accuracy of the enemy aircraft information obtained by the aircraft is not the same, so a confidence factor c (confidence) is added to indicate the accuracy of the enemy information. The larger c, the more accurate the information. It meets the conditions: $0 \leq c \leq 1$.

The air confrontation state can be represented by a 10-dimensional vector:

$$s = (\alpha, \beta, d, v_r, v_b, \tau_r, \tau_b, \gamma_r, \gamma_b, c) \tag{16}$$

3.2. Maneuvering Decision Action Space

In the over-the-horizon air confrontation maneuver decision problem, establishing a reasonable maneuver library is the key to air confrontation intelligent decision-making [16]. Generally, air confrontation maneuver library design is divided into two types: One is the "25 typical tactical actions" based on the classic tactics of pilots in air confrontation, including straight-flat, fixed-height, slow-speed Yo-Yo; The other is a "basic manipulation action library" based on common air confrontation control methods, including maximum acceleration/deceleration, maximum load climb/deep, maximum load left/right turn, stable flight, etc.

As shown in Figure 6, the air confrontation maneuver library is built according to the "Basic Manipulation Action Library", including nine maneuver directions: left climb, climb, right climb, horizontal left turn, horizontal forward fly, horizontal right turn, left dive, dive, and right dive. In this paper, assuming that both sides move on a horizontal plane, there are only three optional actions: horizontal left turn, horizontal forward fly, and horizontal right turn. In these three directions, it can be divided according to the change of speed: increase, hold and decrease. Therefore, there are a total of nine optional maneuvers, that is $|A| = 9$.

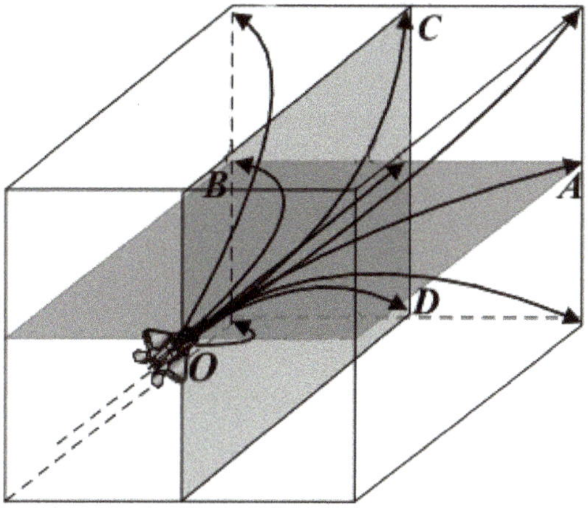

Figure 6. Maneuver direction schematic.

According to the previous description of the aircraft model, the motion of the aircraft is mainly controlled by the three quantities of n_x, n_f and γ_x, which respectively represent tangential overload, normal overload, and track roll angle. n_x is used to control speed, n_f and γ_x are used to control speed direction. According to these three quantities, action space A can be defined as follows (Table 4):

Table 4. Action collection.

Action	γ_x, n_x, n_f	Speed Direction	Speed Size
a_1	$(-1, -1, 0)$	turn left horizontally	Decrease
a_2	$(-1, 0, 0)$	turn left horizontally	Maintain
a_3	$(-1, 1, 0)$	turn left horizontally	Increase
a_4	$(0, -1, 0)$	fly forward horizontally	Decrease
a_5	$(0, 0, 0)$	fly forward horizontally	Maintain
a_6	$(0, 1, 0)$	fly forward horizontally	Increase
a_7	$(1, -1, 0)$	turn right horizontally	Decrease
a_8	$(1, 0, 0)$	turn right horizontally	Maintain
a_9	$(1, 1, 0)$	turn right horizontally	Increase

3.3. Discount Factor and Objective Function

In the application of reinforcement learning, the discount factor has two main functions: (1) the reward and punishment signal decays with time, indicating that it is less important in the far-away time; (2) it can prevent accumulation due to the excessive length of the episode. The reward is too large, and the cumulative reward value can be bounded by the attenuation factor. It is often set to 0.9, so this article also follows this setting.

The optimization objective function uses a state limited discount type objective function, in which it estimates the function V only based on the reward value of the state of the next n moments at the current moment:

$$V^{\pi}(\mathrm{s}_t) = \mathrm{E}_{\pi}(\mathrm{R}(\mathrm{s}_t)) = \mathrm{E}_{\pi}(\sum_{k=0}^{n-1} \gamma^k r_{t+k}) \tag{17}$$

4. Heuristic Q-Network

In view of the over-the-horizon air confrontation maneuver decision problem, this paper adopts an indirect strategy, which is to generate a strategy by obtaining a behavior value function $Q(s, a)$. For the decision-making of maneuvering, this paper also uses the Q-Network algorithm [17].

The reinforcement learning algorithm [18] solves the strategy by interacting with the environment, which is a process of sensing the unknown environment and learning related knowledge. According to the utilization of current knowledge, the learning process of the algorithm can be divided into two kinds of behaviors: exploration and exploitation. Exploitation is based on the currently learned strategy, which enables the agent to obtain many rewards. In addition, exploration is to try new actions in order to find better strategies to get more rewards in the future. In the process of solving practical problems, it is necessary to find a suitable compromise between exploitation and exploration, which will make the algorithm more efficient.

The often-used exploitation strategy is a strategy, which can be expressed as follows (Algorithm 1):

Algorithm 1. $\varepsilon - greedy$ strategy.

Input: control parameter ε
Process:
1: **if** random() < ε
2: action←random from set A
3: **else**
4: action←$\underset{a}{\mathrm{argmax}} Q(s, a)$
5: **end if**

Under the exploration strategy of $\varepsilon - greedy$, the algorithm can converge to an effective strategy through repeated training, but this way of exploring is very inefficient, because in the process of exploration, it randomly selects an action from the action set each time. The randomly selected actions are often useless, which leads to a lot of invalid exploration.

For the over-the-horizon air confrontation maneuver decision problem, we can introduce and use expert knowledge as a heuristic signal to guide the exploration process. This algorithm is called Heuristic Q-Network, which is shown as follows(Algorithm 2):

Algorithm 2. The exploration process of heuristic Q-Network.

Input: control parameter ε
Process:
1: **if** random() < ε
2: action← heuristic_strategy(s)
3: **else**
4: action←$\underset{a}{\mathrm{argmax}} Q(s, a)$
5: **end if**

5. Air Confrontation Strategy Learning

For the two-dimensional over-the-horizon air confrontation problem, according to the previous MDP model, heuristic Q-Network is used. The Q-Network structure used in this paper is an MLP with two hidden layers, which is shown in Figure 7. Its input is the air confrontation states, and the output is the behavior value function $Q(s, a_i)$ corresponding to nine maneuvers. The number of hidden layer nodes can be selected by contrast experiment. The number of nodes in the network hidden layer is determined by experiments—64 in the first hidden layer and 128 in the second layer.

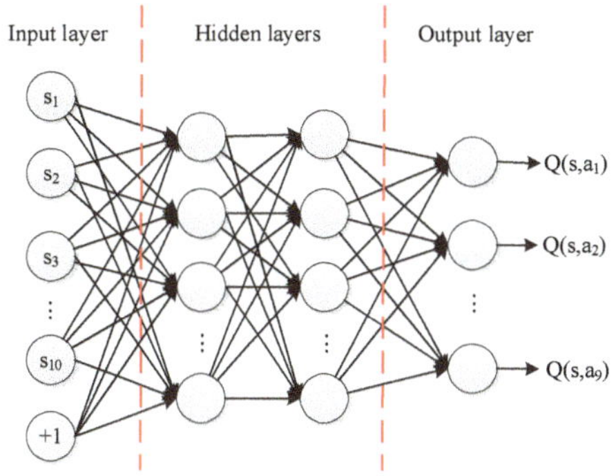

Figure 7. 2D Confrontation Q-Network.

Using the heuristic + random exploration method, the score curve of the Q-Network training process is shown in Figure 8.

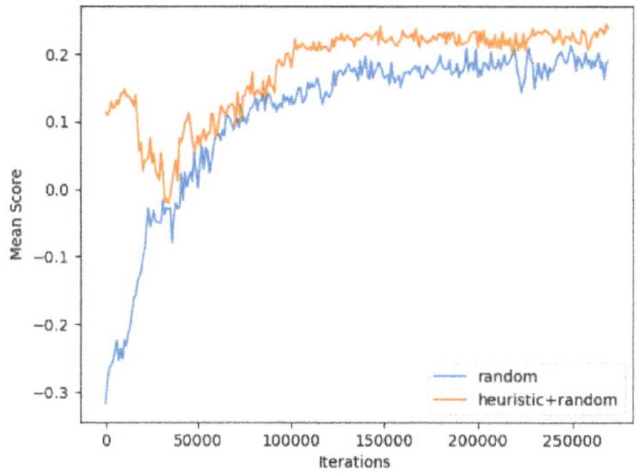

Figure 8. The score curve of 2D Q-Network.

The blue curve indicates the exploration strategy of $\varepsilon - greedy$, and the orange curve indicates the heuristic + random exploration strategy. It can be slightly seen that the heuristic search strategy can obtain a higher expected score, which further validates the effectiveness of the heuristic Q-Network.

6. Simulation Result Verification

The heuristic Q-Network learning strategy is used in air confrontation simulation, and several typical air confrontation cases are selected for analysis. According to the initial air confrontation situation, the initial state of the local aircraft can be divided into: advantage, balance and disadvantages.

- Advantage

Figure 10 records the changes in the relevant data of the aircraft during the above air confrontation process, including: the threat capability (Figure 10a), detection capability (Figure 10b), speed (Figure 10c) and relative advantage of the aircraft to the enemy aircraft (Figure 10d).

As can be seen from Figure 10, when the local aircraft is at an advantage, the local aircraft further increases its advantages from several aspects. In Figure 10a, the local aircraft increases the azimuth threat advantage of the enemy aircraft by changing its own heading, changing the speed to increase the energy advantage of the enemy aircraft, and reducing the distance between the two sides to increase the distance threat advantage. Through these three factors, the overall threat capability of the aircraft to the enemy aircraft is greatly enhanced. In Figure 10b, by changing the heading to increase the azimuth detection advantage and the entry angle detection advantage of the enemy aircraft, by narrowing the distance between the two sides to increase the distance detection advantage, the three factors can improve the comprehensive detection capability of the aircraft to the enemy aircraft. As shown in Figure 10d, the overall relative advantage of the aircraft against the enemy aircraft is on the rise. The fluctuation is due to the change of the enemy's entry angle, which causes the oscillation of the entry angle. This factor is difficult to control for the aircraft. The heading has a greater impact, and the local aircraft mainly enhances the angle advantage by changing the azimuth.

Figure 9 shows the air confrontation process when the unit is initially in an advantageous position, where the unit is indicated in red, and the enemy aircraft is shown in blue.

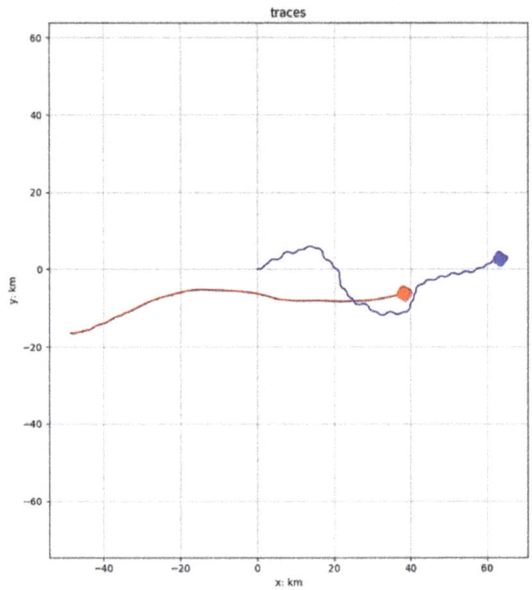

Figure 9. 2D advantage: air confrontation process.

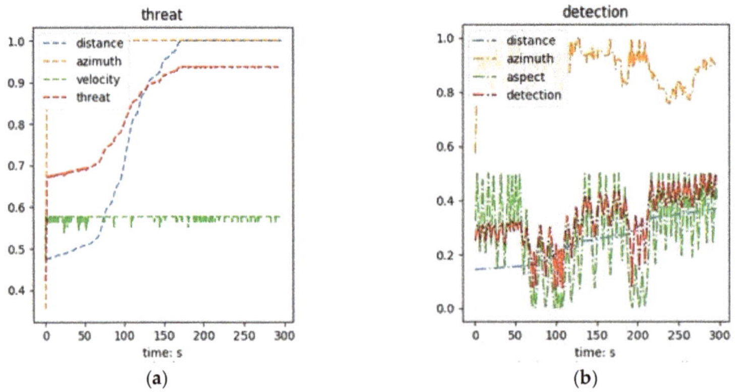

Figure 10. *Cont.*

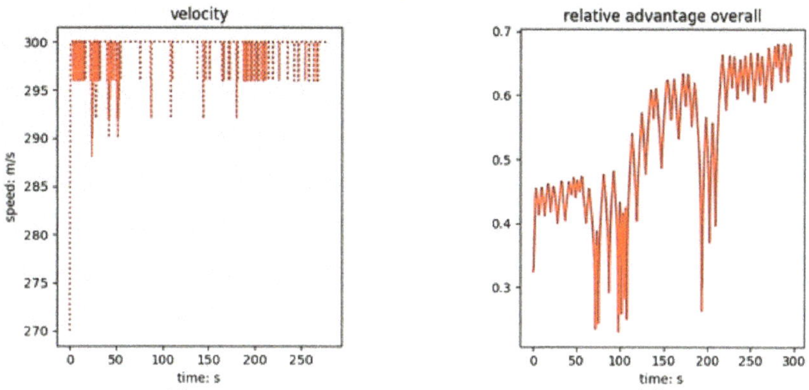

(c) (d)

Figure 10. 2D advantage: native data change: (**a**) Threat ability; (**b**) Detection ability; (**c**) Speed; (**d**) Relative advantage.

- Balance

Figure 11 shows the air confrontation process when the unit is initially in balance. The unit is indicated in red and the enemy aircraft is shown in blue.

Figure 12 shows the data changes of the local aircraft during the above air confrontation. Figure 12a shows changes in threat capabilities and comprehensive threat capabilities in all aspects, Figure 12b shows changes in detection capabilities and comprehensive detection capabilities, and Figure 12c shows changes in local speed. Figure 12d shows the overall advantage of the local aircraft relative to the enemy aircraft changes; from the figure, it can be seen that the initial relative advantage is zero, and the local aircraft, through a series of maneuvering decisions, can improve the relative advantage, so that it is in a higher position.

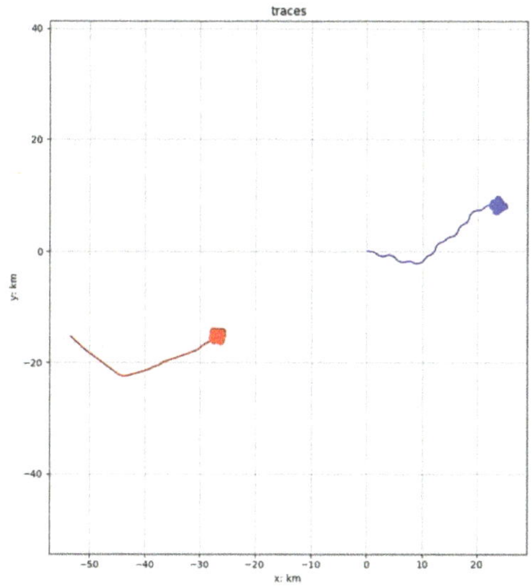

Figure 11. 2D balance: air confrontation process.

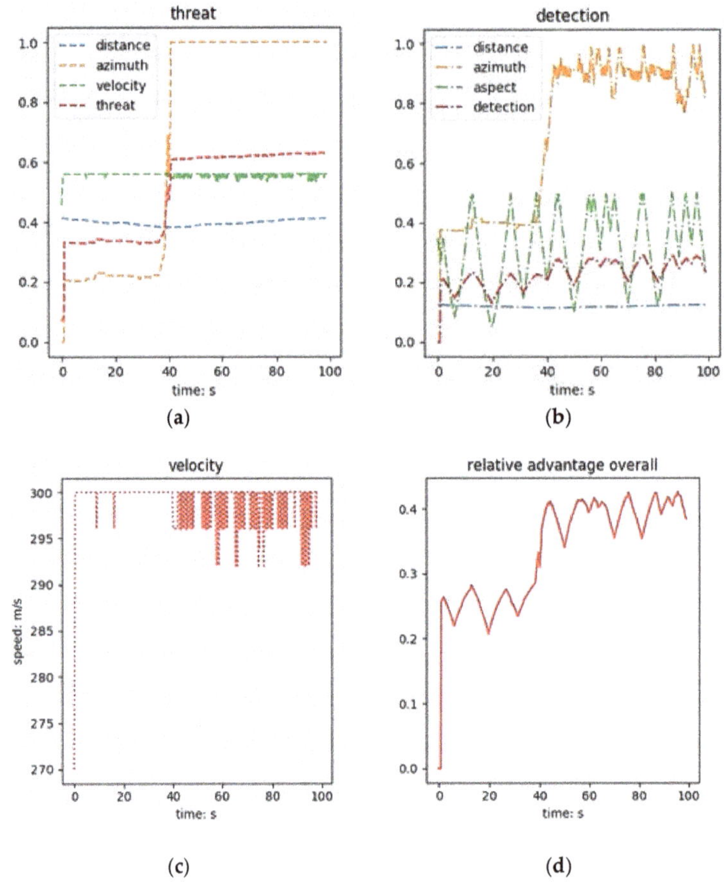

Figure 12. 2D balance: native data change: (**a**) Threat ability; (**b**) Detection ability; (**c**) Speed; (**d**) Relative advantage.

- Disadvantage

Figure 13 shows the air confrontation process when the unit is initially at a disadvantage. The unit is indicated in red and the enemy aircraft is shown in blue.

Figure 14 shows the data changes of the local aircraft during the above air confrontation. Figure 14a,b respectively show the changes in the threat capability and detection capability of the local aircraft to the enemy aircraft. Although there are some fluctuations, the overall trend is correct. And Figure 14c shows changes in local speed. It can also be seen from Figure 14d that the overall advantage of the local aircraft relative to the enemy aircraft increases from the initial negative value to a positive value, and there are some oscillations in the middle, but in the end, it can be stably maintained at a positive value, that is, in an advantageous position.

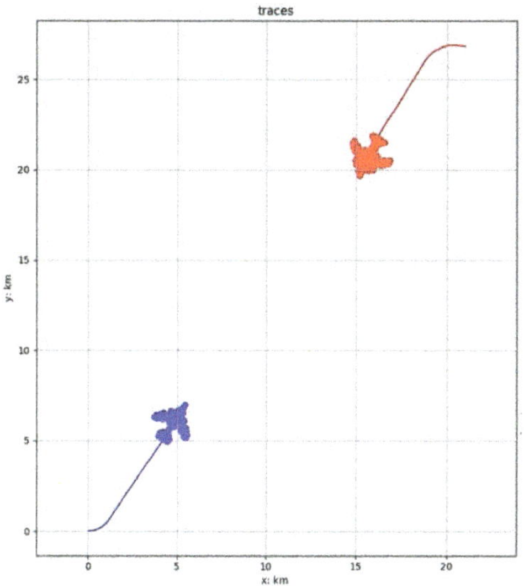

Figure 13. 2D disadvantage: air confrontation process.

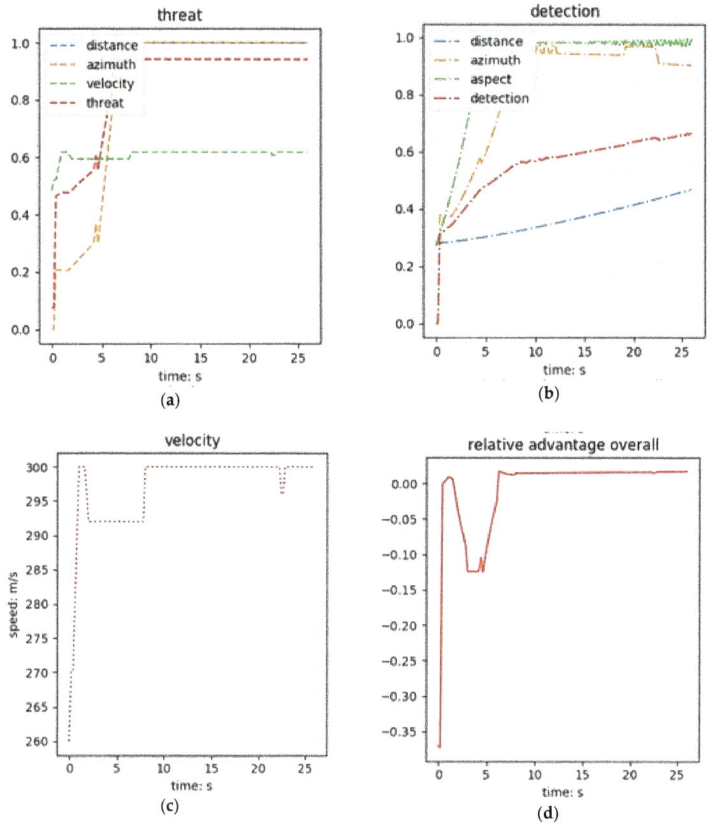

Figure 14. 2D disadvantage: native data change: (**a**) Threat ability; (**b**) Detection ability; (**c**) Speed; (**d**) Relative advantage.

- Other cases

In order to further demonstrate the air confrontation maneuver strategy learned through training, this paper presents a two-dimensional air confrontation simulation process under different initial conditions. As shown in Figure 15, the local aircraft can make better maneuvering decisions in the battle between the two sides and gain a greater advantage in confrontation.

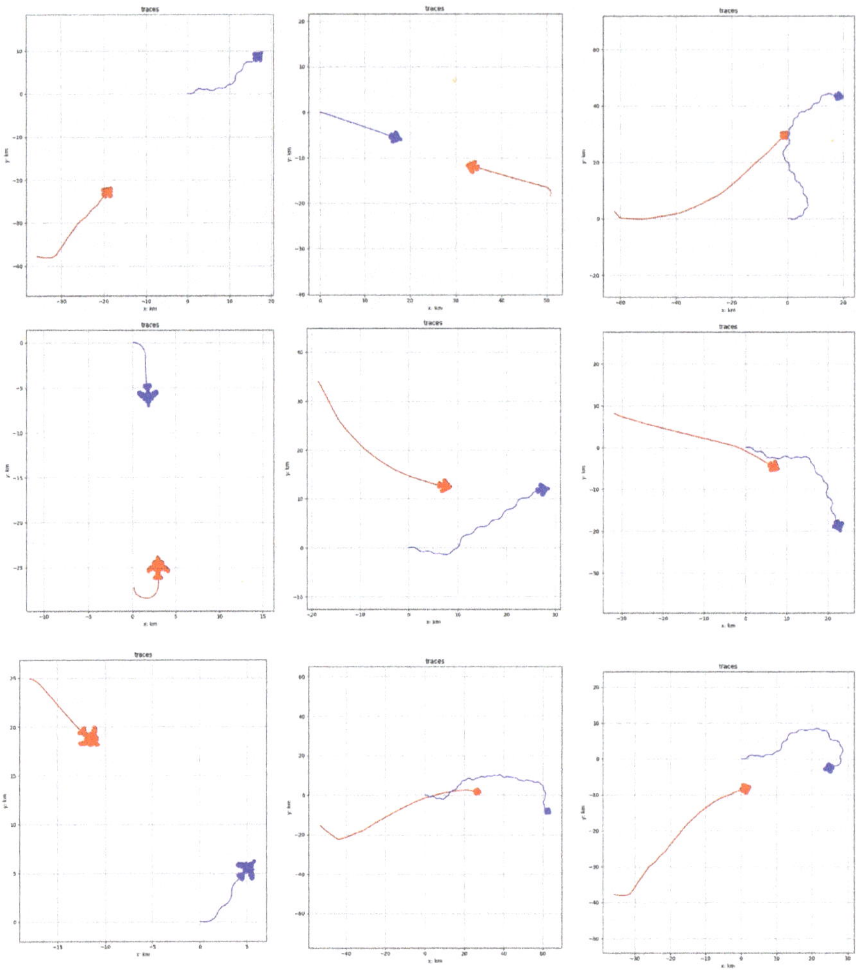

Figure 15. Other cases of 2D air confrontation.

7. Conclusions

In the process of over-the-horizon air confrontation, automated and reasonable maneuver decision-making is the premise of independent decision-making such as weapon attack, sensor use, electronic countermeasures, etc. It is accompanied by the entire air confrontation process and is an extremely important part of the automated air confrontation system/air confrontation assisted decision-making. This paper mainly studies the maneuvering decision-making method of intelligent fighters in this environment based on the single-to-single air confrontation in super-horizon air confrontation. The maneuvering decision algorithm based on reinforcement learning realizes the self-learning of the air confrontation maneuver strategy, and finally helps the fighters make reasonable maneuver decisions independently under different air confrontation situations. However, due to time and condition constraints, this work needs further research. For example, height information can be added to make the air confrontation more realistic, and the air confrontation training environment and enemy aircraft maneuver strategy need to be further improved.

Author Contributions: Conceptualization, X.Z. and C.Y.; Methodology, X.Z. and C.Y.; Software, X.Z. and G.L.; Validation, G.L.; Formal Analysis, X.Z.; Investigation, G.L.; Resources, J.W.; Data Curation, G.L; Writing-Original Draft Preparation, C.Y.; Writing-Review & Editing, X.Z. and C.Y.; Visualization, C.Y.; Supervision, J.W.; Project Administration, J.W.; Funding Acquisition, J.W.

Funding: This research was funded by [Pre-research Shared Technology and Field Fund] grant number [41401010204].

Acknowledgments: Thank my partners for their trust and collaboration, facilitating our work's completion. Thank Senior Shuandao Li for his technical inspiration and technical supports. Thank the editors of the journal MDPI for their patient guidance and correction. Thank my teacher Jiang Wu who supported us in this work all the time.

Conflicts of Interest: The authors declare no conflicts of interest.

References

1. Ehtamo, H.; Raivio, T. On Applied Nonlinear and Bi-level Programming or Pursuit-Evasion Games. *J. Optim. Theory Appl.* **2001**, *108*, 65–96. [CrossRef]
2. Fu, L.; Xie, F.; Wang, D.; Meng, G. The Overview for UAV Air-combat Decision Method. In Proceedings of the Chinese Control and Decision Conference, Changsha, China, 31 May–2 June 2014; pp. 3380–3384.
3. Austin, F.; Carbone, G.; Hinz, H.; Lewis, M.; Falco, M. Game Theory for Automated Maneuvering During Air-to-Air Com-bat. *J. Guidance* **1990**, *13*, 1143–1147. [CrossRef]
4. McManus, J.W.; Chappell, A.R.; Arbuckle, P.D. Situation Assessment in the Paladin Tactical Decision Generation System. In Proceedings of the Air Vehicle Mission Control and Management, CA:AGARD Conference, Paris, France, 1 March 1992; pp. 1–10.
5. Virtanen, K.; Raivio, T.; Hamalainen, R.P. Modeling pilot's sequential maneuvering decisions by a multistage influence diagram. *J. Guid. Control. Dyn.* **2004**, *27*, 665–677. [CrossRef]
6. Ernest, N.; Cohen, K.; Kivelevitch, E.; Schumacher, C.; Casbeer, D. Genetic fuzzy trees and their application towards autonomous training and control of a squadron of unmanned combat aerial vehicles. *Unmanned Syst.* **2015**, *03*, 185–204. [CrossRef]
7. Ernest, N.; Carroll, D. Genetic fuzzy based artificial intelligence for unmanned combat aerial vehicle control in simulated air combat missions. *J. Déf. Manag.* **2016**, *06*. [CrossRef]
8. Krishna Kumar, K.; Kaneshige, J.; Satyadas, A. Challenging Aerospace Problems for Intelligent Systems. In Proceedings of the Von Karman Lecture Series on Intelligent Systems for Aeronautics, Brussels, Belgium, 13–17 May 2002; pp. 1–15.
9. Krishnakumar, K.; Kaneshige, J. Artificial Immune System Approach for Air Combat Maneuvering. *Proc. SPIE Intell. Comput. Theor. Appl. V* **2007**, *6560*, 1–12.
10. Schvaneveldt, R.W.; Goldsmith, T.E.; Benson, A.E.; Waag, W.L. *Neural Network Models of Air Combat Maneuvering*; New Mexico State University: Las Cruces, NM, USA, 1992.
11. McGrew, J.S.; How, J.P.; Williams, B.; Roy, N. Air Combat Strategy using Approximate Dynamic Programming. *J. Guidance Control Dyn.* **2010**, *33*, 1641–1654. [CrossRef]
12. Yang, X.; Wang, X.H.; Shen, G.X.; Wen, C.Y. Modeling and Simulation Research of Six-Degree-of-Freedom Fighter. *J. Syst. Simul.* **2000**, *12*, 210–213.
13. Lachner, R.; Breitner, M.H.; Pesch, H.J. Three-dimensional air combat: Numerical solution of complex differential games. *New Trends Dyn. Games Appl.* **1995**, 165–190. [CrossRef]
14. Howard, R.A. Dynamic Programming and Markov Process. *Math. Gaz.* **1960**, *3*, 120.
15. Li, S.; Liu, G.; Wu, J. A Self-learning Terrain-following Method for Aircrafts. In Proceedings of the China Control Conference, Dalian, China, 26–28 July 2017; pp. 3437–3442.
16. Dong, X.L.; Tong, Z.X.; Wang, B.N. Design of the BVRAC Maneuver Library and Visualization of Movements. *Flight Mech.* **2005**, *23*, 90–93.

17. Touzet, C.F. Neural networks and Q-learning for robotics. In Proceedings of the IJCNN'99 (International Joint Conference (IEEE INNS) on Neural Networks), Washington, DC, USA, 10–16 July 1999.
18. Sutton, R.S.; Barto, A.G. *Reinforcement Learning: An Introduction*; MIT Press: Cambridge, UK, 1998.

electronics

Article

An Efficient SC-FDM Modulation Technique for a UAV Communication Link

Sukhrob Atoev [1], Oh-Heum Kwon [1], Suk-Hwan Lee [2] and Ki-Ryong Kwon [1,*]

1 Department of IT Convergence and Application Engineering, Pukyong National University, Busan 48513, Korea; sukhrobreus@pukyong.ac.kr (S.A.); ohheum@gmail.com (O.-H.K.)

2 Department of Information Security, Tongmyong University, Busan 48520, Korea; skylee@tu.ac.kr

* Correspondence: krkwon@pknu.ac.kr; Tel.: +82-51-629-6257

Received: 25 September 2018; Accepted: 21 November 2018; Published: 25 November 2018

Abstract: Since the communication link of an unmanned aerial vehicle (UAV) and its reliability evaluation represent an arduous field, we have concentrated our work on this topic. The demand regarding the validity and reliability of the communication and data link of UAV is much higher since the environment of the modern battlefield is becoming more and more complex. Therefore, the communication channel between the vehicle and ground control station (GCS) should be secure and provide an efficient data link. In addition, similar to other types of communications, the data link of a UAV has several requirements such as long-range operation, high efficiency, reliability, and low latency. In order to achieve an efficient data link, we need to adopt a highly efficient modulation technique, which leads to an increase in the flight time of the UAV, data transmission rate, and the reliability of the communication link. For this purpose, we have investigated the single-carrier frequency division multiplexing (SC-FDM) modulation technique for a UAV communication system. The results obtained from the comparative study demonstrate that SC-FDM has better performance than the currently used modulation technique for a UAV communication link. We expect that our proposed approach can be a remarkable framework that will help drone manufacturers to establish an efficient UAV communication link and extend the flight duration of drones, especially those being used for search and rescue operations, military tasks, and delivery services.

Keywords: UAV communication system; data link; SC-FDM; peak-to-average power ratio (PAPR); modulation

1. Introduction

With the rapidly advancing technology, unmanned aerial vehicles (UAVs), widely known as drones, are becoming increasingly effective and significantly less costly during recent years. These vehicles can be controlled either under remote control (RC) by a pilot operator or autonomously by onboard computers. A UAV communication channel is a key factor that can affect the performance of the data link in terms of high data rate and reliable transmission of information. In other words, ensuring the efficiency of a UAV communication link represents one of the great challenges of the current works regarding a UAV communication system.

The UAV communication system has the following major requirements:

- Efficient data link
- Long-range operation
- Bidirectional communication
- Low latency
- Long flight time
- Operational capabilities

- Reliable communication

As the transmission of control commands and gathered data, which can be recorded video and photos, is achieved through the communication channel between the UAV and the ground control station (GCS), a UAV data link requires the highest reliability in data transmission as well as a high data transfer rate. Nowadays, in order to provide an efficient communication link, many drones use the spread spectrum technology that allows many different pilots to operate in the same 2.4 GHz band without conflicts. Receivers in this band are virtually immune against interference issues.

Essentially, two types of spread spectrum technology are used for the UAV communication link. The first one is the frequency-hopping spread spectrum (FHSS), which unceasingly changes its narrowband frequency on several occasions a second within the 2.4 GHz frequency range. In this process, the receiver recognizes the patterns of frequency that are utilized by a transmitter. Because the transmitter changes frequency from one to another, the receiver can adopt a suitable frequency. Unlike FHSS, a direct-sequence spread spectrum (DSSS) system uses a much wider bandwidth to transmit the signal on a single selected frequency. The transmitter sends an original narrowband signal via a spreading code generator that multiplies the narrowband data signal using a much higher frequency. Anyway, both spread spectrum modes (FHSS and DSSS) transmit the signal within the 2.4 GHz frequency band. In practice, orthogonal frequency division multiplexing (OFDM) modulation has been considered more efficient than FHSS and DSSS due to its greater tolerance of multipath distortion, higher throughput, and potential data rate [1].

Nowadays, the demand on the long-range operation and long flight time is increasing in UAV communication systems and the currently used modulation techniques have the fundamental constraint to meet this demand. Therefore, we need to adopt a potential modulation technique that fulfills this demand. In fact, it is essential to adopt the most effective technique that can provide a highly efficient data link between the vehicle and GCS. The main contribution of this paper is the investigation of the SC-FDM modulation technique in a UAV communication system in order to provide an efficient UAV communication link and extend the flight time (battery life) of drones. The performance of the adopted modulation technique is analyzed by comparing it with the OFDM modulation. Additionally, SC-FDM has been considered to transfer the data using different kinds of modulation schemes such as M-ary phase-shift keying (BPSK, QPSK, and 8-PSK) and M-ary quadrature amplitude modulation (16-QAM and 64-QAM) in this work.

The remaining part of our paper is structured as follows. In Section 2, some related works are discussed. Section 3 demonstrates the UAV communication link and communication system components. The comparison between the proposed modulation technique and the OFDM modulation is presented in Section 4. We show the experimental setup and performance measures of the system in Section 5. Afterwards, Section 6 illustrates the comparative results obtained from experiments. At the end of the paper, Section 7 presents our conclusions and future work.

2. Related Works

In recent years, extending the battery life and flight time of quadcopters has become a crucial task, since most of these vehicles are used for delivery services and military tasks. To maintain flight time, quadcopter power modeling is a basic technology because the limitation of the flight time actually comes from the battery capacity constraint. Maekawa et al. [2] proposed a simple model of power consumption for delivery quadcopters by testing the Parrot AR. Drone 2.0 on a horizontal flight. The power consumption of the drone was measured by a current logger and light weighted voltage. However, their proposed power model is based only on average power consumption data obtained during the horizontal flight. The work by Sowah et al. [3] presents a rotational energy harvester powered by rotors using a brushless dc (BLDC) generator to increase the flight time of quadcopters. A printed circuit board (PCB) and Eagle PCB design software (EAGLE 6.4.0 Light) were used to build the physical model. The harvester interface circuit consumed 1.5 to 3.2 V as an input from the rectifier circuit and produced an adjustable output of 18 V with 2.7 W output power for each generator at

82% efficiency. A 600 mA of current from the four generators was utilized to provide extra flight time for the quadcopter, thus, gaining about 42% in flight duration. Moreover, in [4], a comparison of BLDC control and field-oriented control (FOC) techniques has been analyzed in order to enhance the flight endurance of multirotor UAVs. According to the power efficiency and output torque quality of propeller electrical drives, FOC has shown better performance by 2–4% in efficiency compared to BLDC control, leading to flight endurance improvement.

A method for improving the efficiency of the UAV communication link can be found in [5]. In this work, the key tasks of a UAV communication system and characteristics of a radio channel between the UAV and ground control unit (GCU) have been analyzed. Considering the various issues associated with a UAV communication link, an author proposed the optimal radio channel construction using the rotary and mounting platform with antennas, power and low-noise amplifiers as well as an OFDM modulation technique to increase the data transfer rate. In [6], Wu et al. proposed OFDM as a transmission system for UAV wireless communications. Initially, to find out the proper OFDM system parameters, the coherence time and Doppler spread have been measured. After obtaining the inter-carrier interference (ICI) coefficients, they evaluated the bit error rate (BER) performance of OFDM technology in a typical UAV communication channel and these performance results were compared with those of OFDM in normal wireless indoor channels. According to their simulation results, an insignificant performance degradation can be seen when the OFDM technology is applied to the UAV communication channel.

In UAV applications, there have been few efforts to adopt the SC-FDM modulation technology for a UAV communication system, while it has been widely accepted in mobile communications. In [7], Miko and Nemeth proposed a hardware architecture which includes a Xilinx field-programmable gate array (FPGA) combined with the software-defined radio (SDR) chip and SC-FDM modulation system to provide a high data transmission rate and radio navigation for the communication link of UAV systems. Their proposed hardware design of the transceiver is shown in Figure 1. They implemented modulation, demodulation, and coding functions in the FPGA. However, there is no indication of the performance of SC-FDM modulation in their work.

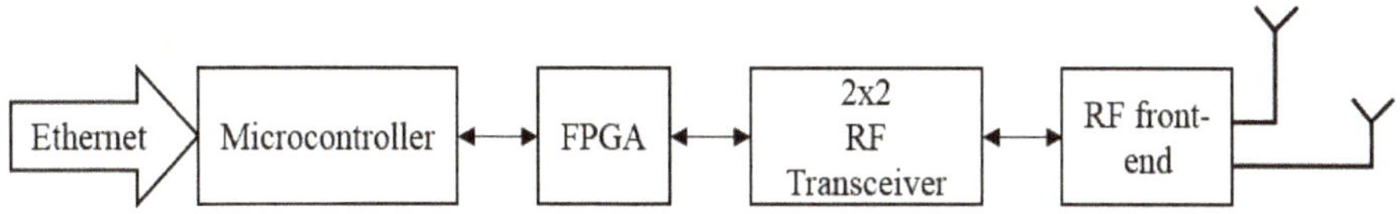

Figure 1. Hardware architecture.

Until now, most of the previous works on the single-carrier frequency division multiple access (SC-FDMA) have been carried out for uplink communications in the long-term evolution (LTE) technology of mobile communication systems [8–13]. As an alternative to the orthogonal frequency division multiple access (OFDMA), SC-FDMA has drawn considerable attention in mobile communications. In [14], Myung gives an overview of SC-FDMA. Another research focuses on PAPR reduction of localized SC-FDMA using a partial transmit sequence (PTS) [15]. Actually, there can be localized and distributed modes of subcarrier mapping in SC-FDMA [14,16]. In localized SC-FDMA, each terminal uses a set of contiguous subcarriers for the transmission of symbols, thereby limiting them to only a portion of the system bandwidth. On the other hand, the subcarriers used by the terminal are propagated throughout the entire bandwidth in the distributed SC-FDMA. Localized SC-FDMA, which is used for uplink transmission of LTE systems, has lower PAPR than distributed SC-FDMA.

Furthermore, Tsiropoulou et al. [17] provided a bargaining model and power optimization framework to solve the problem of subcarrier and power allocation in multiuser SC-FDMA wireless networks. The obtained numerical results and key features of their proposed approach demonstrate that the introduced framework can be a foundation for the supporting heterogeneous services and the implementation of different users' priorities to access the available resources. Towards this direction, a similar work can be found in [18]. In 2016, Tsiropoulou et al. [19] studied and examined the various state-of-the-art resource allocation algorithms and frameworks developed to allocate the subcarriers and transmission power of users in the uplink of SC-FDMA wireless networks. Luo and Xiong [20] proposed the SC-FDMA-IDMA system model, which is the combination of SC-FDMA and interleaved division multiple access (IDMA) and studied the effect of carrier frequency offset (CFO) on the BER performance of this system model.

3. UAV Communication System

In general, a UAV communication link can both send control commands from the GCS to the vehicle and receive data about the flight on downlink, as shown in Figure 2. A bidirectional link can be established in order to provide a communication between the drone and GCS [21]. A communication link between these two components has to provide long-range operations as well as a continuous and stable link. Therefore, the establishment of a channel model that is suitable for UAV characteristics plays an important role in improving the data link of the UAV [22,23]. Furthermore, for improving reliability of the data link, an adaptive information rate method is presented in [24].

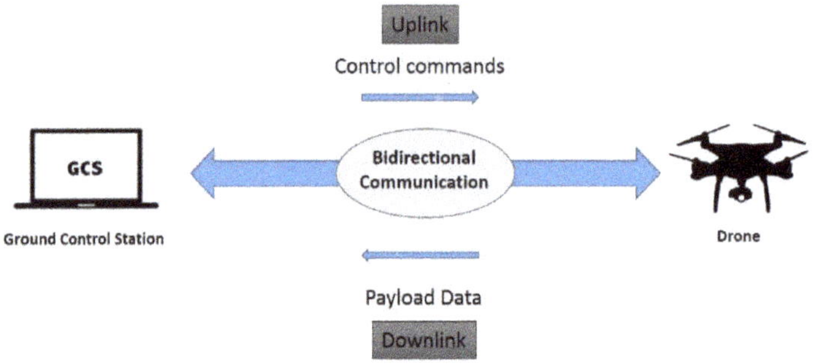

Figure 2. Drone communication link.

Figure 3 illustrates the components of the UAV communication system. The main component is the microcontroller, also referred to as a flight controller, which is the core for all functioning of a UAV. It manages failsafe, autopilot, waypoints, and many other autonomous functions. This microcontroller interprets input from the receiver, global positioning system (GPS) module, battery monitor, inertial measurement unit (IMU) that include an accelerometer and a gyroscope which can be used for providing stability or maintain a reference direction in navigation system [25], and other onboard sensors. The GCS provides relevant data about the vehicle such as speed, attitude, altitude, location, yaw, pitch, roll, warnings, and other information [26]. As shown in Figure 3, a UAV has two links: data link and communication link. To ensure the transmission of data between the vehicle and GCS, a UAV uses a data link that operates in the frequency range from 150 MHz to 1.5 GHz. On the other hand, a 2.4 GHz frequency band that determines the communication link between the transmitter and receiver is used in order to control the vehicle. It should be noted that the transmitter and receiver must both be on the same frequency. In point of fact, drones have exclusive use of their own frequency allocation due to the longer range and potentially worse consequences of radio interference. Initially, drones scan the range of frequencies within the 2.4 GHz band and use only the narrowband frequency that is not in use by another drone. As a result of this, many drones can utilize a 2.4 GHz frequency band simultaneously. This feature of drones can be noticeable when a number of drones are used as flying base stations in wireless cellular networks to serve an arbitrarily located set of users [27–29]. Moreover, typical UAVs use multiple radio interfaces to maintain a continuous connection with essential links to GCSs, other UAVs, and satellite relays.

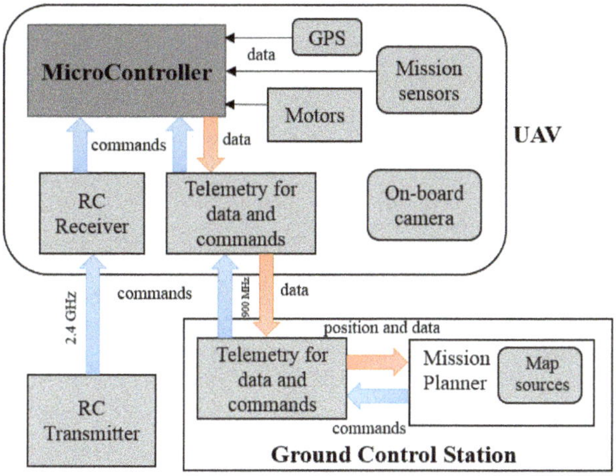

Figure 3. Unmanned aerial vehicle (UAV) communication system components.

The modulation and demodulation process of a transmitted signal through a wireless channel can respectively be done in the transmitter and receiver of the UAV communication system. Figure 4 shows the typical transmitter of a UAV communication system that transmits the control commands and telemetry data. Initially, the input data is stored in a data storage module, then, the channel coding can be used for error correction encoding. After that, the data streams are mapped into the frames and ready for channel modulation. The baseband modulation of each carrier can be selected among BPSK, QPSK, 8-PSK, 16-QAM, and 64-QAM depending on the channel condition. The modulated signals are then directly converted into the radio frequency band for wireless transmission. Before transmitting a signal, the radio frequency (RF) amplifier can be used to convert a low-power frequency signal into a higher one. Finally, the RF signal transmission is done at the transmitter antenna.

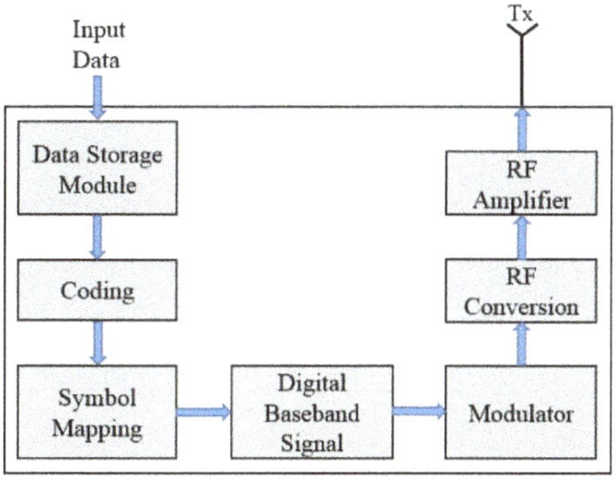

Figure 4. Block diagram of a UAV transmitter.

4. The Proposed SC-FDM and OFDM

In general, SC-FDM behaves like a single-carrier system with a short symbol duration compared to OFDM. To achieve this, SC-FDM introduces an N-point discrete Fourier transform (DFT) block right after the serial to parallel converter in the OFDM structure. The DFT block converts parallel sequences of symbols in the time domain to different frequency points. The major disadvantage of the OFDM is its high PAPR. This is a consequence of the fact that the transferred signal is the amount of all the modulated subcarriers and some of them are in a phase with high amplitudes that cannot be avoided [30]. Due to this reason, the power structure of the transmitter is characterized by relatively low average- and high-power peaks. SC-FDM allows a symbol to be transmitted in parts over multiple subcarriers, but in OFDM, we have one to one mapping between symbol and subcarrier. For example, in OFDM one symbol occupies one subcarrier of 15 kHz, but in SC-FDM, the same symbol is distributed among multiple subcarriers of 15 kHz, as described in Figure 5 [31].

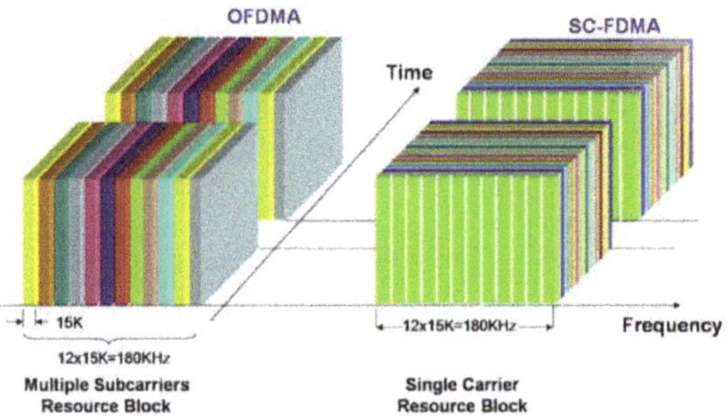

Figure 5. Orthogonal frequency division multiplexing (OFDM) and single-carrier frequency division multiplexing (SC-FDM).

4.1. OFDM

In Figure 5, shown above, the data can be transferred by parallel subcarriers of 15 kHz in OFDM. On the time axis, the further divided subcarriers represent blocks of one symbol duration. This basic unit is known as a resource element, and one symbol is carried by one resource element. In addition, a number of resource elements are used to make a resource block that is the basic unit of scheduling. At the beginning of the modulation process, the data is modulated by a particular modulation scheme in order to transfer the data over these resource elements. This modulation scheme depends on the physical channels mapped on the resource grid. Then, M-point inverse discrete Fourier transform (IDFT) transforms the signals of the parallel frequency domain into samples of a composite time domain signal, which are much easier to generate at the transmitter side. All we need to do is to send these time domain samples at radio frequencies. In wireless channels, due to multipath propagation, there can be delay spread and intersymbol interference (ISI) [32]. This interference may cause a given transmitted symbol to be distorted by other transmitted symbols. Since OFDM uses composite IDFT samples, a cyclic prefix is added by taking some samples from the end of a symbol period and placing them at the beginning. It provides orthogonality between the subcarriers by keeping the OFDM symbol periodic in the duration of the extended symbol and for that reason, avoiding intercarrier and interchannel interference simultaneously. In the next step, a sampled signal is converted into an analog wave by a digital to analog converter (DAC). A further composite waveform is modulated at the desired RF for transmission. The noticeable advantage of OFDM over SC-FDM is that the frequency domain representation of signals simplifies the signal error correction at the receiver [33].

4.2. SC-FDM

SC-FDM is a hybrid modulation technique that combines the frequency allocation flexibility and multipath resistance of OFDM and other important characteristics of a single-carrier system. The crucial characteristic of the SC-FDM signal generation is that the PAPR of finite frequency shifted signal ends up being the same as that of the original modulating data symbols and this is very different from that of OFDM for the same occupied bandwidth in the same data rate. However, if the channel bandwidth is wider, the link between the symbol length and channel bandwidth can be considered as the disadvantage of SC-FDM in comparison with OFDM [33]. Figure 6 presents the block diagram for a signal processing chain of SC-FDM. The first step is the same as for OFDM, modulating the data with one of the modulation schemes for data transmission over the resource elements. The data is placed over the resource elements by adjusting the phase and amplitude of the subcarrier to those derived for the data stream. Mathematically, it means multiplying the complex modulation symbol to the corresponding subcarrier frequency. After this process, there is another block in the signal processing chain. To convert the data symbols from the time domain into the frequency domain, the DFT is performed after the serial to parallel conversion. As an OFDM, afterwards, there is subcarrier mapping and an IDFT to transform the signal in the frequency domain into the time domain signal. The cyclic prefix can be inserted when the parallel to serial conversion is done. Before modulating the signal with a high frequency, a pulse shaping filter can be used to get the desired spectrum [34]. The original values of the symbols can be completely recovered if the transmitted signal is properly sampled at the receiver. The last two steps are the same as for OFDM. Thus, there is no difference in the downlink signal generation chain. In the same way, the reverse of what was done at the transmitter can be accomplished at the receiver. As mentioned in Section 4.1, each subcarrier carries only one particular modulation symbol in OFDM, while the DFT takes a symbol and spreads it via an available subcarrier in SC-FDM. For that reason, SC-FDM is also referred to as DFT-spread OFDM [35]. Since the PAPR is proportional to the square of the number of subcarriers involved, SC-FDM reduces the PAPR by reducing the number of subcarriers.

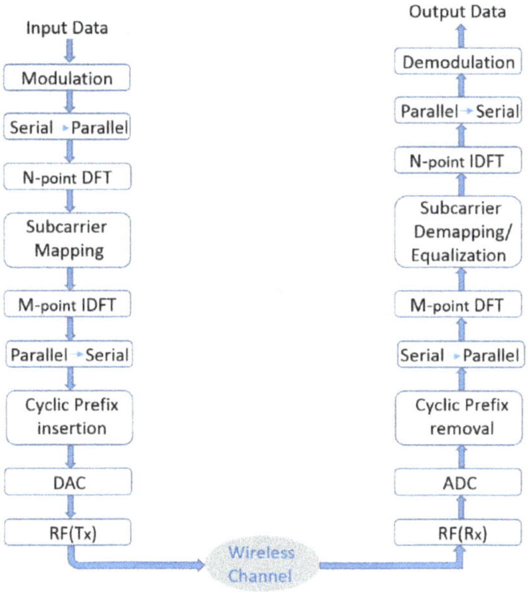

Figure 6. Block diagram of the SC-FDM modulation process.

5. Experiments and Performance Measures

5.1. Experiments

In general, our work is divided into two parts. In the first part of our research, we have performed the experimental setup by using the ArduPilot Mega (APM) 2.8 microcontroller and Mission Planner, which is the suitable GCS application for this microcontroller. The APM 2.8 microcontroller uses the micro air vehicle link (MAVLink) protocol [36] to maintain a connection between the APM 2.8 and GCS. During our experiments, the APM 2.8 board was tested instead of drone in order to obtain the flight data that were used to analyze the SC-FDM modulation for a UAV communication system. As we mentioned in Section 3, the UAV has a data link that utilizes the frequency range from 150 MHz to

1.5 GHz. In our work, two 3DR 915 MHz telemetry radios were used to provide this data link. To set up a connection between the APM 2.8 and Mission Planner GCS, one of the 3DR 915 MHz telemetry radios was plugged into the APM 2.8, as shown in Figure 7, while a second telemetry radio was connected to a PC using a USB cable. Since we have the communication link between two telemetry radios, we can check the link status in terms of the number of transmitted packets, frequency range, number of channels, and T_x power, as shown in Figure 8a. We configured the connection settings using Mission Planner. After establishing a successful connection, the telemetry data were obtained from the GCS. It should be noted that these data can be represented as signals when they are carried by the MAVLink through a wireless link between the APM 2.8 and GCS. All relevant information about APM 2.8 such as altitude, yaw, pitch, roll, attitude, vertical speed, ground speed, etc., are shown in the "Flight Data" screen of the Mission Planner, see Figure 8b. When we move the APM 2.8 microcontroller from one place to another, the "Flight Data" screen will display the telemetry data according to the new position of APM 2.8. Afterwards, the telemetry data collected from the GCS were imported into the MATLAB for the modulation process.

Figure 7. Layout of the experimental setup.

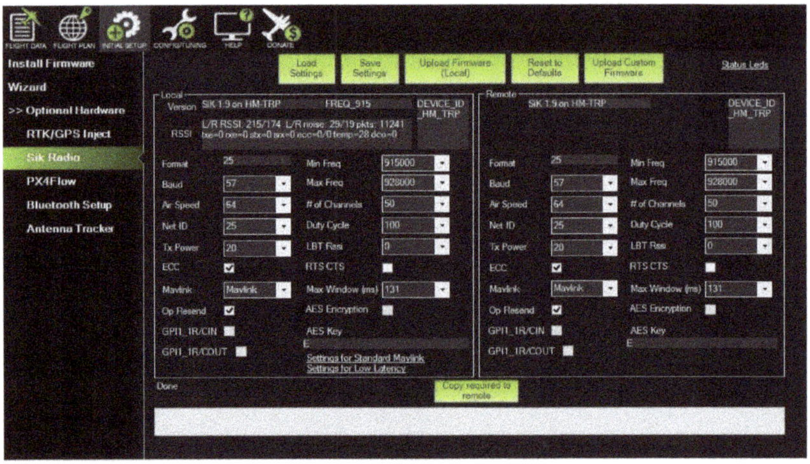

(a)

Figure 8. *Cont.*

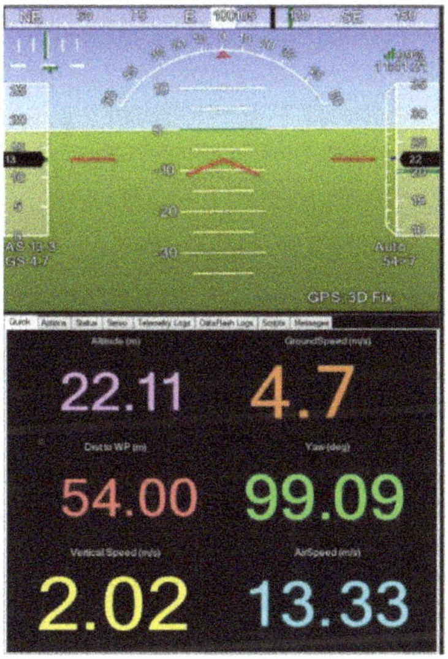

(b)

Figure 8. Mission Planner ground control station (GCS): (**a**) Radio link status; (**b**) "Flight Data" screen.

In the second part of our work, we have used the data that were gathered from the GCS and static simulation parameters shown in Table 1 for the modulation process. For this part, MATLAB was used to perform the modulation process.

Table 1. Simulation parameters.

Parameter	Value
System bandwidth	10 MHz
Frequency	2.4 GHz
Modulation scheme	BPSK, QPSK, 8-PSK, 16-QAM, 64-QAM
Number of subcarriers	1024
Data block size	32
Cyclic prefix	64
Pulse shaping filter	Raised-cosine (RC)/Root raised-cosine (RRC)
Roll-off factor	0.3
Oversampling factor	5
Number of iterations	10^4
Subcarrier spacing	10 kHz
Equalization	Zero forcing/Minimum mean square error (MMSE)

5.2. Performance Measures

The input data symbols x_n for $0 \leq n \leq N-1$ are modulated by one of the modulation schemes using an N-point DFT to generate a representation of the frequency domain of the input symbols and then, the SC-FDM output sequence X_k for $0 \leq k \leq N-1$ is given by:

$$X_k = \sum_{n=0}^{N-1} x_n e^{-j\frac{2\pi nk}{N}}, \quad 0 \leq k \leq N-1. \tag{1}$$

The SC-FDM symbol $X[k]$ is a complex number that consists of real and imaginary parts. According to the central limit theorem, as the number of subcarriers N gets larger, the real and imaginary parts of the SC-FDM symbols follow the normal (Gaussian) distribution and the probability density function (PDF) of $X[k]$ can be shown as [30]:

$$f_{X_k}(x) = \frac{x}{\sigma^2} e^{\frac{-x^2}{2\sigma^2}}, \tag{2}$$

where σ is the standard deviation.

5.2.1. PAPR

One of the key parameters for analyzing the performance of the transferred signal is its peak-to-average power ratio (PAPR), which indicates how extreme the peaks are in a waveform. This can be determined as the ratio of peak power to the average power of the transmitted signal [37]. In a multicarrier system, PAPR occurs when the different subcarriers do not correspond to a phase with each other at each point. It means that these subcarriers are different relative to each other for different phase values. The value of PAPR depends on the number of subcarriers involved as well as on the modulation scheme. It should be mentioned that high PAPR requires a high power consumption for transmitting a signal. In other words, it can be said that the efficiency of the power amplifier will be very low. On the other hand, a lower value of PAPR results in an increase in the flight time (battery life) of the vehicle. The PAPR of the transmitted signal is defined in the units of dB and it can be expressed as follows:

$$PAPR_{dB} = 10\log_{10}\left(\frac{|X_{max}|^2}{X_{ms}{}^2}\right), \tag{3}$$

where X_{max} is the maximum value and X_{ms} is the mean square value of the signal. Here, PAPR is equivalent to the crest factor, as it is defined in decibels. Now, if we consider that X_{max} denotes the crest factor, it can be written by:

$$X_{max} = \underset{k=0,1,\ldots,\ N-1}{max} X_k. \tag{4}$$

The cumulative distribution function (CDF) of X_{max} is the probability that X_{max} will take a value less than or equal to x. The CDF of X_{max} is described by [38]:

$$\begin{aligned} F_{X_{max}}(x) &= P(X_{max} \leq x) \\ &= \int_0^x f_{Y_k}(y)dy \\ &= \int_0^x \frac{y}{\sigma^2} e^{\frac{-y^2}{2\sigma^2}} dy, \end{aligned} \tag{5}$$

$$CDF = 1 - e^{\frac{-x^2}{2\sigma^2}}. \tag{6}$$

Since we have the Equations (5) and (6), we can characterize the PAPR by using a complementary cumulative distribution function (CCDF). The CCDF of the PAPR is the probability of the PAPR which is higher than a certain PAPR value and it can be calculated as:

$$CCDF = 1 - P(PAPR \leq x) = 1 - \left(1 - e^{\frac{-x^2}{2\sigma^2}}\right)^N. \tag{7}$$

5.2.2. BER

Another important parameter for measuring the performance of a wireless channel of a UAV communication system is the bit error rate (BER). When data is transmitted over a wireless link between the vehicle and GCS, the BER specifies the number of errors that appear in the received data. The environmental conditions and changes to the propagation path are the fundamental reasons for the communication channel degradation and the respective BER. To achieve an acceptable BER, i.e., for the transmission of control commands on the uplink, a typically acceptable BER is around 10^{-6}–10^{-9}, while that acceptable value for the transmission of the payload data on the downlink is 10^{-3}–10^{-4} [39]. All the available factors must be balanced. Usually, it is difficult to achieve all the requirements and some compromises are required. Moreover, a higher level of error correction is needed in order to recover the original data. This can help to fix the effects of any occurred bit error. It results in the fact that the overall BER can be improved. If the bidirectional communication between the transmitter and receiver is established very well and the signal-to-noise ratio (SNR) is high, then the BER will be potentially insignificant and will not have an observable effect on the whole UAV communication system. The BER is equal to the number of bit errors (N_E) divided by the total number of transmitted bits (N_T), as expressed by the following equation:

$$BER = \frac{N_E}{N_T}. \tag{8}$$

The N_E can be computed by comparing the transmitted signal with the received signal. The BER is most often expressed in terms of SNR. The SNR can be defined as the ratio of bit energy (E_b) to the noise power spectral density (N_o), which is a power per Hz and it is expressed as follows:

$$SNR = \frac{E_b}{N_o}, \tag{9}$$

where E_b is a measure of energy and can be defined by dividing the carrier power by the bit rate.

The probability of bit error (P_b) represents the probability that the error rate arises in the received signal. The P_b for M-ary PSK can be defined as:

$$P_b \cong 2Q\left[\sqrt{\frac{2E_{av}}{N_o}}\sin\left(\frac{\pi}{M}\right)\right], \tag{10}$$

where E_{av} is the average energy of transmitted symbol. Q represents the scaled form of the complementary error function (erfc) and it is given by:

$$Q(x) = \frac{1}{2}erfc\left(\frac{x}{\sqrt{2}}\right). \tag{11}$$

It is necessary to note that each different kind of modulation scheme has its own value for the error function. This is due to the fact that each type of modulation scheme executes in different ways in the presence of noise.

Finally, we can calculate the P_b for M-ary QAM by using the following equation:

$$P_b \cong 4\left(1 - \frac{1}{\sqrt{M}}\right)Q\left[\sqrt{\frac{3E_{av}}{(M-1)N_o}}\right]. \tag{12}$$

5.2.3. Pulse Shaping

The pulse shaping can be used to make the transmitted signal more suitable for the communication channel. In general, pulse shaping is important to ensure the correspondence of a signal in its frequency band. In this paper, a raised-cosine (RC) and root raised-cosine (RRC) pulse shaping filters are used to get the desired spectrum.

To perform the frequency response of the RC filter, we use the following equation:

$$H_{RC}(f) = \begin{cases} T, & 0 \le |f| \le \frac{1-\alpha}{2T} \\ \frac{T}{2}\left\{1 + \cos\left[\frac{\pi T}{\alpha}\left(|f| - \frac{1-\alpha}{2T}\right)\right]\right\}, & \frac{1-\alpha}{2T} \le |f| \le \frac{1+\alpha}{2T} \\ 0, & otherwise \end{cases} \tag{13}$$

where f is the frequency, T is the symbol period, and α is the roll-off factor and its value can be between 0 and 1. The representation of the time domain of this filter is given by:

$$h_{RC}(t) = \frac{\sin\left(\frac{\pi t}{T}\right) \cos \frac{\pi \alpha t}{T}}{\frac{\pi t}{T}\left[1 - \left(\frac{2\alpha t}{T}\right)^2\right]}, \quad (14)$$

where t is the time. Equations (13) and (14) are used to realize the frequency and impulse responses of the RC filter. Then, the frequency domain transfer function of the RRC filter can be written as:

$$G_{RRC}(f) = \sqrt{H_{RC}(f)}. \quad (15)$$

The impulse response of the RRC filter is given by:

$$g_{RRC}(t) = \frac{\sin\left(\frac{\pi t}{T}(1-\alpha)\right) + \frac{4\alpha t}{T}\cos\left(\frac{\pi t}{T}(1+\alpha)\right)}{\frac{\pi t}{T}\left[1 - \left(\frac{4\alpha t}{T}\right)^2\right]}. \quad (16)$$

Finally, the performance of each modulation scheme for SC-FDM and OFDM can be measured by calculating the different values of the above-mentioned parameters. To achieve the comparative results that are presented in the following section, the simulation parameters and the data that were collected from the GCS were utilized as the inputs of the system performed with MATLAB.

6. Results Analysis

In this research, MATLAB was used to perform the PAPR simulations of SC-FDM and OFDM as well as the probability of bit error simulation for SC-FDM using different kinds of modulation schemes such as BPSK, QPSK, 8-PSK, 16-QAM, and 64-QAM. Of course, it is essential to select the optimal number of subcarriers, types of the pulse shaping filters and channel equalization when simulation parameters are inputted. The number of subcarriers and symbols depends on the cyclic prefix and the subcarrier spacing. When RC and RRC pulse-shaping filters are used to filter a symbol stream, they can minimize an ISI. Half of this filtering can be done on the transmitter and the second half can be done on the receiver.

The impulse and frequency responses of the RC filter based on different roll-off factors have been plotted and shown in Figure 9. Furthermore, the equalizer can be used to get the recovery of the transmit symbols by reducing an ISI. By removing all ISI, the zero forcing equalizer can be the optimal choice in the noiseless channel. On the other hand, when a channel is noisy, this equalizer significantly amplifies the noise at frequencies. In this case, the minimum mean square error (MMSE) equalizer can be more efficient than zero forcing. The main function of an MMSE equalizer is minimizing the ISI components and the entire power of the noise in the output instead of eliminating ISI completely.

Figures 10 and 11 plot the peak-to-average power ratio (PAPR) measurements of SC-FDM and OFDM against the complementary cumulative distribution function (CCDF). According to the simulation results, which are shown in Figures 10 and 11, the SC-FDM has an advantage over OFDM because of its lower PAPR that leads to an increase in battery performance for a UAV. As shown in Figure 10a,b, it can be seen that the PAPR value of SC-FDM for a BPSK modulation scheme (5.6 dB) is almost the same as that for QPSK (5.7 dB) when the CCDF is 10^{-4}. At the same point, the PAPR value of OFDM for BPSK is 8.8 dB and, for QPSK, it equals 8.6 dB. In addition, the simulation results show that 8-PSK has a slightly lower PAPR value than 16-QAM for both SC-FDM and OFDM. From Figure 11, it can be observed that increasing the order of QAM modulation scheme (16-QAM and 64-QAM) results in increasing the PAPR values of SC-FDM and OFDM from 8 dB and 10 dB to 10 dB and 12.2 dB, respectively. From Figures 10 and 11, we can conclude that the abrupt change in the CCDF value comes from the fact that PAPR values are expressed in logarithmic scale for both SC-FDM and OFDM modulation techniques. The PAPR values of SC-FDM and OFDM for different modulation schemes are presented in Table 2.

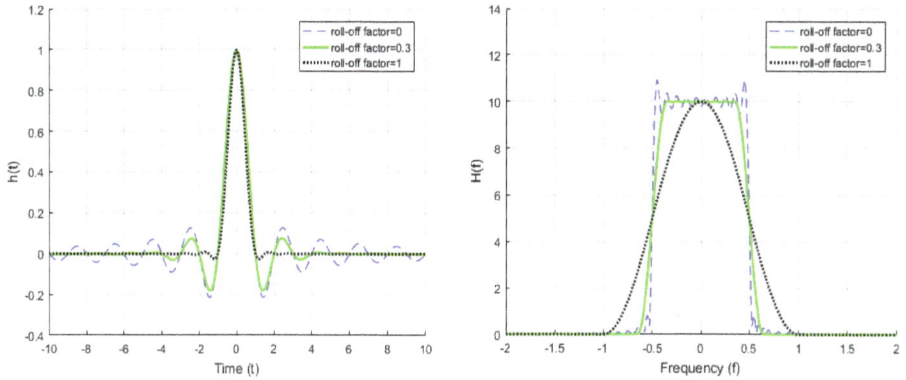

(a) (b)

Figure 9. The impulse and frequency responses of the remote control (RC) filter: (**a**) The impulse response; (**b**) the frequency response.

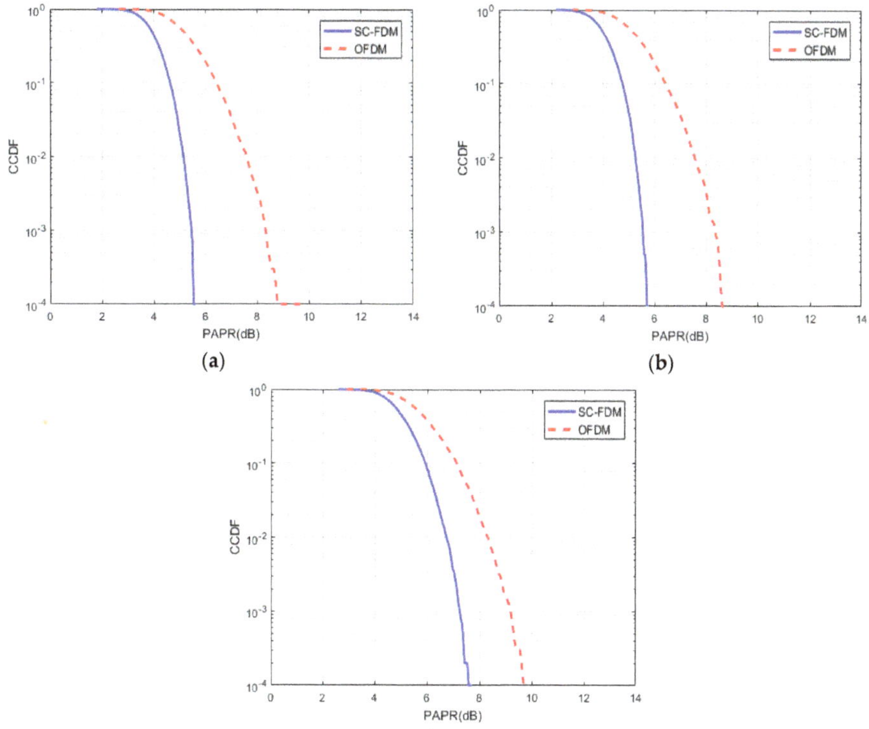

(a) (b) (c)

Figure 10. The PAPR performances of SC-FDM and OFDM for M-ary PSK modulation schemes: (**a**) BPSK; (**b**) QPSK; (**c**) 8-PSK.

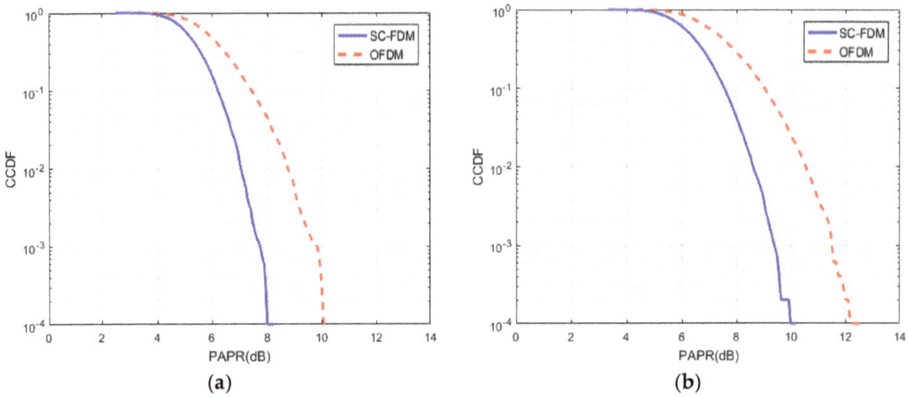

Figure 11. The peak-to-average power ratio (PAPR) performances of SC-FDM and orthogonal frequency division multiplexing (OFDM) for M-ary QAM modulation schemes: (**a**) 16-QAM; (**b**) 64-QAM.

Table 2. PAPR results for SC-FDM and OFDM.

Modulation Scheme	PAPR (dB)	
	SC-FDM	**OFDM**
BPSK	5.6	8.8
QPSK	5.7	8.6
8-PSK	7.6	9.7
16-QAM	8	10
64-QAM	10	12.2

The probability of bit error of different modulation schemes for SC-FDM is shown in Figure 12. From this figure, we can observe that the BPSK and QPSK modulation schemes achieve better performances than other modulation schemes. It is evident that BPSK and QPSK modulation schemes are very suitable for the SC-FDM modulation technique, according to their PAPR and the probability of bit error performances, see Table 2 and Figure 12.

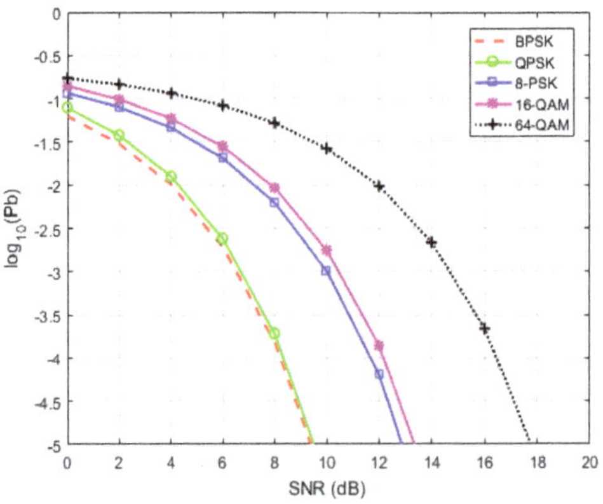

Figure 12. The probability of bit error of different modulation schemes for SC-FDM.

In our work, in order to evaluate the complexity of the proposed framework in terms of computation time, we used a computer with an Intel(R) Core(TM) i7-4790 CPU 3.60 GHz; RAM: 16.0 GB; Operating System: Windows 8.1 Pro 64-bit and MATLAB R2016a. As demonstrated in

Figure 13, the computation time of SC-FDM and OFDM modulation techniques depends on the number of subcarriers.

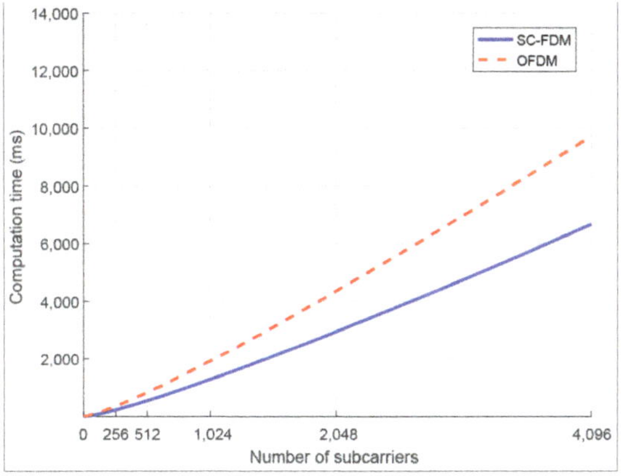

Figure 13. Computation time comparison of SC-FDM and OFDM.

In Table 3, we have shown the computation time of SC-FDM and OFDM modulation techniques. The computation time of SC-FDM ranged from 222 to 6653 ms with $|N| \in [256, 4096]$, while the range for OFDM was between 350 and 9723 ms. We can conclude that the computation time of SC-FDM is lower than that of OFDM.

Table 3. Computation time for SC-FDM and OFDM.

Number of Subcarriers	Computation Time (ms)	
	SC-FDM	OFDM
256	222	350
512	542	829
1024	1278	1916
2048	2941	4348
4096	6653	9723

7. Conclusions

In this paper, we have analyzed the SC-FDM modulation technique in order to provide an efficient communication link between the vehicle and GCS, since the communication link is an essential part of the UAV. The main purpose of this work was to analyze the SC-FDM modulation technique by comparing it with the OFDM. In general, this paper is divided into two parts. In the first part, a UAV communication link, UAV communication system components, as well as some related works have been discussed. On the other hand, the second part provided a brief overview of SC-FDM and OFDM modulation techniques and discussed the simulation results obtained by using various types of modulation schemes such as BPSK, QPSK, 8-PSK, 16-QAM, and 64-QAM. The comparative results show that SC-FDM is more effective than OFDM for the UAV communication system, leading to a noticeable improvement in terms of efficiency of the UAV communication link and the flight time of quadcopters. Moreover, by analyzing the results of our work, we have found that BPSK and QPSK are optimal modulation schemes for the SC-FDM modulation technique.

Due to the fact that the currently used 2.4 GHz band for UAV communication link is utilized by many different communication systems leading to the generation of interferences, some drones have already been designed to operate in the 5.8 GHz band. Thus, our further research will focus on analyzing the 5.8 GHz band for a UAV communication system.

Author Contributions: Conceptualization, S.A. and S.-H.L.; Software, S.A.; Formal Analysis, S.-H.L. and O.-H.K.; Writing-Review & Editing, S.A. and O.-H.K.; Project Administration and Funding Acquisition, K.-R.K.

Funding: This research was supported by the MSIT (Ministry of Science and ICT), Korea, under the Grand Information Technology Research Center support program (IITP-2018-2016-0-00318) supervised by the IITP (Institute for Information & communications Technology Promotion), Basic Science Research Program through the National Research Foundation of Korea (NRF) funded by the Ministry of Science and ICT (No. 2016R1D1A3B03931003, No. 2017R1A2B2012456) and the Korea Technology and Information Promotion Agency for SMEs (TIPA) grant funded by the Korea government (Ministry of SMEs and Startups) (No. C0407372).

Acknowledgments: The authors would like to thank the editors and the reviewers for their valuable time and constructive comments.

Conflicts of Interest: The authors declare no conflict of interest.

References

1. Ehichioya, D.; Golriz, A. Performance comparison of OFDM and DSSS on aeronautical channels. In Proceedings of the 45th International Telemetering Conference, Las Vegas, Nevada, USA, 26–29 October 2009.
2. Maekawa, K.; Negoro, S.; Taniguchi, I.; Tomiyama, H. Power measurement and modeling of quadcopters on horizontal flight. In Proceedings of the 5th International Symposium on Computing and Networking (CANDAR 2017), Aomori, Japan, 19–22 November 2017.
3. Sowah, R.A.; Acquah, M.A.; Ofoli, A.R.; Mills, G.A.; Koumadi, K.M. Rotational energy harvesting to prolong flight duration of quadcopters. *IEEE Trans. Ind. Appl.* **2017**, *53*, 4965–4972. [CrossRef]
4. Bosso, A.; Conficoni, C.; Tilli, A. Multirotor UAV flight endurance and control: The drive perspective. In Proceedings of the 42nd Annual Conference of the IEEE Industrial Electronics Society (IECON 2016), Florence, Italy, 23–26 October 2016; pp. 1839–1845.
5. Kuzmenko, A.O. Optimal choice of technical means of UAV communication link. In Proceedings of the IEEE 3rd International Conference on Actual Problems of Unmanned Aerial Vehicles Developments (APUAVD 2015), Kyiv, Ukraine, 13–15 October 2015; pp. 149–152.
6. Wu, Z.; Kumar, H.; Davari, A. Performance evaluation of OFDM transmission in UAV wireless communication. In Proceedings of the 37th Southeastern Symposium on System Theory (SSST 2005), Tuskegee, AL, USA, 20–22 March 2005; pp. 6–10.
7. Miko, G.; Nemeth, A. SCFDM based communication system for UAV applications. In Proceedings of the 25th International Conference Radioelektronika, Pardubice, Czech Republic, 21–22 April 2015; pp. 222–224.
8. Myung, H.G.; Lim, J.; Goodman, D.J. Single carrier FDMA for uplink wireless transmission. *IEEE Veh. Technol. Mag.* **2006**, *1*, 30–38. [CrossRef]
9. Lande, S.B.; Gawali, J.D.; Kharad, S.M. Performance evolution of SC-FDMA for mobile communication system. In Proceedings of the 5th International Conference on Communication Systems and Network Technologies (CSNT 2015), Gwalior, India, 4–6 April 2015; pp. 416–420.
10. Thomas, P.A.; Mathurakani, M. Effects of different modulation schemes in PAPR reduction of SC-FDMA system for uplink communication. *Int. J. Adv. Res. Electron. Instrum. Eng.* **2014**, *3*, 8531–8539.
11. Kaur, N.; Gupta, N. Simulation and analysis of OFDM and SC-FDMA with STBC using different modulation techniques. *Int. J. Adv. Res. Comput. Eng. Tehnol.* **2015**, *4*, 4184–4189.
12. Luo, Z.; Xiong, X. Performance comparison of SC-FDMA-CDMA and OFDM-CDMA systems for uplink. In Proceedings of the International Conference on Consumer Electronics, Communications and Networks (CECNet 2011), Xianning, China, 16–18 April 2011; pp. 1475–1479.
13. Xiong, X.; Luo, Z. SC-FDMA-IDMA: A hybrid multiple access scheme for LTE uplink. In Proceedings of the 7th International Conference on Wireless Communications, Networking and Mobile Computing (WiCOM 2011), Wuhan, China, 23–25 September 2011.
14. Myung, H.G. Introduction to single carrier FDMA. In Proceedings of the 15th European Signal Processing Conference (EUSIPCO 2007), Poznan, Poland, 3–7 September 2007; pp. 2144–2148.
15. Ahmed, J.J. PAPR reduction of localized single carrier FDMA using partial transmit sequence in LTE systems. *Int. J. Comput. Newt. Techol.* **2017**, *5*, 21–26. [CrossRef]

16. Rana, M.M.; Kim, J.; Cho, W.K. Performance analysis of subcarrier mapping in LTE uplink systems. In Proceedings of the 9th International Conference on Optical Internet (COIN 2010), Jeju, Korea, 11–14 July 2010.
17. Tsiropoulou, E.E.; Kapoukakis, A.; Papavassiliou, S. Energy-efficient subcarrier allocation in SC-FDMA wireless networks based on multilateral model of bargaining. In Proceedings of the 2013 IFIP Networking Conference, Brooklyn, NY, USA, 22–24 May 2013; pp. 1–9.
18. Tsiropoulou, E.E.; Ziras, I.; Papavassiliou, S. Service differentiation and resource allocation in SC-FDMA wireless networks through user-centric Distributed non-cooperative Multilateral Bargaining. In Proceedings of the International Conference on Ad Hoc Networks (ADHOCNETS 2015), San Remo, Italy, 1–2 September 2015; pp. 42–54.
19. Tsiropoulou, E.E.; Kapoukakis, A.; Papavassiliou, S. Uplink resource allocation in SC-FDMA wireless networks: A survey and taxonomy. *J. Comput. Netw.* **2016**, *96*, 1–28. [CrossRef]
20. Luo, Z.; Xiong, X. Analysis of the effect of carrier frequency offsets on the performance of SC-FDMA-IDMA systems. In Proceedings of the 2nd International Conference on Consumer Electronics, Communications and Networks (CECNet 2012), Yichang, China, 21–23 April 2012; pp. 889–893.
21. Crespo, G.; Rivera, G.G.; Garrido, J.; Ponticelli, R. Setup of a communication and control systems of a quadrotor type unmanned aerial vehicle. In Proceedings of the 29th Conference on Design of Circuits and Integrated Systems (DCIS 2014), Madrid, Spain, 26–28 November 2014.
22. Li, B. Study on modeling of communication channel of UAV. *J. Procedia Comput. Sci.* **2017**, *107*, 550–557. [CrossRef]
23. Ma, J.; Liu, H. Performance analysis and simulation of UAV data link and communication link. In Proceedings of the International Conference on Network, Communication, Computer Engineering (NCCE 2018), Chongqing, China, 26–27 May 2018.
24. Li, J.; Ding, Y.; Fang, Z. Key techniques research on UAV data link. *J. Procedia Eng.* **2015**, *99*, 1099–1107. [CrossRef]
25. Petritoli, E.; Leccese, F.; Ciani, L. Reliability and maintenance analysis of unmanned aerial vehicles. *Sensors* **2018**, *18*, 3171. [CrossRef] [PubMed]
26. Lidbom, A.; Kiniklis, E. Providence—UAV Support for Search and Rescue. Ph.D. Thesis, Department of Signals and Systems, Chalmers University of Technology, Gothenburg, Sweden, 2015.
27. Mozaffari, M.; Saad, W.; Bennis, M.; Debbah, M. Unmanned aerial vehicle with underlaid device-to-device communications: Performance and tradeoffs. *IEEE Trans. Wirel. Commun.* **2016**, *15*, 3949–3963. [CrossRef]
28. Kalantari, E.; Yanikomeroglu, H.; Yongacoglu, A. On the number and 3D placement of drone base stations in wireless cellular networks. In Proceedings of the IEEE 84th Vehicular Technology Conference (VTC-Fall 2016), Montreal, QC, Canada, 18–21 September 2016.
29. Mozaffari, M.; Saad, W.; Bennis, M.; Debbah, M. Efficient deployment of multiple unmanned aerial vehicles for optimal wireless coverage. *IEEE Commun. Lett.* **2016**, *20*, 1647–1650. [CrossRef]
30. Ramavath, S.; Kshetrimayum, R.S. Analytical calculations of CCDF for some common PAPR reduction techniques in OFDM systems. In Proceedings of the IEEE International Conference on Communications, Devices and Intelligent Systems (CODIS 2012), Kolkata, India, 28–29 December 2012; pp. 405–408.
31. LTE Uplink and SC-FDMA. Available online: https://www.exploregate.com/Video.aspx?video_id=55 (accessed on 4 September 2018).
32. Kristensen, F.; Nilsson, P.; Olsson, A. A generic transmitter for wireless OFDM systems. In Proceedings of the IEEE 14th Conference on Personal, Indoor and Mobile Radio Communications (PIMRC 2003), Beijing, China, 7–10 September 2003; pp. 2234–2238.
33. Rumney, M. *3GPP LTE: Introducing Single-Carrier FDMA*; Technical Report; Agilent Technologies: Santa Clara, CA, USA, January 2008.
34. Nigam, H.; Patidar, M.K. Performance evaluation of CFO in single carrier-FDMA. *Int. J. Electr. Electron. Comput. Eng.* **2014**, *3*, 104–110.
35. Girdhar, I.; Singh, C.; Kumar, A. Performance analysis of DFT spread OFDM systems. *Int. J. Adv. Comput. Sci. Techol.* **2013**, *2*, 21–26.
36. MAVLink Protocol Overview. Available online: https://mavlink.io/en/protocol/overview.html (accessed on 15 September 2018).

37. Pervej, M.F.; Sarkar, M.Z.I.; Roy, T.K.; Koli, M.N.Y. Impact analysis of input and output block size of DCT-SCFDMA system. In Proceedings of the IEEE 17th International Conference on Computer and Information Technology (ICCIT 2014), Dhaka, Bangladesh, 22–23 December 2014; pp. 440–445.
38. Cho, Y.S.; Kim, J.; Yang, W.Y.; Kang, C.G. *MIMO-OFDM Wireless Communications with MATLAB*; John Wiley & Sons: Singapore, 2010.
39. Baiotti, S.; Scazzola, G.L.; Battaini, G.; Crovari, E. Advances in UAV data links: Analysis of requirement evolution and implications of future equipment. In Proceedings of the RTO SCI Symposium on Warfare Automation: Procedures and Techniques for Unmanned Vehicles, Ankara, Turkey, 26–28 April 1999.

MDPI
St. Alban-Anlage 66
4052 Basel
Switzerland
Tel. +41 61 683 77 34
Fax +41 61 302 89 18
www.mdpi.com

Electronics Editorial Office
E-mail: electronics@mdpi.com
www.mdpi.com/journal/electronics

www.ingramcontent.com/pod-product-compliance
Lightning Source LLC
Chambersburg PA
CBHW051723210326
41597CB00032B/5587
* 9 7 8 3 0 3 9 2 1 0 3 0 5 *